Dialogues in Urban and Regional Planning 5 is a selection of some of the best scholarship in urban and regional planning from around the world. The internationally recognized authors of these award-winning papers take up a range of salient issues from the theory and practice of planning.

The topics they address include the effects of globalization on world cities, metropolitan planning in France and Australia, and new research in pedestrian and traffic design. The breadth of the topics covered in this book will appeal to all those with an interest in urban and regional planning, providing a springboard for further debate and research. The papers focus particularly on themes of inclusion, urban transformation, metropolitan planning, and urban design.

The *Dialogues in Urban and Regional Planning* (*DURP*) book series is published in association with the Global Planning Education Association Network (GPEAN) and its member national and transnational planning schools associations.

Editors: **Michael Hibbard** (ACSP) is Professor Emeritus in the Department of Planning, Public Policy & Management and the Director of the Institute for Policy Research & Innovation at the University of Oregon, USA. **Robert Freestone** (ANZAPS) is Professor of Planning in the Faculty of the Built Environment at the University of New South Wales, Australia. **Tore Øivin Sager** (AESOP) is Professor in the Department of Civil and Transport Engineering at the Norwegian University of Science and Technology, Norway.

Contributors: **Anne Aguiléra, Danièle Bloy, Raymond Bunker, David Caubel, Louise David, Eric Dumbaugh, Pierre Filion, Sergio A. Flores Peña, Robert Freestone, Jill L. Grant, Ludovic Halbert, Michael Hibbard, Amin Y. Kamete, Karina Landman, Jean-Loup Madre, Lai Choo Malone-Lee, Paula Meth, Dominique Mignot, Robert Rae, Sonia Roitman, Tore Øivin Sager, Glen Searle, Saint-Clair Cordeiro da Trindade Júnior, Chris Webster, Patricia Woo.**

Prize-winning papers from the world's planning school associations

Dialogues in Urban and Regional Planning offers a selection of the best urban and regional planning scholarship from the world's planning school associations. The award-winning papers presented here illustrate some of the concerns and discourse of planning scholars and provide a glimpse of planning theory and practice around the world. Everyone with an interest in urban and regional planning will find this collection stimulating in opening avenues for research and debate.

This book is published in association with the Global Planning Education Association Network (GPEAN) and its nine member planning schools associations, which have nominated these papers through regional competitions. These associations represent over 360 planning schools in nearly 50 countries around the world:

Association of African Planning Schools (AAPS)
Association of Collegiate Schools of Planning (ACSP)
Association of Canadian University Planning Programs (ACUPP)
Association of European Schools of Planning (AESOP)
Association of Latin American Schools of Planning and Urbanism (ALEUP)
National Association of Urban and Regional Postgraduate and Research Programs (ANPUR) in Brazil
Australia and New Zealand Association of Planning Schools (ANZAPS)
Association for Development of Planning Education and Research (APERAU)
Asian Planning Schools Association (APSA)

International Editorial Board

Heloisa Costa (ANPUR), Federal University of Minas Gerais, Brazil
Christophe Demazière (APERAU), University of Tours, France
Sergio Flores Peña (ALEUP), National Autonomous University of Mexico
Thomas L. Harper (ACUPP), University of Calgary, Canada

Daniel K. B. Inkoom (AAPS), Kwame Nkrumah University of Science and Technology, Ghana
Neema Kudva (ACSP), Cornell University, USA
Lik Ming Lee (APSA), University Science Malaysia
Awais Piracha (ANZAPS), University of Western Sydney, Australia
Elisabete A. Silva (AESOP), University of Cambridge, UK

Editors

Michael Hibbard (ACSP), University of Oregon, USA
Robert Freestone (ANZAPS), University of New South Wales, Australia
Tore Øivin Sager (AESOP), Norwegian University of Science and Technology, Norway

Dialogues in Urban and Regional Planning 5

Edited by Michael Hibbard,
Robert Freestone, and Tore Øivin Sager

Routledge
Taylor & Francis Group

LONDON AND NEW YORK

First edition published 2013
by Routledge
2 Park Square, Milton Park, Abingdon, Oxon, OX14 4RN

Simultaneously published in the USA and Canada
by Routledge
711 Third Avenue, New York, NY 10017

First issued in paperback 2018

Routledge is an imprint of the Taylor & Francis Group, an informa business

British Library Cataloguing in Publication Data
A catalogue record for this book is available from the British Library

Library of Congress Cataloging-in-Publication Data
Dialogues in urban and regional planning / [edited by] Michael Hibbard, Robert Freestone, Tore Øivin Sager.
    p. cm.
    "Volume 5."
    Includes bibliographical references and index.
    1. City planning—Research. 2. City planning—Case studies. 3. Regional planning—Research. 4. Regional planning—Case studies. I. Hibbard, Michael.
    II. Freestone, Robert. III. Sager, Tore.
    HT165.5.D52 2013
    307.1'216—dc23
    2012046108

Typeset in Galliard by
Swales & Willis Ltd, Exeter, Devon

ISBN 13: 978-1-138-59517-0 (pbk)
ISBN 13: 978-0-415-68077-6 (hbk)

Contents

Rethinking metropolitan frameworks

161

Designing cities

233

Contributors

Anne Aguiléra is Researcher at the Laboratory of Cities, Mobility and Transport (LVMT) belonging to the Department of Planning, Mobility and Environment of the French Institute of Science and Technology for Transport, Development and Networks (IFSTTAR). She earned a PhD in Transportation Economics from the University of Lyon (France) in 2001. Her research deals with (1) the sustainability of land use and everyday travel behaviour phenomena, (2) the trends and drivers of short and long-distance business travel, and (3) the relationships between ICTs (in particular mobile phones) and travel behaviour.

Danièle Bloy is a retired economist and engineer. She studied at the University of Lyon and is a member of LET (Transport Economic Laboratory) in Lyon. She participates in numerous research projects concerning transport and land use.

Raymond Bunker retired from full-time employment in 1994, but has continued to research and write about Australian cities. He is Adjunct Associate Professor at the City Futures Research Centre, University of New South Wales (UNSW). He first worked as a town planner in local government in England before moving to the Department of Town and Country Planning at the University of Sydney in 1959, where he became an Associate Professor and Acting Department Head. In 1975 he left to head the Strategy Division of the national Department of Urban and Regional Development. He then became head of the Research Directorate of the national Department of Environment, Housing and Community Development. In 1978 he moved to the then South Australian Institute of Technology as Head of the School of Planning. In Adelaide he was also Chair of the State Planning Commission, acted as advisor to the Premier on the Multi-Function Polis, and conducted research on urban consolidation.

David Caubel earned a PhD in Economics and specializes in public policy evaluation, sustainable urban development, transportation, and territorial economics. His aim in his research is to enhance his critical thinking and analysis to use his skills to help local authorities achieve sustainable urban development policies.

Louise David is a PhD student in Urban Planning and Regional Studies at the LATTS, a research center at the Université Paris-Est. She studies the financialization of real estate in the metropolis of Mexico. She has taught urban sociology at the Université Paris-Est.

Eric Dumbaugh is Associate Professor in the School of Urban and Regional Planning at Florida Atlantic University. He holds a PhD in Civil and Environmental Engineering from Georgia Tech, and Masters degrees in Civil Engineering and City Planning, also from Georgia Tech. His research areas include transportation systems planning, design, and community livability. His paper "The Design of Safe Urban Roadsides" received the Transportation Research Board's 2006 award for Outstanding Paper in Geometric Design and "Safe Urban Form" (in this volume) received the 2009 award for Best Paper from the *Journal of the American Planning Association*.

Pierre Filion is Professor in the School of Planning at the University of Waterloo, Canada. His research interests include metropolitan-scale planning, the implementation of smart growth strategies, and traditional and suburban downtowns. He has been co-editor of four editions of *Canadian Cities in Transition* (Oxford University Press) and has published over one hundred journal articles and book chapters. He has served on the Planning Advisory Committee of the National Capital Commission and on the Scientific Advisory Committee of the International Joint Commission on the Great Lakes.

Robert Freestone is Professor of Planning in the Faculty of the Built Environment at the University of New South Wales (UNSW), Sydney, Australia. His main research interests are in metropolitan planning history and heritage. He has worked in the private and public sector including appointments in the New South Wales Department of Planning, University of Melbourne, and the Australian National University. He is a council member of the International Planning History Society and was President 2002–2006. He was elected a Fellow of the Academy of the Social Sciences in Australia in 2008 and the Institute of Australian Geographers in 2009.

Jill L. Grant is Professor in the School of Planning at Dalhousie University, Halifax. She has a doctorate in planning from the University of Waterloo, and Masters degrees in Planning (Waterloo) and Anthropology (McMaster University). Her recent research focuses on the design and planning of residential environments (with a special interest in development trends such as new urbanism and gated communities), the application of creative cities ideas in local economic development, and the relationship between youth health and the built environment. She is the author or editor of four books: *The Drama of Democracy: Contention and Dispute in Community Planning* (1994, University of Toronto Press), *Planning* *the Good Community: New Urbanism in Theory and Practice* (2006, Routledge), *Towards Sustainable Cities: East Asian, North American and European Perspectives on Managing Urban Regions* (2004, Ashgate), and *A Reader in Canadian Planning: Linking Theory and Practice* (2008, Thomson Nelson).

Ludovic Halbert is an economic geographer and planner. After graduating from the Ecole Normale Supérieure de Fontenay Saint-Cloud, he successfully passed his PhD in Urban and Economic Geography at the Paris-1 Panthéon-Sorbonne University. He was given a full-time permanent tenure position at the French National Centre for Scientific Research (CNRS) in 2005 and subsequently joined LATTS, a joint research unit at the Université Paris-Est. His research interest focuses on the interdependencies between the transformations of productive systems (globalization, financialization) and new urban geographies, with particular attention to large-scale city-regions. His research fields cover European cities, with a particular interest in the Paris region, and Indian cities with dedicated fieldwork in Bangalore. He has published two books, co-edited two special journal issues (*Built Environment*, 2006; *Regional Studies*, forthcoming) and authored 13 international peer-reviewed journal articles and book chapters. He has also co-authored two documentary films.

Michael Hibbard (ACSP) is Professor Emeritus in the Department of Planning, Public Policy & Management and the Director of the Institute for Policy Research & Innovation at the University of Oregon, USA. He is also a participating faculty member in the University's Environmental Studies, International Studies, and Historic Preservation programs. His research focuses on community and regional development, with a special interest in the social impacts of economic change on small towns, indigenous communities, and rural regions, both in the US and internationally. He has published and consulted widely on those issues. He has served as president of the Association of Collegiate Schools of Planning and ACSP's Global Planning Educators Interest Group; as editor of the *Journal of Planning Education and Research*; and as US representative on the steering committee that organized the first World Planning Schools Congress in Shanghai, in 2001. Before entering academia full-time he worked for more than ten years in rural community and regional development. He received his PhD in regional planning from UCLA.

Amin Y. Kamete is Senior Lecturer in Urban Studies in the School of Social and Political Sciences at the University of Glasgow. Previously he was Lecturer in Planning in the School of Environment, Natural Resources and Geography at Bangor University. He launched his academic career in the Department of Rural and Urban Planning at the University of Zimbabwe. In 2003, he joined the Nordic Africa Institute in Sweden as Senior Researcher and Research Programme

Coordinator for the urban program. His research interests are planning theory and practice with special emphasis on governmentality, cities, space, and power in the context of development planning and management practice vis-à-vis informality, marginality, resistance, (in)security, and sustainability in sub-Saharan Africa. His recent publications focus on planning, informality, power, spatialized resistance, and urban governance in the contested urban spaces of Zimbabwe.

Karina Landman is Senior Lecturer in the Department of Town and Regional Planning at the University of Pretoria, South Africa. She has a background in architecture, urban design and city planning, with a PhD in Urban Design. She has been involved in extensive research related to urban and spatial transformation, crime prevention in the built environment, housing, and sustainable development. Other areas of research also include the privatization of urban space, services, and local governance; gated communities, urban segregation, medium-density mixed housing developments, and affordable housing. She has published widely in these fields, including published articles, book chapters, and conference papers presented at national and international conferences and symposia.

Jean-Loup Madre is the Head of the Department of Transport, Economics and Sociology (DEST) and works with the French Institute of Science and Technology for Transport, Development and Networks (IFSTTAR, formerly INRETS). He earned his PhD in Mathematical Statistics and in Economics. His current research interests include motorization and mobility in metropolitan areas, longitudinal approaches in travel behavior, forecasting methods (based on demography and econometrics), equity and distribution effects, survey methods (responsible for the Border Survey 1989 and the French National Personal Transportation Survey 1993–1994, European projects MEST and TEST), and international comparisons (European projects SCENARIOS, SCENE, TRANSFORUM, KITE, etc.).

Lai Choo Malone-Lee is Director of the Centre for Sustainable Asian Cities, an urban research center in the School of Design and Environment at the National University of Singapore. She received her Masters in Urban Planning from the University of Sydney and her PhD at the Tokyo Institute of Technology, Japan, with the benefit of a Ronpaku Fellowship. Her current research interest is in the areas of urban planning policies/strategies for sustainable cities, focusing on issues of city benchmarking, densification, livability, and resource optimization.

Paula Meth is Senior Lecturer in the Department of Town and Regional Planning at the University of Sheffield, UK. She completed her BA and Masters in Human Geography and Town and Regional Planning respectively at the

University of KwaZulu Natal, South Africa. She completed her PhD in Geography at the University of Cambridge in 1998. Her work focuses on "urban lives in the global South," focusing on interconnections between violence, fear, gender, and place, focusing on urban South Africa in particular. She also has a strong interest in qualitative methodologies and the use of diaries in social research. She co-authored *Geographies of Developing Areas* (2009, Routledge) and has published a range of journal articles and book chapters.

Dominique Mignot has served as Deputy Scientific Director of the French Institute of Science and Technology for Transport, Development and Networks (IFSTTAR) since January 2011. He is a member of the COST domain Transport and Urban Development Committee and a member of the board of the Forum of European Road Safety Research Institutes (FERSI). He is engineer of the French Bridges, Waters and Forests Corps, Director of Research in Economic Sciences (2000), and Doctor of Economic Sciences (University of Lyon, 1992). He was President of the Association de Science Régionale de Langue Française, the French section of the Regional Science Association International (ASRDLF) from 2006 to 2008. He was Deputy Scientific Director of INRETS in charge of evaluation from 2007 to 2010, Deputy Director of LET (Transport Economic Laboratory) in Lyon, and Director of the ENTPE team of the laboratory from 1997 to 2007. His research focuses on transportation and land use, transport and inequalities, transport economics, and road safety economics.

Sergio A. Flores Peña was originally an architect from the Universidad Nacional Autónoma de México (UNAM, 1973). He holds Masters degrees in Landscape Architecture (1978) and City and Regional Planning (1979) from the University of California, Berkeley. In his career he has been able to combine professional practice and academic work, covering a diverse number of planning–related fields. He has had articles published on topics such as: land economics, property tax and cadastral management; land value modeling; transportation and urban change; risk management in cities; urban services provision; plan making; urban design; and urban restructuring. He has recently finished conducting the preparation of the Urban Strategic Plan for Mexico City's Metropolitan Area, and a Manual for Sustainable Urban Interventions in Poverty Areas of Mexico for the Inter-American Development Bank (IDB). Currently he leads a research team on the Transformation of Central Areas in Great Cities, case study Mexico City.

Robert Rae currently works as a transportation planner for Kimley-Horn and Associates in Dallas, Texas. He focuses on helping cities and metropolitan planning organizations provide safer and more efficient transportation systems. This also includes integrating multi-modal improvements in locations and corridors that have the most need. The work completed for the co-authored paper in this volume was done while he was a Masters student at Texas A & M University.

Sonia Roitman is Lecturer in Development Planning at the University of Queensland (Australia), and was educated as an urban sociologist and planner with a PhD in Urban and Regional Planning (DPU, UCL). She formerly worked as Research Fellow at CONICET (Argentina), Postdoctoral Research Associate at the Bartlett School of Planning (London), and Postdoctoral Research Fellow in the Latin American Institute at the Free University Berlin (Germany). She also worked for the Provincial Spatial Planning Agency in Mendoza (Argentina) in the elaboration of the Provincial Development Plan 2030. Her research interests and publications focus on urban planning and development; segregation, social theory, and urban inequalities; gated communities and private governance; land and housing policies; and "glocalization," growth and planning. Her areas of study are Latin America, Europe, and East Africa.

Tore Øivin Sager is Professor in the Department of Civil and Transport Engineering at the Norwegian University of Science and Technology, Trondheim, Norway, where he teaches transport economics and transport planning. His research is mostly directed at the interfaces between institutional economics, decision processes in transport, and planning theory. He has published on communicative planning theory in a dozen international academic journals. His main publications in English are *Communicative Planning Theory* (1994, Ashgate), *Democratic Planning and Social Choice Dilemmas* (2002, Ashgate), and *Reviving Critical Planning Theory* (2013, Routledge). He has recently been researching the moral responsibilities of planning theorists, the tensions between communicative planning theory and neo-liberalism, and the use of cost–benefit analysis in participatory planning.

Glen Searle is Associate Professor in Planning at the School of Geography, Planning and Environmental Management at the University of Queensland, Australia. He was previously Director of the Planning Program at the University of Technology, Sydney. He has also held urban and regional policy positions in the Inner Cities Directorate of the United Kingdom Department of Environment, and the New South Wales Departments of Decentralisation and Development, Treasury, and Planning, where he was Deputy Manager for Policy. His principal research focus concerns metropolitan strategic spatial planning, seeking an institutionally-based understanding of strategic planning outcomes in Australian cities. A second focus has been planning and the contemporary economy of Australian cities, with *Sydney as a Global City* being published in 1996 and subsequent publications on the economic geography of advanced producer services in Sydney and Melbourne. He is Chief Editor of the *Urban Policy and Research* journal.

Saint-Clair Cordeiro da Trindade Júnior is a Brazilian geographer. He received his PhD in Human Geography at the Universidade de São Paulo (USP) in 1998 and his Masters degree in Development Planning in 1986 from the Universidade Federal do Pará (UFPA) in Brazil. Since 1989, he has worked at UFPA as a teacher, researcher, and consultant. He specializes in urban geography, and urban and regional planning, in the context of the Amazon. He also is a researcher of the Conselho Nacional de Desenvolvimento Científico e Tecnológico (CNPq) in Brazil.

Chris Webster is Dean Designate of the Faculty of Architecture at Hong Kong University, and formerly Professor of Urban Planning and Development at Cardiff University. He has degrees in Urban Planning, Computer Science, Economics, and Economic Geography, and is a leading urban theorist, writing and speaking about how cities develop at the boundary of market and government order. He has published over 150 scholarly papers on the idea of spontaneous urban order. Recent co-authored or co-edited books include *Property Rights, Planning and Markets* (2003, Edward Elgar), *Private Cities* (2006, Routledge), *Urban Poverty in China* (2010, Edward Elgar), and *Marginalisation in Urban China* (2010, Palgrave Macmillan). Recent funded research includes two UK government funded projects on urban property rights and poverty in Chinese cities; and collaborative work with Cardiff Medical School investigating the relationship between individual morbidity and evolved and designed urban morphology.

Patricia Woo spent 15 years in the formulation and implementation of urban policies in the Hong Kong Government. She obtained a Masters of Science (Environmental Management) from the National University of Singapore. She was a graduate student researcher in the Centre for Sustainable Asian Cities, National University of Singapore from 2010 to 2011, and is now working on a project of rekindling awareness of urban culture and connectivity in Hong Kong.

Preface

The book series *Dialogues in Urban and Regional Planning* offers a selection of the best urban planning scholarship from every region of the world. It provides both a lens into the contemporary issues attracting concern and at the same time helps bridge language and planning culture among planning researchers on different continents. Selected through a process of peer review, the papers provide glimpses into approaches to planning theory, methodology, and practice by planning academics around the world. Inaugurated following the first World Planning Schools Congress in Shanghai in 2001, the *Dialogues* series is produced by the nine member planning school associations of the Global Planning Education Association Network (GPEAN). International distribution through the nine associations and wider professional channels helps make this a truly global initiative. The hope is that it will make intellectual connections and deliver theoretical and policy insights which will further help integrate and enrich planning scholarship around the world.

This fifth *Dialogues* volume includes 12 papers originating from 6 continents. They draw on local concerns but also reflect international issues – problems, challenges, and opportunities presented by a variety of planning institutions, intellectual traditions, and approaches to practice. They are organized into four broad themes: inclusion and exclusion, urban transformations, rethinking metropolitan frameworks, and designing cities.

The papers included here were chosen through a rigorous selection process. Each of GPEAN's nine member planning school associations nominated one or more papers. The editors then worked with an International Editorial Board with representation from all the associations to make the final selections. Taken together, the papers convey a set of issues seen as important by planning educators today as well as the diversity of approaches to research and practice which distinguish the world's planning scholarship communities. We hope all those with an interest in urban and regional planning will find this collection valuable not only in its critical approach to important topical, theoretical, and practical issues but also in expanding perspectives and approaches pertaining to their own research.

The *Dialogues* series could not be sustained without the efforts of a diverse group of stakeholders drawn from planning school associations, publishing houses, and universities. For this fifth volume, members of the International Editorial Board played a vital role in selection of the papers and the final editing of the volume. We acknowledge their input: Heloisa Costa (Associação Nacional de Pós-graduação em Planejamento Urbano e Regional, Brazil); Christophe Demazière (Association pour la Promotion de l'Enseignement et de la Recherche en Aménagement et Urbanisme, France); Sergio Flores (Asociación Latinoamericana de Escuelas de Urbanismo y Planeación); Thomas L. Harper (Association of Canadian University Planning Programs); Daniel K. B. Inkoom (Association of African Planning Schools); Neema Kudva (Association of Collegiate Schools of Planning, USA); Lik Ming Lee (Asian Planning Schools Association); Awais Piracha (Australian and New Zealand Association of Planning Schools); and Elisabete A. Silva (Association of European Schools of Planning).

The *Dialogues* series has benefited greatly from the vision and support first of Alex Hollingsworth and more recently of Nicole Solano at Routledge, as well as Grace Chang at the University of Oregon.

Michael Hibbard
University of Oregon, USA
Robert Freestone
University of New South Wales, Australia
Tore Øivin Sager
Norwegian University of Science and Technology, Norway

Chapter 1

Introduction

Issues and options for the twenty-first century

Robert Freestone, Michael Hibbard, and Tore Øivin Sager

This is the fifth volume in the *Dialogues in Urban and Regional Planning* (*DURP*) book series. The purpose of the *DURP* series is to make widely available a selection of the best scholarship in urban and regional planning from the world's scholarly planning communities. *DURP* is one outgrowth of a phenomenon that emerged in the mid 1980s – the coming together of planning scholars in various ways but primarily through conferences and scholarly journals – to present and critically discuss planning research and educational issues unconstrained by the concern for immediate application which preoccupies planning practitioners. The tacit aim has thus been to lay the foundations for a *discipline* of urban and regional planning. The discipline of planning does not ignore professional practice; on the contrary, theoretically informed and empirically rigorous planning scholarship fosters advances in professional practice. But the primary ambition is to facilitate interchanges between planning academics worldwide and disseminate their research to a wider audience.

We begin this opening chapter by examining the institutional and intellectual background of the *DURP* series. Next we discuss the current state of thinking about planning to construct a conceptual framework for this volume. We use that framework to distill the issues and options posed by each of the selected papers, and then close with a brief final commentary.

The globalization of planning scholarship

The first truly global meeting of planning scholars was the initial World Planning Schools Congress (WPSC) hosted by Tongji University in Shanghai in 2001. It was the climax of a ferment for greater interchange among and between planning scholars that began in the 1980s (Stiftel and Watson 2004) and was organized by four regional associations – the US-based Association of Collegiate Schools of Planning (ACSP), the Association of European Schools of Planning (AESOP), the Asian Planning Schools Association (APSA), and the Australia and New Zealand Association of Planning Schools (ANZAPS). More than 650 planning scholars from more than 250 planning schools in 60 countries presented papers at WPSC 1. In conjunction with WPSC 1, leaders of ten planning school associations from around the world met to discuss ways to promote ongoing cooperation for advancing the quality of planning scholarship and education. Nine of the associations endorsed the Statement of Shanghai which, among other things, established the Global Planning Education Association Network (GPEAN).

GPEAN provides an institutional framework for integration, cross-fertilization, and critical analysis in the global planning academy. The leadership of the member associations meets to exchange information at least annually, usually at one of the annual meetings of one of the member associations. There are also regular meetings of people in counterpart positions in the member associations such as journal editors and conference conveners so that opportunities for intellectual exchange have continued to emerge. Through GPEAN, WPSC 2 was held in Mexico City in 2006, hosted by the National University of Mexico, and WPSC 3 met in Perth in 2011, hosted by the University of Western Australia. A fourth international conference is projected for 2016.

Recognizing its potential for promoting scholarly cross-fertilization, GPEAN conceived the *DURP* series as an ongoing anthology of the best planning scholarship globally. The ambition of the series is to provide a research lens on the range of problems with which planners deal. A range of questions arises from this endeavor: what new issues have attracted the interest of the planning academy and what advances in understanding have been made? Are there universal trends in planning theory and practice? If so, how do they connect to specific local issues? How do planning institutions and regulations contribute to the health and prosperity of cities and regions? Is there a convergence of planning ideas? And do we learn across borders, across cultures, across political and economic systems?

As a prelude to the organization of this volume, it is timely to revisit the themes which have surfaced through the first four volumes of the *DURP* series. Contributions in *DURP 1* (Stiftel and Watson 2004) were organized around three major international issues: the relationship between planning and economy, concerns over environment and conservation, and the nature of the planning process and decision-making. In *DURP 2* (Stiftel et al. 2006) these were refined and expanded to six main strands: economy, urban space and planning; environment and conservation of heritage; planning processes and the nature of decision-making; the development of planning ideas; planning and transport; and planning and gender. *DURP 3* (Harper et al. 2008) identified three main research vectors: the forces shaping urbanization and the different planning responses to them; evolving theoretical and conceptual ideas about urban and regional planning; and analysis of specific planning practices. The editors of *DURP 4* (Harper et al. 2010) drew on *Planning Sustainable Cities* (UN-Habitat 2009), which itself constituted a major achievement for global planning scholarship with many leading GPEAN members involved as expert external advisers and authors of background papers. The UN-Habitat report delineated five major challenges for planning that resonated through the selected papers: environmental, demographic, economic, socio-spatial, and institutional.

The *DURP* series to date has assembled approximately 50 contributions which had previously run rigorous peer-reviewed gauntlets in their first appearance as journal articles, book chapters, and conference papers, with some translated into English for their appearance in the book series. Despite the diversity of topics covered, and additional to the thematic coherence noted above, other qualities unify this body of planning scholarship. The editors of *DURP 4* identified three elements of convergence in planning scholarship common to all the volumes (Harper et al. 2010):

1 Planning scholarship is grounded in practice. While there is a broad range of methodological approaches, scholarship addresses current issues and options facing various social groups, communities, cities, and regions.
2 Empirical work is embedded in broader debates in the social sciences and environmental design that both position planning and build its own theoretical base.
3 There is evidence of growing mutual influence of planning scholarship across the world as locally embedded findings are "globally appropriated, adapted and transformed."

DURP 5 sustains these same qualities but through its own organizational matrix and prompts further reflections about the nature of and challenges in global planning scholarship.

Thinking about planning

In approaching the production of *DURP 5*, we were certainly mindful of the extraordinary diversity of challenges confronted by planning practitioners and educators. In sympathetic self-portraits, planning describes itself as seeking to address some of the most vexing and potentially catastrophic issues facing humanity. For example, in their call to the WPSC 3 Congress in 2011, the organizers declared that "with policy makers globally bracing themselves for the wide-ranging ramifications of climate change, future population growth and social diversity, amongst other concerns, urban and regional planning issues are being given unprecedented attention by governments worldwide." This is consistent with the notion put forward by UN-Habitat in the program for their sixth World Urban Forum, held in Naples, Italy in September 2012: "Urban Planning plays a key role in creating and sustaining prosperity in cities. Cities that have the tools in place to develop in a coordinated manner and with good urban design have a head start toward prosperity and urban planning should have an important place in any strategy towards prosperity. Urban planning and functional urban design creates the necessary conditions for cities to prosper by effectively responding to the demands of urban growth and rapid urbanization. The institutional and regulatory framework defines the role of different actors and modality of their engagement. On the whole, the planning framework determines the promotion of livable spaces and improvement in the quality of life."

The primary goals of urban and regional planning have evolved in tune with the times. The origins of planning at the turn of the last century lay firmly in improving the built environment and this focus remained the major calling for decades. This dominant paradigm was not really challenged until the 1960s when broader social, economic, and environmental dimensions were factored into the mix, a process that was far from seamless (Hall 2002). The sustainability agenda from the late 1980s has helped provide a more considered and welcome foundation for an integrative approach. Healey (2010) attempts to capture the contemporary "planning project" a decade into the twenty-first century as an amalgam of five major social and policy impulses:

1 an orientation to the future and a belief that action now can shape future potentialities;
2 an emphasis on livability and sustainability for the many, not the few;
3 an emphasis on interdependences and interconnectivities between one phenomenon and another, across time and space;
4 an emphasis on expanding the knowledgeability of public action, expanding the intelligence of a polity;
5 a commitment to open, transparent government processes, to open processes of reasoning in and about the public realm.

The themes which help order the selection of papers in this volume all relate in some way to this declaration of intent and how it restates the broader and often competing objectives scripted for planning by government, the community, and the market. The papers capture the challenges and constraints in moving forward in practice. Planning, for all its enemies and ambiguities, remains a necessary and resilient survivor (Gleeson 2003; Tewdwr-Jones 2011).

Following the same carefully considered protocols of review and consultation with the editorial board that were established from the commencement of the *DURP* series, we have been forced to reduce a total of 24 nominated papers through the planning associations to exactly half that number for this volume. We were mindful all the way through this process of the need to secure not only a globally representative but also coherent selection which was more than the mere sum of its parts. We see the 12 selected papers as divided equally across a quartet of themes which link into wider discourses of scholarship and policy. These do not define the scope of planning research nor do they capture all its leading edges, but they are representative of the range and depth of current concerns:

1 inclusion and exclusion
2 urban transformations
3 rethinking metropolitan frameworks
4 designing cities.

This first theme captures an enduring but intensifying planning concern with social inequality. It has been variously expressed through different binaries: rich and poor, haves and have-nots, developed and developing, first and third worlds, north and south. The scale of concern thus ranges from the differentiation of neighborhoods to divide world regions. Planning's engagement has been similarly captured through different emphases such as social justice, spatial equity, social polarization, and social inclusion which all inject a normative dimension into spatial planning. This was brilliantly captured by Harvey (1973) for the late modernist city and must still be reiterated as a central concern for the post-millennium city (Fainstein 2010). Many urban and regional policies are underpinned by at least a rhetoric of redistribution but an emerging imperative, frustrated at inaction and political barriers, talks of "the right to the city" (Purcell 2002) and links to wider social movements for greater democratic access in determining life chances. The notion of the "public interest" which uncritically informed the mission of planning for decades is now a concept rightfully contested from every dimension: not only socio-economic status but multiculturalism, age, disability, and gender. Sandercock's (2003) search for a more just postmodern urbanism captures some of the challenges of the age in which planning is now inextricably implicated: marginalization, racism, fundamentalism, empowerment, fear, and dealing with difference.

The second theme of "urban transformations" picks up an ongoing disciplinary obligation to not only monitor continuing evolving urban forms but unpack their drivers, reflect on the implications, and consider interventions which might restabilize economic, social, and environmental dysfunctionality. At the global scale, the inescapable preoccupation is with the relentless rate of urbanization which is defining urban space as "the central terrain on which the critical issues of human development will play out over the course of the twenty-first century" (Birch and Wachter 2011: 3). Planning research disag-gregates urbanization processes which recur globally but often shape distinc-tive local forms – from sprawl to renewal, peri-urbanization to regeneration, centralization to poly-nucleation, growth and decline. The concern at the top of urban hierarchies is not just the transformation of cities by bigness but their transmutation by function; this is the shift from world cities to global cities and their connectedness within what Manuel Castells (1996) memorably called the "spaces of flows." At the bottom it is often about the same thing but at this level come other concerns more sharply focused on regional development, quality of life, competitiveness, and sheer survival. In between come a range of fortunes needing to be teased out at both the inter-urban and intra-metropolitan levels and a whole taxonomy of places present: transition towns, new towns, eco-towns; inner suburbs; outer suburbs; edge cities, slow cities, toxic cities, tran-sitional cities, shrinking cities. The descriptors are multiplying, perhaps even to the point of obfuscation (Taylor and Lang 2004).

The third immanent theme in the papers crystallized an interest in the metropolis, the pre-eminent urban form developing its own populist literature of urbanology (Glaeser 2011; Kotkin 2006) but for planners one of the primary manifestations of globalization (Jonas and Ward 2007). Since the 1960s "the metropolitan explosion" has been the focus of planning research. The theory and practice of metropolitan planning encompasses several major themes: global competition and integration; environmental quality; urban structure; balancing private and public transport; infrastructure funding; governance; the time horizon and the spatial range of intervention (Gleeson et al. 2004). Like every significant contemporary planning problem, defining optimal metropolitan planning regimes defies simplistic one-type-fits-all strategies because of different environmental, economic, institutional, and social settings. Divergent planning cultures long ago set in train path dependencies (Sanyal 2005). But good governance and integration across multi-jurisdictional regimes remain pressing concerns everywhere.

Finally, a fourth theme of "designing cities" in our selection gravitates toward issues of urban design which are becoming, at least at the precinct scale, almost indivisible from planning proper (Banerjee and Loukaitou-Sideris 2010; Carmona et al. 2010). Urban design targets place-making through valuation of the public realm and intertwines other manifestos of livability, sustainability, com-munity safety, public health, and mobility. A central role in enabling transition of urban spaces to a post-carbon era has been flagged (The Penn Resolution 2011). While urban design has been critiqued as physicalist and lacking in any socio-the-oretic grounding (Cuthbert 2006), its ability to channel and consolidate progres-sive ideas into practical place-based initiatives is increasingly recognized. Design and regeneration are closely linked (Punter 2010), and this nexus increasingly provides the most visible manifestation of delivering integration between land use planning and transport provision and hence the goal of "sustainable accessibility" (Curtis 2008).

Issues and options

Although we have tagged each of the papers in this volume to one of these themes, it is by no means a mutually exclusive categorization. The papers deserve to be read in depth because, like all the best products of planning research, they have nuances, implications, and interdependencies which defy identification with any singular theme.

Inclusion and exclusion

Roitman, Webster, and Landman (Chapter 2) examine gated communities which are conventionally depicted as fortified and fragmented spaces of privilege, retreating from the chaos and crime of the failing public city, and reinforcing the trend towards social exclusion (Atkinson and Blandy 2005). They represent a global trend which has received intense scholarly attention through case studies without these coalescing within a rigorous theoretical scaffold. Here the authors argue that clearer frameworks for empirical investigations are needed, both for specific disciplines and also for providing an interdisciplinary perspective. The intent is thus unabashedly methodological rather than densely empirical. The authors highlight three different approaches (social, spatial, and institutional) to the analysis of urban fragmentation en route to outlining a new conceptual framework for interdisciplinary analysis of gated communities and related phenomena of private urban governance.

Kamete (Chapter 3) addresses a radically different aspect of exclusion. Many youths in urban Zimbabwe have taken up "precarious modes of survival" – working in the informal sector at "illegal" sites. These practices contest officialdom and resist often-violent retaliation by the authorities. The youths' utilization of urban places contravenes both property and land use controls. In democratic governance systems, these spatial conflicts can be resolved without overt violence. However, according to Kamete this is not necessarily the case when an authoritarian government embraces planning as the path to urban modernity. He identifies three modes of resistance that he terms "docility," "fighting fire with fire," and "resistance at the margins." An uncomfortable exploration of the "dark side" of planning regulation (Yiftachel 1998), the paper highlights the appropriation of planning as a spatial technology of domination.

Meth (Chapter 4) also takes up the issue of exclusion in Africa. She uses two case studies from Durban, South Africa to tell stories of marginalized women living precariously who participate through insurgency in shaping their city. Their contributions to resolving unmet housing and employment needs represent acts against a state which has, in part, retreated from the provision of shelter, employment, and other service provision through commitment to neoliberal ideology. The women manage crime and violence in their local areas using a range of strategies directly challenging the authority of the state. Their crime management practices are in turn a form of vigilantism which can turn violent. This contradiction "unsettles" the conventional representation of community insurgency (Sandercock 2003). It raises questions for planning theory about dealing with this contradiction of repressive practices emanating from below and highlights the everyday ambiguities of living in challenging urban settings.

Urban transformations

David and Halbert (Chapter 5) consider how globalized financial investment can profoundly shape the urban landscape. Bangalore in India and Mexico City serve as points of reference, the former emerging with a key role in information technology services and the latter through investment in commercial, services, and logistics corridors. With the globalization of financial transactions, real estate has been financialized and has become an investment vehicle like any other financial asset. Investment and development practices need to bridge international drivers and local circumstances, with new international property standards spreading to local real estate markets. The implications for planning lie not only in the transformation of urban and regional property markets and their spatial outcomes, but in grasping and intervening in the international real estate investment process. Like Tiesdell and Adams (2012) they see planners as "market actors" with the power to transform outcomes for the better.

Flores Peña (Chapter 6) looks more closely at the economic transformation of Mexico City, focusing on the 1950s suburb of Naucalpan on the northern outskirts. While linked to the broader literature on "shrinking cities" (Pallagst et al. 2012), the author describes a divided pattern of development, with an affluent western oriented sector benefitting from public infrastructure endowments and poorer, lower middle-class settlements supplying labour for local industries. Dual future pathways are sketched, one based on the further "informalization" of industry constructed through local networks; the other a more global pull toward logistics and services of the kind sketched in the previous chapter. Within a particular setting of economic restructuring, it thus revisits the earlier trope of inclusion/exclusion.

Trindade (Chapter 7) analyzes yet another kind of transformation in Brazil's Amazonia region where company towns (*cidades-empresa*) were developed in the second half of the twentieth century to support large-scale mineral extraction, value adding processing, and supportive power generation projects. The exploitation of the Amazon by corporate interests began much earlier (e.g. Grandin 2009) but the new wave of development is creating a distinctive class of urban spaces where local production meets the world market. The analysis draws from the spatial theorizing of Brazilian geographer Milton Santos. The new "towns *in* the forest" contrast markedly with the more traditional and organic "towns *of* the forest" in their functional connectivity to command and control centers far from the region, and pose challenges for integrated regional planning and development.

Rethinking metropolitan frameworks

Mignot, Aguiléra, Bloy, Caubel, and Madre (Chapter 8) study three metropolitan areas in France – Lille, Lyon, and Marseille – to discern the impact of divergent spatial structures on commuting patterns and intra-urban segregation. This research links into a wider interest in polycentric urban structures within Western European spatial planning (Hall and Pain 2006). The primary purpose is to evaluate the sustainability of different urban forms. Like other chapters which take broad normative positions and then seek to drill down into empirical reality, the outcomes prove ambivalent or at least complex with each of the three cities exhibiting different patterns of physical differentiation and social fragmentation. Nevertheless the importance of appropriate planning and transport policies to guide development rather than rely completely on market forces of dispersion and recentralization is apparent.

Searle and Bunker (Chapter 9) explore cross-national differences in approaches to metropolitan strategic planning. The renaissance of large-scale integrated regional strategies is explained by various factors, some common across various jurisdictions such as global pressures to be competitive. However there are significantly different historical and institutional differences between the three main regions surveyed: Europe, the United States, and Australia. European metropolitan planning is characterized as more conceptual but underpinned by complex and dynamic governance structures. The US presents a far more heterogeneous set of responses driven by local plan-shaping forces. By contrast, the Australian experience, the central concern of the authors, is depicted as a distinctive paradigm marked by a uniformly high degree of spatial specificity in plan detail virtually akin to old-style "blueprint" plans of the 1940s and 1950s. The authors relate this to an historical perpetuation of the importance of state governments in coordinating local planning and metropolitan-scale infrastructure provision.

Filion (Chapter 10) considers the evolution of trends in metropolitan planning and development in Toronto, Canada since the mid 1950s as a representative North American metropolis. The central aim here is to examine the evolution and balance of forces promoting an earlier urban form ideal of low density/auto-dependence versus the called-for transition to more high density/less auto-dependence "smart growth" structure. The key problematic is to identify the relevant mix of conservative versus transformative factors. Developing a conceptual framework drawing upon structuration theory, five key conditions tending to promote the status quo are identified – legitimacy, familiarity, efficiency, self-interest, and the absence of credible alternatives – and then mapped at different stages in metropolitan Toronto's history. Like the previous chapter, the importance of path dependency of inherited structural conditions is highlighted in impeding transition toward alternative and more sustainable models of urban development.

Designing cities

Dumbaugh and Rae (Chapter 11) work at the interface of community design, planning history, and road safety literatures in presenting an empirical analysis of the relationship between community design and the incidence of car accidents. Noting the absence of such studies, they revisit some of the critical assumptions inherited from early twentieth-century planning experts such as Clarence Stein and Clarence Perry. A GIS-based study of crash incidence in neighborhoods in San Antonio, USA questions several received wisdoms linking road safety and particular urban forms. The interaction of design theories, market-led development, vehicle speed, traffic calming, and changing attitudes towards arterial and neighborhood models of mobility produce a mosaic of road–land use configurations with variable findings. Overall, more modern road hierarchy forms are found "to substitute one set of safety problems for another" while more traditional, pedestrian-oriented configurations were found to be associated with fewer crashes. Major implications for professional design practice are discussed.

Woo and Malone-Lee (Chapter 12) assess the usage and user perceptions of skywalk systems in central Hong Kong with a view to evaluating their advantages and disadvantages in high-density urban environments. Skywalks, in segregating pedestrians from vehicular traffic, hark back to modernist visions of efficient cities where the pathways of different transport modes were separated for functional efficiency. Their popularity in many cities in the 1960s was variously a response to cold weather climates and as agents of downtown regeneration particularly because of competition from suburban shopping malls. While the skywalk system in Hong Kong is said to be "thriving," this study looks more critically at their design and impacts. It draws from traffic counts, user questionnaire surveys, and interviews with experts to offer a planning "balance sheet" of their positive and adverse impacts. While an imperfect substitution for the vitality and social space of the traditional street, they nonetheless perform a key role in pedestrian circulation that could be enhanced through better attention to their "design details."

Grant (Chapter 13) deals with design more expansively in exploring some of the reasons behind the international standing of Vancouver, Canada as a socially progressive city. This reputation derives not only from its insertion of urban design into place-making but its related pursuit of other liveable city goals such as downtown residential development, prioritization of active transport, and commitment to affordable housing as well as participatory and more transparent decision-making processes. The empirical foundation is an interview with Larry Beasley, co-director of city planning responsible for development activities between 1994 and 2006. Beasley's philosophy of "experiential planning" was attuned to securing consensus through facilitating "meaningful visions based on the everyday experiences and aspirations of residents." Carrying a broader message like each of the other papers, the success of this people-centered, human-scaled planning with its valuation of the public realm carries valuable lessons for approaches in other cities.

Conclusion

The papers assembled as chapters in this book affirm the continuing development of planning as an academic discipline. They converge toward the three key dimensions already established in the *DURP* series and cited earlier (Harper et al. 2010): policy relevance, theoretical contextualization, and cross-cultural fertilization. Moreover they capture another outstanding quality of the series to date, the "glocal" refraction of global themes within diverse local contexts, or what the editors of *DURP 2* termed the intersection of "global commonality and regional specificity" (Stiftel et al. 2006).

Cross-cutting the four major themes into which we have organized this collection is an acknowledgement of three deeper ideologies which currently animate an even wider set of concerns in contemporary planning scholarship and practice: the all-pervasive stranglehold of neoliberalism with its call to market sensitivity and pro-business thinking (Sager 2011); the quest to effect transitioning towards resilient and biophilic urban places usually under the banner of sustainability (Riddell 2004); and the commitment to participatory democracy, often articulated as a desire for inclusive, non-repressive governance (Friedmann 2011).

Where to from here? Research agendas for planning are constantly being revisited (e.g. Blanco et al., 2009a, 2009b). Contained within this volume are the seeds for further innovative work in several directions, although it is the possibilities for transnational comparative research as sampled by several papers here (Chapters 2, 5, and 9) that would clearly advance the GPEAN agenda of better dialogue between planning research communities. Comparative work in planning research is admittedly "something of a minefield" but nevertheless offers prospects of dividends in both widening the scope of research in urban and regional planning as well as deepening our understanding of policy processes (Booth 2011: 26). Indeed, greater collaborative work across the planning school associations and collegially of the kind that helped produce *Planning Sustainable Cities* (UN-Habitat 2009) is also an enticing prospect in moving forward. We would expect *DURP 6* to not only revisit some of the themes, issues, and options canvassed in this book but also to be confidently taking on new intellectual challenges.

References

Atkinson, R. and Blandy, F. (2005) "Introduction: international perspectives on the new enclavism and the rise of gated communities," *Housing Studies*, 20(2): 177–186.

Banerjee, T. and Loukaitou-Sideris, A. (eds) (2010) *Companion to Urban Design*, New York: Routledge.

Birch, E. L. and Wachter, S. M. (2011) "World urbanization: the critical issue of the twenty-first century," in Birch, E. L. and Wachter, S. M. (eds), *Global Urbanization*, Philadelphia: University of Pennsylvania Press, pp. 3–23.

Blanco, H., Alberti, M., Forsyth, A., Krizek, K. J., Rodríguez, D. A., Talen, E. and Ellis, C. (2009a) "Hot, congested, crowded and diverse: emerging research agendas in planning," *Progress in Planning*, 71(4): 153–205.

Blanco, H. M., Alberti, R., Olshansky, S., Chang, S. M., Wheeler, J., Randolph, J. B., London, J. B., Hollander, K. M., Pallagst, T., Schwarz, F. J., Popper, S., Parnell, E., Pieterse and Watson, V. (2009b) "Shaken, shrinking, hot, impoverished and informal: emerging research agendas in planning," *Progress in Planning*, 72(4): 195–250.

Booth, P. (2011) "Culture, planning and path dependence: some reflections on the problems of comparison," *Town Planning Review*, 82(1): 13–28.

Carmona M., Heath, T., Oc, T. and Tiesdell, S. (2010) *Public Places, Urban Spaces: The Dimensions of Urban Design*, 2nd edition, Oxford: Butterworth Heinemann.

Castells, M. (1996) *The Rise of the Network Society*, Oxford: Blackwell.

Curtis, C. (2008) "Planning for sustainable accessibility: the implementation challenge," *Transport Policy*, 15(2): 104–112.

Cuthbert, A. R. (2006) *The Form of Cities: Political Economy and Urban Design*, Oxford: Blackwell.

Fainstein, S. (2010) *The Just City*, Ithaca, NY: Cornell University Press.

Friedmann, J. (2011) *Insurgencies: Essays in Planning Theory*, New York: Routledge.

Glaeser, E. (2011) *Triumph of the City: How Our Greatest Invention Makes Us Richer, Smarter, Greener, Healthier, and Happier*, New York: Penguin.

Gleeson, B. (2003) "The difference that planning makes," *Environment and Planning A*, 35(5): 765–770.

Gleeson, B., Darbas, T., Johnson, L. and Lawson, S. (2004) *What Is Metropolitan Planning?*, Research Monograph 1, Brisbane: Griffith University, Urban Policy Program.

Grandin, G. (2009) *Fordlandia: The Rise and Fall of Henry Ford's Forgotten Jungle City*, New York: Metropolitan Books.

Hall, P. (2002) *Cities of Tomorrow: An Intellectual History of Urban Planning and Design in the Twentieth Century*, Oxford: Blackwell.

Hall, P. and Pain, K. (2006) *The Polycentric Metropolis: Learning from Mega-City Regions in Europe*, London: Earthscan.

Harper, T. L., Yeh, A.G.O. and Costa, H. (eds) (2008) *Dialogues in Urban and Regional Planning 3*, London: Routledge.

Harper, T. L., Hibbard, M., Costa, H. and Yeh, A.G.O. (eds) (2010) *Dialogues in Urban and Regional Planning 4*, London: Routledge.

Harvey, D. (1973) *Social Justice and the City*, London: Edward Arnold.

Healey, P. (2010) *Making Better Places: The Planning Project in the Twenty-First Century*, London: Palgrave Macmillan.

Jonas, A. E. G. and Ward, K. (2007) "Introduction to a debate on city-regions: new geographies of governance, democracy and social reproduction," *International Journal of Urban and Regional Research*, 31(1): 169–178.

Kotkin, J. (2006) *The City: A Global History*, New York: Random House.

Pallagst, K. M., Wiechmann, T. and Martinez-Fernandez, C. (eds) (2012) *Shrinking Cities: International Perspectives and Policy Implications*, London: Routledge.

The Penn Resolution: Educating Urban Designers for Post-Carbon Cities (2011), Philadelphia: Penn Design and Penn Institute for Urban Research.

Punter J. (ed.) (2010) *Urban Design and the British Urban Renaissance*, Oxford: Routledge.

Purcell, M. (2002) "Excavating Lefebvre: the right to the city and its urban politics of the inhabitant," *GeoJournal*, 58(2–3): 99–108.

Riddell, R. (2004) *Sustainable Urban Planning: Tipping the Balance*, Oxford: Blackwell.

Sager, T. (2011) "Neo-liberal urban planning policies: a literature survey 1990–2010," *Progress in Planning*, 76(4): 147–199.

Sandercock, L. (2003) *Cosmopolis II: Mongrel Cities of the 21st Century*, New York: Continuum.

Sanyal, B. (ed.) (2005) *Comparative Planning Cultures*, New York: Routledge.

Stiftel, B. and Watson, V. (eds) (2004) *Dialogues in Urban and Regional Planning 1*, London: Routledge.

Stiftel, B., Watson, V. and Acselrad, H. (eds) (2006) *Dialogues in Urban and Regional Planning 2*, London: Routledge.

Taylor, P. J. and Lang, R. E. (2004) "The shock of the new: 100 concepts describing recent urban change," *Environment and Planning A*, 36(6): 951–958.

Tewdwr-Jones, M. (2011) *Urban Reflections: Narratives of Place, Planning and Change*, Bristol: Policy Press.

Tiesdell, S. and Adams, D. (2012) *Shaping Places: Urban Planning, Design and Development*, London: Routledge.

UN-Habitat (United Nations Human Settlements Programme) (2009) *Planning Sustainable Cities: Global Report on Human Settlements 2009*, London: Earthscan.

Yiftachel, O. (1998) "Planning and social control: exploring the dark side," *Journal of Planning Literature*, 12(4): 395–406.

Inclusion and exclusion

Chapter 2

Methodological frameworks and interdisciplinary research on gated communities

Sonia Roitman, Chris Webster, and Karina Landman

This chapter examines gated communities as an object of study that has received intense scholarly attention from diverse disciplines over the last ten years. The many conference presentations and published papers on the subject have not, however, always contributed to a cohesive body of knowledge. We suggest in this chapter that clearer frameworks for empirical investigations are needed; not only for specific disciplines, but also for providing an interdisciplinary perspective. The chapter focuses on methodology: first highlighting three different approaches to the analysis of urban fragmentation (social, spatial and institutional); and second, outlining a framework for interdisciplinary analysis. In the latter part, we illustrate the connections that may be made between the analyses of the social, spatial and institutional fragmentation effects and causes of gated communities and suggest ways of handling phenomenological as well as linguistic complexity in this multidisciplinary area of urban scholarship.

Introduction

Since 1999, an international group of scientists has been tracking the global emergence and spread of gated communities, private neighbourhood governance and urban territorial enclosure. Its findings have so far been published in a series of books and special issues of journals (Environment and Planning B 2002, Geographica Helvetica 2003, Ciudades 2003, Housing Studies 2005,

Submitted by the Association of European Schools of Planning (AESOP).

Originally published in *International Planning Studies*, 15:1: 3–23 (2010). © 2010 By kind permission of Taylor & Francis Ltd. www.informaworld.com.

Trialog 2006, Geojournal 2006, European Journal of Geography 2007, Housing Policy Debate 2007, Urban Design International 2008) and discussed in nine conferences (Hamburg 1999, Mainz 2001, New York 2003, New Orleans 2003, Glasgow 2004, Pretoria 2005, Paris 2007, Santiago de Chile 2009, Istanbul 2011)[1]. The network's conferences embody an implicit methodology that is consciously multidisciplinary. The disciplines represented on the scientific committee include law, geography, anthropology, economics, sociology, architecture, urban planning and political science. Many papers are presented in plenary, where presenters have to defend their positions before researchers from other disciplines. This has the effect of shaping a shared knowledge base in a distinctive way, with a rich flow of ideas and an intensified mixing and competition between them.

The three big constraints and challenges for a multidisciplinary or interdisciplinary approach to any scientific endeavour are differences in values, underlying theory and methodology. These three dimensions of the morphology of knowledge are closely connected and present significant barriers to learning across scholastic or disciplinary traditions. Nevertheless multidisciplinary conferences devoted to understanding specific urban issues are rare and we believe it is worth persisting, especially in the study of gated communities. In writing this chapter we set ourselves the challenge of presenting three contrasting theoretical and methodological approaches to the study of gated communities and then drawing connections between them in a way that adds value to each. The research questions, conceptual framework and empirical methods of each are different and we present them alongside each other to encourage researchers, particularly young researchers, to understand the interconnectedness of different lines of enquiry into this urban phenomenon and to search for useful knowledge across disciplinary boundaries. Interdisciplinary research can be considered as the mixing together of disciplines, where contributions are integrated to provide a holistic or systemic outcome (Lawrence 2004; Bruce et al. 2004). In order to do this, interdisciplinary research requires a common conceptual framework and analytical methods based on shared terminology, mental images and common goals (Lawrence 2004). Developing these interdisciplinary frameworks, however, flows from in-depth multidisciplinary research that identifies cross-border issues and interconnectedness between disciplines or issues related to each.

The focus of this 'panel' chapter is methodology. It is clear from the many conference presentations and published papers on gated communities that greater clarity is needed about viable frameworks for empirical investigations. It would also help if those frameworks were robust across, or at least translatable across, boundaries in disciplines, paradigms, values and theories. In the chapter we look at one particular type of complex urban dynamic – socio-spatial interaction – from three different perspectives and present one common framework for the interdisciplinary research of gated communities. We propose it as one that allows for a coherent and comparable investigation of the urban complexities related to the three perspectives.

The first three sections of the chapter report, respectively, a sociological study of the social consequences of gating; an urban designer's study of the effects of gating on spatial systems; and an institutional study. While the first study is specifically of gated communities, the other two could be applied to other objects of study or specific urban phenomena. The three sections focus respectively on three specific themes found in the gated community literature: *social fragmentation*; *spatial fragmentation*; and *institutional fragmentation*. A fourth section places these within an overarching research framework, which is offered both by way of example and as a challenge to researchers to be more explicit about their methodological models and to think about how these might contribute to interdisciplinary research. It might also constitute something of a guide to an empirical research agenda. A final section concludes with a call for more interdisciplinary research of gated communities.

Gated communities and social fragmentation

Gated communities are defined in different ways (Roitman 2004, 2008). However, there are agreed elements in most definitions that refer to their social and physical features. Generally, they can be understood to be closed urban residential schemes voluntarily lived in by a homogeneous social group where public space has been privatised, restricting access through the implementation of security devices. They are designed with the intention of providing security for their residents and preventing penetration by non-residents. These compounds usually possess high-quality housing as well as services and amenities that can be used only by their residents and have a private government that enforces internal rules concerning social behaviour and construction (Roitman 2008).

There have been repetitive arguments in the gated communities literature about the alleged impact of this style of development on socio-spatial segregation (Blakely and Snyder 1997; Castell 1997; Caldeira 2000; Svampa 2001; Low 2003; Campos and García 2004; Sellés and Stambuk 2004). Gated communities, it is argued, encourage segregation through physical and institutional barriers that inhibit non-resident access. There is a smaller group of scholars, however (Manzi and Smith-Bowers 2005; Sabatini and Cáceres 2004; Salcedo and Torres 2004; Álvarez 2005; Sabatini and Salcedo 2005; Webster 2002) who disagree with the argument that gated communities produce more segregation and say they may even foster integration. When gated communities are located in poor neighbourhoods (often the case in developing countries with high income inequalities), the closer territorial proximity reduces the geographical scale of segregation and produces opportunities for functional integration. Álvarez points out that gated community residents belong to homogeneous social groups with already closed social circles before moving within gates and therefore it is "existing social and residential segregation (that) facilitates the move to gated communities" (Álvarez 2005: 11). If higher income social groups can be coaxed (by gating and private services) to live nearer to the poor, then there may be at least more economic integration between the two groups as gated communities can be sources of employment for low-skilled outside neighbours. This link may not always be strong, or there at all, and is under-researched.

In the remainder of this section a methodological framework is described, which is designed to shed light on this disagreement. It is based on four key concepts: gated communities, segregation, social practices and viewpoints. It is elaborated in greater detail in Roitman (2008).

Social practices and viewpoints: a methodological framework

Two scales of analysis can be identified when studying urban segregation. At a city scale, we can study the socio-spatial features of different groups and their distribution in the urban space describing patterns and inferring processes. At a micro scale we can study a social group and the individual preferences, attitudes and behaviour of its members. The processes that give rise to patterns at both scales are interrelated.

The study of segregation at the macro scale is often concerned with structural effects, or put another way, the effects of constraints on individual behaviour. Income is the principal constraint on an individual's freedom to locate but there are others such as race, religion, degree of education and other personal attributes. These effects, particularly income, can create socio-spatial groups that are not formed by choice. Micro-scale segregation studies often focus more on choice, looking at the separation of one social group (for example, gated community residents) from other social groups (those living in the surrounding areas) on the basis of particular interests, needs and values. The process of separation of one particular social group from other social groups living in spatial proximity can be called *urban social group segregation*.

Another important concept to consider is therefore that of *social practices*. These are regular, conscious and recognised actions carried out by individuals to satisfy their needs and interests (Roitman 2008). Individuals need to interact with others in order to consume and produce, work and play. Some of these actions will involve more social interaction than others. All will, one way or another, contribute (or not) to the evolving pattern of socio-spatial segregation – but some will do so more directly.

A third useful concept is *viewpoint*. *Viewpoint* refers to the expression of values, beliefs, feelings, attitudes, perceptions and knowledge held by individuals and these can be positive or negative with respect to other individuals and groups with whom an individual interacts or with respect to the social practices performed by other individuals and groups (Roitman 2008).

The relationship between living in gated communities and urban social group segregation can be analysed through examining the social practices and viewpoints of gated community residents and members of surrounding communities. Segregation can be said to result from a lack of social interaction between two social groups living in spatial proximity and is the consequence of social practices and viewpoints. Segregation can be said to be *intended* where there is an absence of social interaction between the gated community residents and outsiders and where negative viewpoints exist between the two groups – in one or both directions. Segregation can be said to be *unintended* when social practices lead to an absence of social interaction but there are positive (or neutral) viewpoints between gated and non-gated groups. An examination of particular social practices can indicate whether they encourage or discourage social interaction and therefore contribute to segregation.

One approach to examining the emergence of segregation at the micro scale is to identify several "ideal types" of social practices that influence the interactions of gated community residents with those outside. Two such social practices are (a) access to gated communities, which refers to how residents and non-residents act when they enter these residential estates; (b) hiring staff to work in gated communities, which refers to how residents manage to hire low-skilled employees to work within the gates.

Access to gated communities

Gated communities are residential schemes with clear limits marked by walls, fences, and sometimes gates. These physical barriers are designed to inhibit free circulation of people and vehicles from inside to outside and vice versa. The method of access (by residents and non-residents) is an otherwise trivial practice that becomes imbued with social significance because of the purpose attached to gates and other entry technology.

Gated community residents might enter through a residents' entrance that might be operated by barriers activated by remote control, digital identification or guards. Exiting the neighbourhood also involves passing through surveillance points and/or the opening of gates and barriers. Access is a practice performed every time residents enter or leave their neighbourhood and understanding how it is established is critical to an understanding of emergent segregation. The special rules or procedures required to negotiate access are as, or more, important than the physical security devices themselves in creating obstacles to social interaction between residents and outside neighbours in everyday activities.

A non-resident performs a different practice from a resident, but also has to follow certain procedures. Many gated communities have one or two entrances for residents and guests and a different one for service staff – builders, maintenance workers, guards, gardeners, home-helps and so on. This complements the analysis of the access practice performed by gated communities' residents.

Access practices form part of and help shape the social relations that develop between the two social groups living in spatial proximity. Getting access may be a determinant of social segregation as it may inhibit the development of new social interactions. In a study of access it will be useful to analyse how both social groups perceive these security devices and how they justify their use. Viewpoints about security measures indicate whether segregation is intended or unintended and can help uncover dynamics that might influence the pattern of segregation. This should be analysed against the general image that each group has of the other and evidence looked for on feedback between access practices and viewpoints.

The analysis of this practice involves the examination of three variables: access practice (which is considered here as restricted access since this is likely to be the case in most gated communities); viewpoints about physical barriers and viewpoints about the other social group. The latter two variables each have two categories: the wall and barrier can be justified as a means of security or as a tool to mark social differences; and the other group can be perceived positively or negatively. The categories can be combined to classify urban social group segregation as intended or unintended.

Some gated communities will be associated with unintended segregation. Residents' preference for gates is in response to security concerns but there is no desire to detach themselves from surrounding communities socially. Intended segregation happens when access is restricted for security reasons and there is a negative image of the neighbouring residents. Alternatively, access may be restricted and although there is a positive image about the other social group, security devices are justified to make social differences between the two groups clear. The separation of the inside residents is pursued as a form of differentiation. The third possibility of intended segregation takes place when access is restricted, security devices are used as tools to show social differences and there are negative viewpoints about the other social group. This is the most severe form of intended urban group segregation since the three categories contribute to inhibit contact between groups.

The analytical framework might be further elaborated by considering the direction of viewpoints (negative–negative; negative–positive; positive–negative;

positive– positive); the degree of accessibility constraint (as discussed); and the social or ethnic differentiation of local neighbours.

Hiring staff to work in gated communities

In low- and middle-income countries where domestic labour is cheap, many gated community residents will hire members of the surrounding local communities to work inside the gated community. These settlements are potentially important job engines for the surrounding local communities, especially in the provision of low-skilled jobs, such as gardeners, home-helps and security guards.

The presence of members of surrounding communities working inside a gated community would mean no segregation, regardless of the type of viewpoints existing from one social group towards the other. It would imply social interaction between the two social groups mediated by labour transactions. The absence of outsiders working inside implies a process of segregation since social interaction is reduced. Segregation would be unintended when despite having no working links, the two social groups living in spatial proximity have positive viewpoints towards each other. Intended segregation would happen when viewpoints about the other social group are negative; this may be one of the reasons for not hiring local members to work inside.

The analytical framework could be extended by examining the degree to which (a) gated community residents work in the surrounding local areas – implying the possibility of social interaction between the two social groups via labour transactions, for example in an employer–employee relationship; and (b) the degree to which they use the services of local communities, implying social interaction through trade transactions.

This framework has been applied in a study of gated communities in Argentina. Detailed analysis, including the elaboration of more social practices, can be found in Roitman (2007, 2008). The framework proved useful for identifying in which situations, and based on what reasons, gated community residents segregate from nearby local communities and could be used for analysing gated communities in other locations. It also emphasises the idea that segregation can be an *unintended* process that is not pursued by gated community residents. The application of the framework in that study provided evidence supporting the argument that gated communities encourage segregation. In the Argentinean case, it was found that the majority of the social practices analysed contributed to intended urban social group segregation of gated community residents from their local surrounding communities. The research also highlighted the role of access practices in shaping social relations and pointed towards the influence of physical measures on social practices.

Gated communities and spatial fragmentation

The gated communities debate has given rise to a growing body of work which portrays the "fortress city" as one of the key characteristics of urban life at the turn of the century, with an overemphasis on fortification and, in extreme cases, militarisation of the urban landscape (Soja 1989; Flusty 1995 and 2000; Dear 1986; Caldeira 2000; Graham and Marvin 2001), or what Ellin (1997) refers to as "defensive urbanism". Davis, referring to the militarisation of public space in Los Angeles, describes a city in which the "defence of luxury has given birth to an arsenal of security systems and an obsession with the policing of social boundaries through architecture" (Davis 1992: 154).

Despite the fact that some of the most distinguishing features of gated communities are their physical elements including walls or electrified fences, ostentatious entrance gates and booms, and a range of other security measures such as CCTV cameras and alarm systems, limited research has been conducted from an urban design/architectural perspective, or at a neighbourhood scale with a spatial focus. Among the little-addressed research issues are the effects gated communities have on segmenting the physical city, preventing access and creating physical and emblematic barriers that shape system and human flows and interactions. A few authors have started to allude to some of the effects, some making explicit, though brief, reference to the impact on the urban form and/or implications for urban design (Davis 1992; Sennett 1995; Ellin 1997 and 2001; Flusty 1995 and 2000; Marcuse 1995 and 2001; Tiesdell and Oc 1998; Giglia 2003; Grant and Mittelsteadt 2004; Fernández-Praxjou 2005; Sheinbaum 2005). Yet, very little has been said about the process of spatial transformation and how gated communities specifically transform the urban landscape.

Ellin (1997, 2001) argues that "form follows fear" in the contemporary city, resulting in people in perceived high-risk areas constructing defensive enclaves of various forms and nature to protect their own interests. The architect's and urban designer's concern with gating as a morphological form inevitably leads to a consideration of the social drivers and built form. Above the building level, this means considering the impact of enclosure on the urban system as a whole. The greatest handicap in taking forward a research agenda focused on the systemic impacts of particular changes in the built environment – global impacts of local actions – is complexity. Everything is related to everything else by relationships that map to space and time. According to Meadows (cited in du Plessis 2008: 12) the "world is a complex, interconnected and finite system". Given this, the main aim of science in the new ecological paradigm of complexity and thus also of the empirical investigation of particular urban phenomena, is to "understand the relationships and interactions within and across different scales of complex, adaptive systems (such as social-ecological systems) [including socio-spatial systems] and the properties that emerge from these interactions" (du Plessis 2008: 13). In the following sections, we describe an analytical framework that provides thinking tools for investigating systemic causes and effects of urban enclosure.

A conceptual framework to analyse and understand spatial transformation

In order to understand the transformation of urban space, one has to understand urban space and the aspects influencing its changes. One way to understand the urban development process is to concentrate on development agencies, the structures they interact with in the form of resources, rules and ideas, and the social and spatial contexts in which they operate (Madanipour 1996a). This emphasises the close relationship between space and society: social drivers influence spatial change, leading to specific social interpretation and response.

This conceptual framework offers a way to understand spatial transformation as a socio-spatial (multidimensional) process. This happens through a process involving space, need, idea, order, form and meaning, and parallel with these, the production and management of the spatial intervention in a specific context (Figure 2.1). Space refers to the unbound natural or existing 'man-made' space. There are many different interpretations and meanings of space (see, for example Madanipour 1996). In this context, 'space' refers to physical form and its function, either natural or built, defined as physical space. Spatial transformation

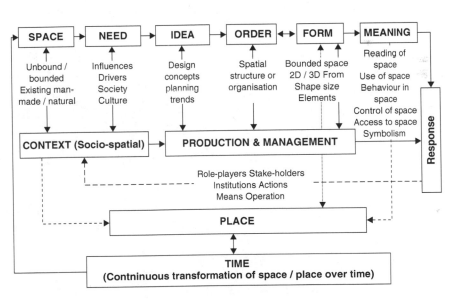

2.1 A framework for investigating the making and changing of urban space

does not happen randomly. It is informed by particular needs at a specific time (related to the context). The need/demand gives rise to a specific idea about how to address this need. This is the beginning of order, of structural organisation to order the idea and guide form. Form is the physical manifestation of the need and idea, and takes on a particular shape, texture and size. It reflects the character of a space and contributes to the creation of a particular sense of place, which in turn can be modified over time. Space and place are not arbitrary. They encompass meaning. Spaces or places can therefore be "read" and "experienced" and can appeal to people's feelings or emotions – for example, feeling comfortable in a place or feeling safe. In such a way it can also influence the use of space and thus people's behavioural patterns and consequently people can react differently to different spaces and places. Their reaction may depend on a number of predispositions, including current feelings and experiences. Places can elicit a number of responses, which in turn can add to the transformation of specific spaces if deemed appropriate by a sufficient number of supporters. This returns the cycle to the beginning, where a need arises to change existing 'man-made' space. This process is influenced by a range of players involved in the production and management of space, which constantly influences the need/demand, idea, form, order and meaning in settlements.

Although this is a highly simplified abstraction of a complex process (Figure 2.1), it offers a way to conceptualise space and place in the urban design process and understand some of the aspects involved in the process of urban transformation. It can also be argued that this is a design process that is based on an individual designer's process, and may therefore prove difficult to apply to a complex social process where many actors are involved and no one designs the outcome. While this is true, a broader interpretation will allow for a wider application, where this framework becomes a representation of multiple processes where need/demand represent a multiple range of needs, and production and management represent a wide range of players involved in many actions that occur simultaneously and on a constant basis. If one takes such a viewpoint, the framework starts to offer a way to interpret spatial transformation and design in urban areas as part of a much broader socio-spatial process, building on previous research (Boyer 1995; Liggett and Perry 1995; Madanipour 1996; Short 1996; Graham and Healey 1999; Massey 1994 and 1999).

It also offers a framework for collecting evidence about the spatial transformation of a city and the various factors influencing urban complexity in this regard. The framework serves as a tool for identifying the main drivers (*need* and *idea*) that influence spatial changes in the form of urban fortification and privatisation; how this generally occurs in cities (*order* and *form*); followed by the impact and implications of these changes (*meaning* and *response*). It also helps to highlight the implications for the transformation of place and the conceptualisation of time within urban environments. The main drivers are directly linked to and influenced by aspects such as resources, capital flows, power struggles and development frameworks (*production and management*), as well as the broader *context.*

Applying the framework and mapping the process of gating in South Africa

This framework has been used to analyse urban transformation through gated communities in South Africa (Landman 2006). Research has indicated that gated communities[2] are a significant mechanism of urban transformation in post-apartheid South Africa, changing both neighbourhoods and cities in many ways. The proposed framework serves as a tool to highlight the actual changes occurring in practice in a specific context (place and time). The framework proved to be very useful in studying the complex urban transformations happening in South African neighbourhoods but is applicable to a broader range of socio-spatial dynamics.

Using the framework, existing urban spaces in South Africa are reinterpreted within the context of post-apartheid transition where a new spatial order has contributed to a range of growing insecurities, leading to a growing need for safer living places (for example, Beall et al. 2002; Durlington 2006; Landman 2007; and Lemanski 2006). Safety is also linked to a greater sense of community and identity and the need to bring the entire community together to support common values. This resulted in a desire for ideal places that are able to fulfil these needs and gave rise to ideas about how to address them. These ideas are based on abstract concepts translated to specific design models or planning trends that are often imported from abroad, combined by local variations or adaptation. In the case of enclosed neighbourhoods, residents combined concepts of Crime Prevention through Environmental Design (CPTED) with ideas to recreate old suburban neighbourhoods. Larger security estates incorporated a wider range of ideas, based on selective notions of the Garden City and Neighbourhood Unit, combined with a partial inclusion of elements from both New Urbanism and the Urban Villages movement.

These ideas started to guide the structural organisation and form through, for example, an organic layout inside a walled neighbourhood with green spaces and a central clubhouse. In the case of South Africa's retro-fitted enclosed neighbourhoods, interventions were restricted to what is possible in an existing neighbourhood within the policy framework of the local authority and national laws. These included interventions such as fencing in the neighbourhood, closing roads through gates or booms, and in some cases upgrading existing neighbourhood spaces such as the park or providing additional infrastructure. Security estates incorporate all the physical elements related to security including perimeter walls and fences, often electrified and enhanced with surveillance cameras or alarm systems, access control measures such as booms and/or gates combined with visitor check-in-booths, and/or PIN numbers or access disc for residents.

The process of physical transformation was influenced by a range of actors involved in the production and management of the physical space. This process differs between the two types of gated communities in South Africa. Enclosed neighbourhoods are established through a process of community mobilisation led by a homeowners' association (HOA) in municipalities that makes provision for road closures for security purposes. The HOA is also responsible for the ongoing management of the neighbourhood. Large security estates are established by private developers, usually on private undeveloped land on the urban periphery. After an initial phase, the neighbourhood management is usually transferred from the developers to the newly established residents' association in the estate, which is responsible for the daily management of the estate.

Space and place also encompass meaning. This has been clearly demonstrated through the manifestation of gated communities in South Africa. The different levels of meaning relate to access and use of space; behaviour in space; and management and control of space, which in turn results in a range of responses from different groups. It is perceived by many outside South Africa's gated communities that those inside are getting or enhancing control over local resources inside, including former public spaces such as roads, sidewalks and parks, as well as a range of facilities and amenities. The tensions between the different opposing groups (those in favour and against) have become so heated that local authorities have often been forced to intervene. In this way spatial changes begin to influence social behaviour and in turn have a direct implication for institutional practices.

Changing physical spaces and places can therefore elicit a number of responses from different actors based on contested meanings, which in turn can add to the transformation of these spaces if deemed appropriate by a sufficient number of supporters. Over time, this re-establishes a need to change existing physical space, for example to remove certain gates and fences that are illegal, as occurred in the City of Johannesburg. This clearly illustrates the need to understand the institutional implications of and influence on the spread of gated communities.

Gated communities and institutional fragmentation

Gated communities represent a reordering of micro society and of space, as we have argued. They also represent a reordering of the rules, norms and customs that affect the allocation of shared civic goods and services. Two extreme cases are to be found in the gated community literature that focuses on institutional, organisational and governance issues. First are the antagonists who make the assumption that the governance of shared space and facilities in cities should be organised by public government. Second, there are the extreme protagonists who argue that gated communities do not go far enough in privatising territorial governance. For the latter, the ideal organisation of the city is through multi-tenant estates in which rental contracts between user and landlord permit the landlord to supply collective goods at efficient levels. As in a shopping mall, science park or hotel, rents finance neighbourhood goods, which can be supplied to meet the specific needs of renters (and their customers if the renter is a business). Sympathy for the *city as a hotel* thesis can be found in the libertarian writings of Spencer Heath MacCallum (1970) and Fred Foldvary (1994), for example. Less extreme protagonists are those who see in gated communities and the homeowners' associations and condominium institutions that organise them, the possibility of neighbourhood-level contractual territorial government (Nelson 1999; Webster 2001, 2002 and 2005; Lee and Webster 2006; Chen and Webster 2005; Le Goix and Webster 2008). Between the extremes is a range of views about the relative efficiency of contractual versus political governance.

Actually, this polarisation is not always sufficiently distinguished: gated communities are contractually governed in the sense that title deeds to homes confer rights and responsibilities in respect of shared and private goods but they are also politically governed in that the homeowner representatives are voted into office. When spatial scale is added as a dimension of analysis, it turns out that the differences between contractual and political might not be that great. Eric Charmes' (2007) work on French peri-urban communes demonstrates this nicely, showing that these small communities of up to about 5,000 people have many similarities to North American-style gated communities. The similarities are due to the tight articulation between finance and spending. More formal analysis of the efficiency of geographical scale and shared consumption arrangements can be found in the economic theory of clubs (see Cornes and Sandler 1996 for a review, Webster and Lai 2003 for an urban analysis and Fisch 1975 for an early statement of the problem). Evan McKenzie considers the efficiency of gated community governance from the perspective of legal and administrative costs (McKenzie 1994, 2005). From his evidence it seems possible that the organisational transaction costs of contractual neighbourhood government (particularly litigation costs) could turn out to be greater than those of public government. Another important consideration in addressing the real costs of contractual governance is the financial sustainability of various private neighbourhood institutions. Many gated communities built in the 1960s and 1970s around the world are now suffering from chronic underinvestment and ageing infrastructure (Webster and Le Goix 2005; Le Goix 2004; Le Goix and Webster 2008). This is less of a fundamental finding, however since modern contracts make better provision for reinvestment funds (often under the influence of government regulation).

So there is a range of issues to be considered when examining the nature of gated community governance institutions. In the remainder of this section we focus on the relationship between scale and function of an institution and inevitably find ourselves talking about the capacity of an institution for taking collective action on what might broadly be called neighbourhood problems.

Consider Table 2.1, which shows scale of collective consumption against scale of collective action. For simplicity, building means a construction on a single plot shared by several individuals or households; neighbourhood means several buildings in a physical arrangement that fosters a degree of shared consumption such as along a street; city a single settlement within which individuals and households share certain metropolitan-level goods and services; nation means the territory within which citizens of many cities share certain goods and services, costs and benefits.

We shall not elaborate on each cell but the reader is invited to extrapolate the logic. For ease of exposition we take examples from China's so called *urban villages* – villages swallowed up by rapidly expanding cities but still in the collective ownership of villagers. These are not gated communities in the popularly understood sense but they are an intriguing form of co-owned neighbourhood. They are, in fact, nearly all physically and institutionally enclosed areas. Most have a clearly defined entrance, typically marked symbolically by an arch; and some have boom gates. Institutionally, they are micro-territories, co-owned by ex-agriculturalist turned urban workers and landlords. Under Chinese land law, villagers retain a collective and inalienable form of rights over the land on which their village houses were built. When village farmland is expropriated and swallowed up by the city, surrounding the village housing land on all sides by high-density commodity housing, industry and commerce, villagers uniformly capitalise on urban land values by redeveloping their housing land. They do so at very high density and low quality, and rent homes to migrant workers and in some parts of the country, rent factories to manufacturers. The examples in the following are drawn from a UK Economic and Social Research Council/Department

Table 2.1 Institutional designs at different spatial scales

	Scale of collective action					
	Room	Floor	Building	Neighbourhood	City	Nation
Scale of collective consumption						
Room	A	B	C	D	E	F
Floor	G	H	I	J	K	L
Building	M	N	O	P	Q	R
Neighbourhood	S	T	U	V	W	X
City	Y	Z	AA	AB	AC	AD
Nation	AE	AF	AG	AH	AI	AJ

For International Development funded research project on property rights and poverty in China (He S. et al. 2009).

Cell A

Imagine a 10-floor building shared by 50 migrant worker families with an average of 15 individuals per floor separated only by makeshift hardboard partitions. On one floor a family of three live in a space of about 6 m², which gives room for only one bed. They bed-share so the teenage son sleeps at night while the parents are out working. Notwithstanding family problems in such conditions, the informal time-sharing rule (institution) governed by household decision-making works efficiently. Collective action and consumption occur at the same scale and the institution of bed time-sharing can be said to be efficient (although the living conditions of this family are quite unacceptable by most people's values).

Cell E

On each floor there is a small space for cooking. Fumes of vaporised cooking oil pervade the partitioned bedrooms and make it difficult to talk and breathe and lead to health problems that keep the father off work and interrupt the bed-sharing arrangement. The son has to be sent home to the village for a while and misses out on months of schooling. The families try to come to an agreement about cooking times but it does not work. The landlady (once a poor farmer herself) has no incentive to intervene – there are plenty of other migrants queuing up to rent. The only organisation with the power to do anything about it is the Municipal Authority, which has the legislative power to enact public health by-laws. Because of the scale of the problem – and the public health costs associated – it is likely that the city will enact such a law in time and the landlord will be forced to install a properly insulated kitchen on every other floor. In this case there are no solutions at B, C or D scales and E seems the most efficient institutional design in the circumstances.

Cell P

An entrepreneur from another city (also a migrant worker but one with access to some capital) organises the enclosure and joining up of several of the better-built village 'houses' (actually 6-storey buildings) by some simple building work and some internal renovations that include installing a roof garden, a 24-hour-manned security gate, painting and fixtures. The village collective, which also operates as a joint stock company, creates another company owned by a subset of villagers to own the new apartment complex – which is rented to students and young professionals attracted to the mixed, quirky bohemian life in the urban village. Environmental and security problems in the streets between the houses are overcome. The informal HOA is well organised and liaises with the owners over the standard and management of the services provided. By and large, the institution works because the numbers are not large and there is a contractual (rental) relationship that gives a degree of power to the renters, who will move out if the company fails to organise neighbourhood goods sufficiently well.

Cell V

Villagers' fields and a large historical refuse dump within the village boundary are expropriated by the municipal government at the time of urban encroachment. The refuse dump is sold cheaply to an entrepreneur who invests a sizeable amount of equity in reclamation, building an elite upper-end gated community of villas situated around a lake stocked with rare water birds. The management of the lake and wildlife – together with a clubhouse with a 50-metre outdoor heated pool and a boutique-style hotel for visitors is under the control of the developer who holds the leasehold of the entire site and lives there too. Because this is an owner-developer-occupied estate and because the population is not large (500 households), the HOA tends to work quite well. Its members are voted into office and sit on a committee that advises on neighbourhood management decisions. The decisions are taken by the owner-developer, however, who owns the company that has title to all shared facilities. Because the owner-developer and the individual homeowners would all lose out if the collective organisation collapses, there is a shared interest that keeps the neighbourhood well managed.

Cell W

Next to this estate is a much larger one in which the development company also owns the collective spaces, which are managed by a property management company (PMC) forced upon it and the homeowners by the municipal government. The PMC is a private company that had its origins as the estates division of a large state-owned enterprise and has close links to the municipal government. There are many complaints among residents. If the estate were in the USA, they would have taken court action but in China, HOAs are not permitted to incorporate in law (which would mean territorial democracy by the back door). So there exists a serious institutional gap. There are conflicts over security, traffic, road maintenance, use of parks, poor landscaping and crowded sports facilities – all conflicts (collective consumption problems) at a neighbourhood scale (neighbourhood externalities). But there is no effective institution of collective action at the same scale. As a result, the housing bureau of the Municipal Government has stepped in to try and manage the problems. The administrative solution is problematic because the housing bureau is not the owner and does not have sufficient powers to coerce the owner on many of the matters. It is also overwhelmed by complainants who are doubly frustrated by the lack of solutions and the wait. The city-scale publicly organised institution is ill-fitted to the scale of the problem. Cell V is, in fact, the most common institutional design in cities. Municipal agencies impose administrative solutions to neighbourhood collective consumption problems. It can work if there is an abundance of municipal funds. But when municipal funds are constrained, Cell V institutions become a zero-sum game. One neighbourhood benefits only at the expense of others.

The framework we illustrate here can be used positively or normatively. Positively, it can be taken to a city to investigate and record the morphology of urban governance. It may be used generally or to analyse particular public goods and collective consumption problems. This can help identify significant institutional gaps. Normatively, it may be used to assess the adequacy (efficiency, fairness, practicality) of particular institutional arrangements. Gated communities are not institutionally homogeneous. In the UK, for example, they may be organised around a company that owns common areas, in which homeowners own shares; a community land trust that holds the common areas in pursuit of legally specified community objectives; or a Commonhold Association. The framework in Table 2.1 suggests that there is an efficient spatial scale for any particular collectively consumed good – closely allied to the ideas of subsidiarity and fiscal federalism. This is true in theory if the costs of organising supply are not considered. When they are, then it becomes clear that certain public goods should probably be supplied together – for economies of scale and complementarities. Certain civic goods should undoubtedly be organised by city-level governments. It is useful to ponder, however, which may be organised through neighbourhood-scale institutions.

A framework for interdisciplinary analysis of enclosed communities

Interdisciplinary research moves beyond multidisciplinary studies – projects where each researcher or specialist remains within his or her discipline and contributes using disciplinary concepts and methods (Lawrence 2004). This can happen without any intention toward interdisciplinary problem-solving. It is research that is thematically orientated, contributing to the theme from different perspectives. Our three research cameos in the preceding sections of the chapter illustrate this approach. In multidisciplinary work, no collaboration is necessary between the different participating research programmes or disciplines (Balsiger 2004). Interdisciplinary research, on the other hand, requires a common framework and analytical methods based on shared terminology, mental images and common goals.

This section aims to draw together the three otherwise parallel but untriangulated perspectives relating to social, spatial and institutional fragmentation in order to explore an interdisciplinary approach.

We assume that research is all about finding answers to questions. It should be possible for a researcher to be explicit about the unknown that she or he is investigating. What do we not yet know about gated communities? In a field of enquiry dominated by qualitative research, that can be surprisingly difficult to answer. When most knowledge is generated by case studies, generalisation is not easy and we run the risk of forever rediscovering. One of the motivations behind this chapter is to encourage the creation and discussion of explicit empirical research frameworks so that we can more easily undertake interdisciplinary analysis and better connect our multidisciplinary studies. We would like it if case study comparisons and generalisation more powerfully came together to identify what it is we do and do not know about gated communities.

We model the three urban dynamic processes (social, spatial and institutional fragmentation) as an initial state (I), an end state (E) and a mechanism of change (M):

Social fragmentation {v1..vn}: Initial State X → Mechanism of change ? → End state Y

Spatial fragmentation {v1..vn}: Initial State ? → Mechanism of change Z → End state Y

Institu. fragmentation {v1..vn}: Initial State X → Mechanism of change Z → End state ?

X and Y represent observations, variables or concepts. Z represents a hypothetical causal mechanism. All I:M:E models are value-specific, represented by a vector of values {v1..vn}. Question marks represent an unknown initial state, end state or causal mechanism, making it clear that empirical research is very difficult if there is more than one unknown. An empirical research project on gated communities with more than one question mark in the I:M:E model of its design is likely to fail to generate any useful knowledge.

Take our first theme – social fragmentation. Research may start from observations about the initial and end state of an urban dynamic process. For example, we may know from previous studies that wealthy gated communities located next to poor communities tend to foster a greater sense of social barrier than wealthy communities located at a distance from poor communities (I = distance between communities; E = sense of social barrier; ? = unknown mechanisms). Research into this relationship may pose a number of hypotheses {Z1..Zn}. Z1 might be, for example, that day-to-day contact through the employment by rich families of domestic services supplied by workers from the poor neighbouring community tends to reinforce awareness of class division. This is a strong research model with two *knowns* {X:?:Y}

As a second example, we may have observed a mechanism of change that we suspect has an effect of some interest. For example, we suspect that the use of separate entrances for residents and for domestic service employees leads to a change in attitude and/or behaviour. We may have a general interest in this without much idea of what effects this mechanism causes, in which case there may be two question marks in the research model: {?:Z1:?}. If we make certain assumptions about the relationship between Z and I then the research model is strengthened. For example, we may suppose that, or know with a degree of certainty from previous studies, that separate access actually reduces ill-feeling between residents and domestic staff because it reduces ambiguity in social relations. It may also have been found that friendships between residents and their staff and charitable giving by residents of the poor neighbouring community increases with separate access as a result of reduced social transactional ambiguity. The model is more easily resolved: {I:Z1:?}, where I = measured degree of friendship (or existence/non-existence of friendship relations) or local charitable giving in a gated community with no separate access; Z1 = the hypothesis about separate access; and ? = the unknown (to be assessed) degree of friendship (or level of local charitable giving) in a gated community with separate access. This might be a before–after study if a separate access is installed to an existing estate or a cross-sectional, possibly matched-pairs study between communities matched by socio-economic and other profiles. The study will cast light on specific socio-spatial relationships, i.e. the effect of physical measures on social relations, and also on the impact of spatial fragmentation on social fragmentation.

If we also suspect that separate access in gated communities with unintended segregation (a positive view of outsiders but with separate access rationalised as a security device) has a different effect to separate access in communities with intended segregation (defined earlier in the chapter), then we may enrich the research framework with two models: {I:Z1:?}, {I:Z2:?}, where Z1 is a hypothesis for gated communities characterised as displaying intended urban social segregation and Z2, a hypothesis for those displaying unintended segregation. Or expressing the model differently: {I:Z3:?}, where I = intended/unintended classification, Z3 = hypothesis relating effect of separate access to intentionality, and ? = difference in change in friendship (or giving) between intended and unintended communities.

Take our second and third themes – spatial and institutional fragmentation. A common allegation about gated communities in the critical academic press is that the fragmentation of territorial governance into contractual neighbourhood units gives rise to unacceptable spillover effects. The greater efficiency in neighbourhood security gives rise to crime displacement. Road closure leads to traffic displacement and longer walking times. A quantitative research design to investigate this could be a before–after study with the following structure: {I:Z:?}, where I = a measure of crime variance across urban neighbourhoods before the implementation of a pro-gating law or policy; Z = an hypothesis about dispersal effects resulting from the new institution; and ? = crime variance after the passing of the law.

For a second example, imagine that we could classify a set of collectively consumed civic goods and services in terms of the scale of co-consumption, following Table 2.1 and then classify the scale at which collective action in a particular city is taken to address congestion/conflict arising from that co-consumption. We could then record evidence of congestion/conflict by good/service and look at the correlation between this pattern and the scale mismatch. This suggests a study structure: {I:?:E}, where I = some measured pattern of institutional spatial mismatch; E = a corresponding pattern of indicators of collective consumption problems; and ? = an unknown set of mechanisms by which the two are related. Again, this would start to show the interconnected relationship between physical modification and consequent fragmentation (through neighbourhood enclosure) and possible institutional fragmentation.

We finish with a framework pitched at a higher level of analysis. Returning to the issue of complexity, imagine a 3 × 3 table with the three types of fragmentation (social, spatial, institutional) in the rows and columns. The diagonals represent research models {I,Z,E} that investigate the ways in which social fragmentation causes more social fragmentation; spatial fragmentation causes more spatial fragmentation; and institutional fragmentation causes more institutional fragmentation. All urban processes are complex and take place over time with a good deal of feedback effects, non-linearity and unpredictability. It may be that the social fragmentation produced by gated communities, possible in part due to spatial fragmentation, leads to more social fragmentation in time – for example, by teaching children who live within gates patterns of alienation or misunderstanding in respect of other groups (social fragmentation → social fragmentation, {I,Z,?}). The non-diagonals of the 3 × 3 matrix represent other dimensions of complex urban dynamics. We might conjecture or observe that gated communities bring groups of diverse incomes in closer spatial proximity with some positive effects (spatial fragmentation → social fragmentation [+ve]), {I,?,E}); thus reversing the effects of social fragmentation. We may view the rise of private neighbourhood governance as a necessary response to an institutional gap in the twentieth century model of municipal government, brought about through social fragmentation, increasing heterogeneity and rising competition for scarce tax-funded civic goods and services (social fragmentation → institutional fragmentation, {I,?,E}). The six non-diagonal cells in the 3 × 3 matrix represent six different models for enquiry on the relationships of cause and effects between the three types of fragmentation:

1. Social fragmentation → Spatial fragmentation → Institutional fragmentation
2. Social fragmentation → Institutional fragmentation → Spatial fragmentation

3. Institutional fragmentation → Spatial fragmentation → Social fragmentation
4. Institutional fragmentation → Social fragmentation → Spatial fragmentation
5. Spatial fragmentation → Institutional fragmentation → Social fragmentation
6. Spatial fragmentation → Social fragmentation → Institutional fragmentation

Plausible research models may be constructed under each of these including, by way of illustration:

1. It is pre-existing and underlying social fragmentation processes that lead to gated enclaves, and institutional fragmentation (the emergence of condominium and other co-ownership laws and related legislation and policy) follows to reduce the social and private costs of contractual urban governance.
2. Fractures in an increasingly heterogeneous society give rise to informal rules and practices such as those that create ethnically based spontaneous housing market areas. This may in turn encourage gating.
3. A breakdown in the civic social contract because of scale and fiscal poverty might induce institutional fragmentation – such as a move to greater localisation in governance. This may induce new local patterns of infrastructure investment and differentiation that leads to social sorting.
4. A change in property law that makes it easier to develop shared-ownership housing estates might lead to greater social sorting. The new pattern of political power resulting leads to increasing differentiation in infrastructure between neighbourhoods and hence more enhanced physical differentiation.
5. An urban redevelopment scheme creates scope for a municipal government to experiment with new forms of collective ownership and micro-governance. Over time this leads to social differentiation, facilitated by a range of spatial measures.
6. A redevelopment scheme socially engineers the city, significantly changing the socio-spatial geography. This gives rise to new patterns of conflict and new rules and new physical or design modifications to accommodate these.

These are, of course, merely different starting points in what are inevitably and always complex webs of cause and effect.

Conclusion

We are arguing in this chapter that clearer frameworks for empirical investigations are needed, not only for specific disciplines, but also for providing an interdisciplinary and international comparative perspective of gated communities. We illustrate the multidisciplinary approach to gated communities with independent research frameworks for investigating social, spatial, and institutional urban fragmentation. Explicitly laid out frameworks such as these will facilitate in-depth single and multidisciplinary studies. They are also the first step towards building a more consciously interdisciplinary approach.

Overcoming the limitations and barriers of single methodological approaches is one of the big challenges of empirical urban research. After more than ten years of academic discussion and research on the nature, causes and consequences of gated communities, we suggest that there is a need to advance the understanding of the subject by being clearer about the knowledge that all this activity has generated. This understanding is likely to be deeper and clearer if robust methodological frameworks are developed and combined. Interdisciplinarity allows for a better grasp of the complexity of urban dynamics where the social, spatial and institutional realms interrelate. This chapter illustrates how three methodological frameworks that each give partial insight can be combined within higher level frameworks. These transcend the qualitative and quantitative divide and provide platforms for triangulated mixed-method studies that cross disciplinary boundaries.

There are always connections to be made between the viewpoints and approaches of researchers from different traditions and perspectives. Every research team necessarily fixes its concepts, variables, values and assumptions about mechanisms, spatial scale of analysis and temporal scale. And within its chosen frames of reference, it fixes its *knowns* and *unknowns*. The community of researchers investigating the dynamics of gated communities, private urban governance, urban territorial governance and socio-spatial development would benefit from making these frameworks more explicit so that multi- and interdisciplinary mapping is made easier and so we can be more certain about what we do know and what we do not know about this fascinating urban phenomenon.

Acknowledgements

The authors wish to acknowledge the support of the UK Economic and Social Research Council and the UK Government Department for International Development for support in the Chinese work reported in this chapter (project RES-167-25-0448, The Development of Migrant Villages under China's Rapid Urbanization: Implications for Poverty and Slum Policies; and project RES-167-25-0005, Urban poverty and property right changes).

Notes

1   For more information on these conferences, visit the Private Urban Governance and Gated Communities Research Network website (www.gated-communities. de).

2   Gated communities in South Africa are broadly categorised into two groups, namely enclosed neighbourhoods and security villages. *Enclosed neighbourhoods* refer to existing neighbourhoods that have been fenced or walled in and where access is controlled or prohibited by means of gates or booms that have been erected across existing public roads. *Large security estates* are private developments where the entire area is developed by a private developer. These areas/buildings are physically walled or fenced off and usually have a security gate or controlled access point, with or without a security guard. They offer an entire lifestyle package, including a secure environment; a range of services (garden services, refuse removal, etc.); and a variety of facilities and amenities such as golf courses, squash courts, cycle routes, hiking routes, equestrian routes and water activities.

References

Álvarez, M. J. (2005) "Golden ghettos: gated communities and class residential segregation in Montevideo, Uruguay." Rep. No. 02/2005, Research and Training Network Urban Europe.

Balsiger, P. W. (2004) "Supradisciplinary research practices: history, objectives and rationale," *Futures*, 36: 407–421.

Beall, J., Crankshaw, O. and Parnell, S. (2002) *Uniting a Divided City: Governance and Social Exclusion in Johannesburg*, London: Earthscan.

Blakely, E. J. and Snyder, M. G. (1997) *Fortress America: Gated Communities in the United States*, Washington, D.C.: Brookings Institution Press.

Boyer, M. C. (1995) "The great frame-up: fantastic appearances in contemporary spatial politics," in Liggett, H. and Perry, D. C. (eds), *Critical Explorations in Social/Spatial Theory*, Thousand Oaks, CA: Sage Publications.

Bruce, A., Lyall, C., Tait, J. and Williams, R. (2004) "Interdisciplinary integration in Europe: the case of the Fifth Framework programme," *Futures*, 36: 357–470.

Caldeira, T. P. R. (2000) *City of Walls: Crime, Segregation and Citizenship in São Paulo*, Berkeley: University of California Press.

Campos, D. and García, C. (2004) "Identidad y sociabilidad en las nuevas comunidades enrejadas: observando la construcción de la distancia social en Huechuraba," in Cáceres, G. and Sabatini, F. (eds), *Barrios Cerrados en Santiago de Chile: entre la exclusión y la integración residencial*, pp. 179–205, Pontificia Universidad Católica de Chile-Instituto de Geografía and Lincoln Institute of Land Policy, Santiago de Chile.

Castell, D. B. (1997) "An investigation into inward-looking residential developments in London." Unpublished MPhil Thesis in Town Planning, University College London.

Charmes, E. (2007) "Carte scolaire et clubbisation des petites communes periurbaines," *Sociétés Contemporaines*, 67: 67–94.

Chen, C. Y. and Webster, C. J. (2005) "Homeowners' associations, collective action and rent seeking," *Housing Studies* 20(2): 205–220.

Cornes, R. and Sandler, T. (1996) *The Theory of Externalities, Public Goods and Club Goods*, 2nd edn, Cambridge: CUP.

Davis, M. (1992) "Fortress Los Angeles: the militarization of urban space," in Sorkin,

M. (ed.), *Variations on a Theme Park: Scenes from the New American City and the End of Public Space,* pp. 154–180, New York: Hill and Wang.

Dear, M. (1986) "Post-modernism and planning," *Environment and Planning D,* 7(4): 449–462.

Du Plessis, C. (2008) "A conceptual framework for understanding social-ecological systems," in Burns, M. and Weaver, A. (eds), *Exploring Sustainability Science: A Southern African Perspective,* Cape Town: Sun Press.

Durlington, M. (2006) "Race, space and place in suburban Durban: an ethnographic assessment of gated community environments and residents," *GeoJournal,* 66: 147–160.

Ellin, N. (1997) *Post Modern Urbanism,* New York: Princeton University Press.

Ellin, N. (2001) "Thresholds of fear: embracing the urban shadow," *Urban Studies,* 38 (5–6): 869–883.

Fernández-Prajoux, V. (2005) "Urban fragmentation and urban development of gated communities in two boroughs of the Great Santiago: Huechuraba and Peñalolen." Paper delivered at the conference "Territory, Control and Enclosure: The Ecology of Urban Fragmentation," Pretoria, 1–2 March 2005.

Fisch, O. (1975) "Optimal city size and the economic theory of clubs and exclusionary zoning," *Public Choice,* 24(1): 59–70.

Flusty, S. (1995) "Building paranoia," in Ellin, N. (ed.) *Architecture of Fear,* New York: Princeton Architectural Press.

Flusty, S. (2000) "Thrashing downtown: play as resistance to the spatial and representational regulation of Los Angeles," *Cities,* 17 (2): 149–158.

Foldvary, F. (1994) *Public Goods and Private Communities,* Aldershot: Edward Elgar.

Frantz, K. (2000–1) "Gated communities in the USA: a new trend in urban development," *Espace, Populations, Sociétés,* 2000–1: 101–113.

Giglia, A. (2003) "Gated communities in Mexico City." Paper delivered at the conference "Gated Communities: Building Social Division or Safer Communities, Glasgow," 18–19 September.

Graham, S. and Healy, P. (1999) "Relational concepts of space and place: issues for planning theory and practice," *European Planning Studies,* 7 (5): 623–646.

Graham, S. and Marvin, S. (2001) *Splintering Urbanism: Network Infrastructures, Technological Mobilities and the Urban Condition,* London: Routledge.

Grant, J. and Mittelsteadt, L. (2004) "Types of gated communities," *Environment and Planning B: Planning and Design,* 31 (6): 913–930.

He, S., Liu, Y., Webster, C. and Wu, F. (2009) "Property rights redistribution, entitlement failure and the impoverishment of landless peasants in China," *Urban Studies* (in press).

Landman, K. (2006) "An exploration of urban transformation in post-apartheid South Africa through gated communities, with a focus on its relation to crime and impact on socio-spatial integration." Unpublished PhD thesis, University of Newcastle upon Tyne, UK.

Landman, K. (2007) "Exploring the impact of gated communities on social and spatial justice and its relation to restorative justice and peace-building in South Africa," *Acta Juridica,* 07: 134–155.

Lawrence, R. J. (2004) "Housing and health: from interdisciplinary principles to transdisciplinary research and practice," *Futures* 36: 487–502.

Lee, S. and Webster, C. J. (2006) "Enclosure of the urban commons," *GeoJournal*, 66 (1–2): 27–42.

Le Goix, R. (2004) "Gated communities: sprawl and social segregation in Southern California," *Housing Studies*, 20(2): 323–344.

Le Goix, R. and Webster, C. J. (2008) "Gated communities," *Geography Compass*, 2(4): 1189–1214.

Lemanski, C. (2006) "Residential responses to fear (of crime plus) in two Cape Town suburbs: implications for the post-apartheid city," *Journal of International Development*, 18 (6): 787–802.

Liggett, H. and Perry, D. C. (eds) (1995) *Critical Explorations in Social/Spatial Theory*, Thousand Oaks, CA: Sage Publications.

Low, S. (2003) *Behind the Gates*, New York and London: Routledge.

MacCallum, S. (1970) *The Art of Community*, Menlo Park, CA: Institute for Humane Studies.

Madanipour, A. (1996) *Design of Urban Space: An Inquiry into a Socio-Spatial Process*, Chichester: John Wiley.

Manzi, T. and Smith-Bowers, B. (2005) "Gated communities as club goods: segregation or social cohesion?," *Housing Studies*, 20 (2): 345–359.

Marcuse, P. (1995) "Not chaos, but walls: postmodernism and the partitioned city," in Watson, S. and Gibson, K. (eds), *Postmodern Cities and Spaces*, Oxford: Blackwell.

Marcuse, P. (2001) "Enclaves yes, ghettos no: segregation and the state." Paper prepared for the International Seminar on Segregation in the City, Lincoln Institute of Land Policy, Cambridge USA, 25–28 July 2001.

Massey, D. (1994) *Space, Place and Gender*, Cambridge: Polity Press.

Massey, D. (1999) "City life: proximity and difference," in Massey, D., Allen, D. and Pile, S. (eds), *City Worlds*, London: Routledge in association with the Open University Press.

McKenzie, E. (1994) *Privatopia: Home Owner Associations and the Rise of Private Residential Government*, Newhaven CT: Yale UP.

McKenzie, E. (2005) "Planning through residential clubs: home owners' associations," *Economic Affairs*, 25(4): 28–31.

Nelson, R. (1999) "Privatising the neighbourhood: a proposal to replace zoning with private collective property rights to existing neighbourhoods," *George Mason Law Review*, 7(4): 827–880.

Roitman, S. (2004) "Urbanizaciones cerradas: estado de la cuestión hoy y propuesta teórica," *Revista de Geografía Norte Grande*, 32: 5–19.

Roitman, S. (2007) "Social practices and viewpoints: a conceptual framework for the analysis of urban social group segregation and gated communities." Paper presented at the 4th International Conference of the Research Network Private Urban Governance & Gated Communities, 6–8 June 2007, Paris.

Roitman, S. (2008) "Urban social group segregation: a gated community in Mendoza, Argentina." Unpublished PhD thesis, Development Planning Unit, The Bartlett, University College London.

Sabatini, F. and Cáceres, G. (2004) "Los barrios cerrados y la ruptura del patrón tradicional de segregación en las ciudades latinoamericanas: el caso de Santiago de Chile," in Cáceres, G. and Sabatini, F. (eds) *Barrios Cerrados en Santiago de Chile: entre la exclusión y la integración residencial*, pp. 9–43, Pontificia Universidad Católica de Chile, Instituto de Geografía and Lincoln Institute of Land Policy, Santiago de Chile.

Sabatini, F. and Salcedo, R. (2005) "Gated communities and the poor in Santiago, Chile: functional and symbolic integration in a context of aggressive capitalist colonization of lower class areas." Paper presented at the conference "Territory, Control and Enclosure: The Ecology of Urban Fragmentation," Pretoria, South Africa, 28 February–3 March 2005.

Salcedo, R. and Torres, Á. (2004) "Gated communities: walls or frontier?" *International Journal of Urban and Regional Research*, 28(1): 27–44.

Sellés, F. and Stambuk, L. (2004) "Asentamiento de grupos medios-altos en sectores populares bajo la forma de comunidades enrejadas: una mirada externa," in Cáceres, G. and Sabatini, F. (eds), *Barrios Cerrados en Santiago de Chile: entre la exclusión y la integración residencial*, pp. 229–255, Pontificia Universidad Católica de Chile, Instituto de Geografía and Lincoln Institute of Land Policy, Santiago de Chile.

Sennett, R. (1995) "The search for a place in the world," in Ellin, N. (ed.) *Architecture of Fear*, New York: Princeton Architectural Press.

Sheinbaum, D. (2005) "Gated communities from the inside: a case study in Mexico City." Paper delivered at the conference "Territory, Control and Enclosure: The Ecology of Urban Fragmentation," Pretoria, 1–2 March 2005.

Short, J. R. (1996) *The Urban Order: An Introduction to Cities, Culture and Power*, Cambridge, MA: Blackwell Publishers.

Soja, E. (1989) "Postmodern geographies and the critique of historicism," in Jones, J. P. III, Natter, W. and Schatzki, T. (eds), *Postmodern Contentions: Epochs, Politics, Space*, pp. 113–136, New York: The Guilford Press.

Svampa, M. (2001) *Los Que Ganaron. La vida en los countries y barrios privados*, Biblos: Buenos Aires.

Tiesdell, S. and Oc, T. (1998) "Beyond 'fortress' and 'panoptic' cities – towards a safer urban public realm," *Environment and Planning B: Planning and Design*, 25: 639–655.

Webster, C. J. (2001) "Gated cities of tomorrow," *Town Planning Review*, 72(2): 149–170.

Webster, C. J. (2002) "Property rights and the public realm: gates, green belts and gemeinschaft," *Environment and Planning B*, 29(3): 397–412.

Webster, C. J. (2005) "Editorial: Diversifying the institutions of local planning," *Economic Affairs*, 25(4): pp. 4–10.

Webster, C. J. and Lai, L. W. C. (2003) *Property Rights, Planning and Markets: Managing Spontaneous Cities*, Cheltenham, UK and Northampton MA, USA: Edward Elgar.

Webster, C. J. and Le Goix, R. (2005) "Planning by commonhold," *Economic Affairs* 25(4): 19–23.

Chapter 3

Defending illicit livelihoods

Youth resistance in Harare's contested spaces

Amin Y. Kamete

In response to incessant assaults by the Zimbabwean state's repressive apparatus, spearheaded by the urban planning system, youth in Harare have
shifted their modes of resistance. The most successful forms of resistance
appear to be those that are multifarious, non-confrontational and less
docile. Empirical material from Harare suggests that the situation is best
understood in the framework of more sophisticated conceptualizations of
human agency and resistance than those proposed by modernist perspectives. It is shown that the resistance of the youth is about localized struggles that disrupt institutions and normalization. Arguing that the youth's
continued occupation of contested urban spaces is a result of abandoning
full-scale confrontation in favour of 'resistance at the margins', this chapter
concludes that a postmodernist analysis best explains the youths' modes of
resistance.

Introduction

Many youth in urban Zimbabwe have taken up precarious modes of survival.
The adoption and continuance of these practices entail defying official land use
planning controls and resisting the often-violent 'restoration of order' by the
authorities. The antagonistic relationship between youth and the planning system is almost inevitable. The youths' (ab)use of urban places clashes with the
official urban order in two important ways. First, their occupation of the spaces
is illegal as it contravenes property laws. Second, through their provisional and

Submitted by the Association of African Planning Schools (AAPS).

Originally published in *International Journal of Urban and Regional Research*, 34:1: 55–75
(2010). © 2010 By kind permission of Wiley-Blackwell Publishing.

improvised survival practices they contravene land use controls, thereby directly defying the planning system which is legitimately tasked with "the public production of space" (Yiftachel and Huxley 2000: 907).

In democratic governance systems, these spatial conflicts can be resolved without overt violence. However, rarely is this the case when an authoritarian government embraces planning as the path to urban modernity (Swilling et al. 2002). From colonial times, the authorities in Zimbabwe have routinely deployed the planning system to create and restore order to urban spaces. In executing its mandate, planning is always backed by state agencies that specialize in repression, social control and penalization. Predictably, the 'restoration of order' is accomplished through violence: destruction of property, eviction and detention. But this bid to restore order has not been executed without resistance.

It is this resistance that is the focus of this chapter. Based on research conducted in 2004 and 2005, the chapter analyses how youth have resisted attempts by the authorities to eradicate livelihood practices that are incongruent with prescriptions of the planning system. This chapter is the second study to come out of this research. The first study examines the different ways in which the authorities have clamped down on "spatial unruliness" (Kamete 2008). The present chapter looks at the other side of the contestation: the *reaction* of the youth to the clampdown. By zeroing in on resistance, the chapter helps us appreciate the agency of youth, not as a static phenomenon, but as expressed through localized mutating practices that are no more than by-products of trial and error, always powered by the desire to defend threatened livelihood practices.

In the next section I discuss planning, violence and resistance in urban space. This is followed by an outline of the research on which the chapter is based, which leads to an examination of the three modes of youth resistance. I also discuss *Operation Murambatsvina/Restore Order*. I conclude by reflecting on planning, power and resistance.

Planning, violence and resistance in urban space

The urban planning system – which consists of discourses, ideologies, techniques and technologies for ordering urban space – is an integral component of what Pile (1997: 3) terms "the spatial technologies of domination [that] need to continually resolve specific spatial problems." The hegemonic production of space is not without challenge. Contestation erupts at that point when planning seeks "to connect [its] forms of knowledge with forms of action in the public domain" (Friedmann 1993: 482). Sometimes planning relies on violence to maintain its mastery of space. Epistemic violence is exercised to suppress contesting knowledges (see Spivak 1988; Castro-Gómez 2002); physical violence is deployed to eliminate deviant spatial practices. It is this 'double violence' that largely explains the tenuous hegemony of planning systems in some African cities.

Lefebvre (1991: 365) asserts that "spatial contradictions 'express' conflicts between socio-political interests and forces." Unsurprisingly, in the face of defiance, planning as "a state-directed activity" (Sandercock 1998: 45) can marshal state resources to forcefully uphold its version of the city (Kamete 2008). In African cities, the state's repressive manoeuvres against informality are a testimony to this fact (see Simone 2004). In Anglophone sub-Saharan Africa, modernist planning has incited and rationalized state repression against informal livelihood practices. This has been the case in Malawi (Jimu 2005), Zimbabwe (Potts 2006), Kenya (Robertson 1997), Lesotho (Setšabi 2006), Tanzania (Tripp 1997), Zambia (Tranberg-Hansen 2004), and Ghana (Clark 1994). However, by inscribing its conception of order on urban space and "designating only specific parcels of land for specific purposes" (Short 1996: 381), planning incites non-compliance, for "disorder only emerges out of a vision of order" (Mooney 1999: 55).

The confrontation between planning and non-compliant livelihood practices is best understood in the framework of Foucault's power analytics (Foucault 1998; see Cato 2002). As an incoherent amalgam of rationalities, discourses, ideologies, techniques and technologies, the planning system relies on the exercise of power in two important ways: in its designation of what counts as normal and acceptable, and in its reaction to non-compliance. Directing attention to the existence of points of resistance in all power relations, Foucault (1998: 95) insists that "where there is power, there is resistance . . . [T]hese points of resistance are present everywhere in the power network." Since planning relies on power to mould and dominate urban spaces, it can be argued that where there is planning there is resistance. Hence, despite the attempts of the totalizing project that is planning – even under totalitarian governance systems – totalization is never fully realized (cf. Amin and Thrift 2002: 108); there are always some points of resistance (Deleuze 1988: 44). As de Certeau (1984) has shown, 'the weak' always come up with tactical everyday practices that challenge dominant spatial orders.

Resistance is essentially "oppositional practice" (Aggleton and Whitty 1985: 62) that is, "collective, directed actions taken by a subordinate group towards a dominant one" (Raby 2005: 151). It is manifested through oppositional, conflicting or contesting actions, attitudes and behaviours (see Fernandes, 1988) against "oppressive or threatening situations, structural arrangements, and ideologies" (Abowitz 2000: 878). The conception of resistance as oppositional practice raises the need for it to "be located within broader theoretical positions on such issues as power and agency" (Raby 2005: 151). It is the locus and exercise of power that marks the attempt to dominate; and it is the agency of the dominated that delineates oppositional practice.

Raby usefully contrasts modernist and postmodernist conceptualizations and theorizations of resistance, explaining that the two perspectives fundamentally differ in terms of their "understandings of power and subjectivity" (Raby 2005: 152). The modernist perspective typically places these out in 'neat' packages. This is helped by a conception of power "in terms of a binary between dominance and submission [where] power is something that is possessed by the dominant group and wielded against the subordinate" (Raby 2005: 152). This power could be "infrastructural power" (power to) or "despotic power" (power over) (Mann 1986, in Tilly 1991: 594). Thus, when the powerless 'resist', their actions are directed against domination and oppression. Scott (1985; 1990) demonstrates that this resistance takes many forms: overt collective action; localized individual action; passive collective or individual action; and the appropriation of symbols of dominance and using them in new ways. Depending on the degree of expression of attitudes, behaviours and actions, resistance can be latent or manifest (Fernandes 1988: 176). In the modernist conceptualization, there is a clear source of agency. It is embodied in "a rational, prediscursive, internally coherent acting subject" (Raby 2005: 155). Consequently, resistance is easily noticeable because its source(s), target(s), and goal(s) are clear. It is the oppressed resisting the oppressors in order to "overcome or change some perceived effect of power" (Rose 2000: 385).

In planning, the modernist perspective provides for a straightforward analysis of resistance. Since "resistance opposes power" (Pile 1997: 1), it follows that the planning system, and ultimately the state, are the targets of resistance (Aggleton and Whitty 1985: 62). The purpose of the oppositional practice is to overcome spatial oppression resulting from the administrative and legal controls imposed by planning. Ultimate success would be the winning of concessions on the occupation and use of space that allow informal livelihood practices not only to continue existing, but also to find a place in planning policy that accommodates the special operational and locational characteristics of informality.

The modernist perspective obviously offers clear prospects of protest, defiance, revolution and emancipation. The fact that the dominated can specifically target those responsible for their plight implies that their situation can be transformed if the oppositional practice directed at the oppressors succeeds. One of the problems in this perspective is its failure to capture subtle resistance intended to bring short-term gains and that "never ventures to contest the formal definitions of hierarchy and power" (Scott 1985: 33). This is addressed in the postmodernist perspective.

The postmodernist view of resistance is complex – and messy. Here, "collective, organized and oppositional resistance" is shunned in favour of a concept that allows for the "complex flows of power relations, fragmented and constricted subjectivities and local and individualized activities" (Raby 2005: 161). The driving influence in this perspective is the Foucauldian conception of power, not as a possession, but as existing in "power relationships" and being present "everywhere" (Foucault 1998: 93). Foucault emphasizes the imbrication of resistance within power, arguing that resistance is nothing more than an integral component of power relations, and is in fact made possible by power. Resistance, therefore, "is never in a position of exteriority in relation to power" (Foucault 1998: 93).

This is because relations of power "envelop or run through even those points that serve as resistance" (Cato 2002: 1). In this regard, "there is no point outside of power, no escaping power" (Cato 2002: 10). Accordingly, as noted by Mills (2000: 162), "resistance and the subject who resists are fundamentally implicated within the relations of power they oppose." This explains why postmodernist researchers argue for the fragmentation and inconsistency of resistance against power. In addition to the fuzzy source of resistance, there is also the problem of agency and the target. Since all who are involved in power relations can exercise power, and since these "occupy multiple subjectivities or locations in relations of power" (Raby 2005: 162), it follows that what is opposed is not so clear. Not surprisingly, postmodernism has been labelled as having "no agenda for social justice" (Atkinson 2002: 74). Those who deploy Foucault's power analytics in conceptualizing resistance, notably Judith Butler (1997a; 1997b), have been accused of "foreclosing the possibility of resistance . . . through an anti-humanist negation of any possibility of agency" (Mills 2000: 269). Somehow reflecting this view, Allen fatalistically laments, "If power is believed to have no real whereabouts to speak of, no obvious landmarks to target, putting oneself against it is a tricky, if not hopeless, option" (Allen, 2003: 196). However, this accusation is not wholly accurate as postmodern writings are replete with "sophisticated conceptualizations of human agency and resistance" (Schutz 2004: 18; see Atkinson 2002).

In the postmodernist perspective, resistance to planning cannot be a revolution – a frontal attack on the planning system and the state that backs and grants it authority and legitimacy. Rather, it is about localized struggles that "disrupt institutions and normalization" (Martin 1988: 9). A Foucauldian conception of resistance treats resistance to planning as diverse and local, characterized by, among other things, the "proliferation of discourses" (Raby 2005: 162) where suppressed knowledges on the designation, ownership and use of urban spaces are unearthed and dominant ones destabilized.

Despite claims made against its efficacy, the postmodernist view is suitable for the analysis of "weapons of the weak" (Scott 1985) where direct revolt against the authorities would bring undue attention and elicit a violent reprisal to crush that revolt. This is the case in authoritarian systems of governance which have little regard for human rights and do not hesitate to use force and violence to suppress direct challenges to dominant spatial orders. In any case, planning itself is not a monolithic, coherent unit; rather it consists of discourses, ideologies, rationalities, techniques and technologies, located in different institutions and mediated by different individuals. It is not possible for people at the bottom rungs of society to successfully openly confront, once and for all, on a large scale, all elements of this matrix. Postmodernists hold that there is "subversive potential" in every activity – including planning – no matter how dominant (Schutz 2004: 19); there is always an 'outside' – such as informality – to systems that constitute such activities. In the case of planning, informality, which stands 'outside' planning, contests the very idea of defining proper and improper forms of livelihood and land use.

Resistance in urban Africa

In focusing on resistance to the dominatory practices of urban planning, this chapter complements studies of resistance in urban Africa, especially with respect to disadvantaged groups in general and youth in particular. Resistance in urban Africa comes in many forms and for many reasons. Implying the multiplicity of these modes of resistance against the "hegemony of political establishments," Lovell (2006: 228) mentions "political activism, dissent, and unbridled use of violence" which she argues "have often been characterised as the prerogative of the young" (2006: 230). A variety of perspectives have been deployed to examine and explain various types of oppositional practice in urban Africa. In all cases, this resistance is always against an agent or a condition. Youth political activists, militants and soldiers are good examples of the former; migrant workers and the unemployed typify the latter.

Dorman's study of Eritrean youth is a good example of a modernist analysis of resistance (Dorman 2005). Here we see university students openly voicing opposition to national service, to which the state, the target of that resistance, responds with brutal repression. Faced with this brutality, the youth adopt another strategy, draft-dodging, which is a form of "resistance to military service recruitment" consisting of "fleeing the country illegally, hiding from soldiers and resisting arrest" (Dorman 2005: 199). Biaya (2005) provides a more detailed and nuanced description of the *shege*[1] of Kinshasa. The typical *shege*, who "thrives in times of crisis" and whose time is devoted to making money, is an integral part of "the popular protest against the power of the rich and influential" (Biaya 2005: 220). Biaya deploys a postmodernist analysis to interpret how the *shege* subvert order by using, among other things, music. Ferguson's exploration of the consequences and character of material decline on the Zambian Copperbelt analyses how beleaguered urbanites cope with adversity. Instead of straightforward modernist theorizing about resistance to the dark side of capitalism, Ferguson (1999: 43) develops "'bushy', nonlinear ways of conceptualizing the ways that different, coexisting strategies . . . have been distributed over time."

The most spectacular modernist narratives of youth resistance are found in analyses of youth violence and political activism. A typical example is Kagwanja's (2005) analysis of youth identity and violence in Kenya. In this analysis, the militant *Mungiki* who espouse "a culture of vigilantism, banditry, urban criminality and general youth violence" are portrayed as openly resisting state authority and power through "bloody confrontation" with security forces (Kagwanja 2005: 82). Everatt (2005) suggests a similar perspective when he describes South Africa's urban youth as leading the anti-apartheid uprisings by taking on security forces and making the country ungovernable. Similarly, Seekings (1996: 104) asserts that "young black South Africans were the so-called 'shock troops' or 'foot soldiers' in the struggle for political change." In this open confrontation, youth "built barricades and fought street battles against the state's security forces, and took action against alleged collaborators" (1996: 104).

As the preceding overview shows, urban resistance in Africa has been a subject of analysis from various perspectives. This chapter complements these studies by exploring oppositional spatial practices of urban youth in an authoritarian environment where the intention of the youth is not to challenge the status quo (cf. Routledge 1997: 69). It demonstrates how, in a bid to *defend* their livelihoods, youth find themselves (unwittingly?) challenging the diktats of the planning system, and through this, the authority of the state.

Resisting the spatial order in Harare

Background

Post-2000 Zimbabwe is plagued by crises. Politically, the country has been beset by what some critics diagnose as a "crisis of governance" (Chikuhwa 2004). Intolerance, repression and human rights violations are some of the ills attributed to the state (HRW 2006). Supported by a sophisticated infrastructure of repression, the state has been accused of presiding over and promoting a breakdown in the rule of law. The breakdown is characterized by a sustained attack on the private media, civil society, and opposition-controlled urban councils; a blatant disregard of the independence of the judiciary; and persistent electoral fraud (ICG 2006; EIU 2006). The repression has been worsened by the militarization of the public sector (Mugabe 2005) and the establishment of party militia.

The embattled government has readily used brutal force to repress dissent and protest. This has been evident in urban centres, which, since 2000, have become "bastions of opposition support" (Maroleng 2005: 1). The successive local and national electoral triumphs of the opposition Movement for Democratic Change (MDC) have resulted in major urban centres having opposition Members of Parliament and opposition-controlled councils. Among the staunchest supporters of the opposition are youth (Kamete 2007).

The crisis of governance is blamed for the severe economic problems that have hit the country particularly since 1997, and worsened dramatically from 2000, thanks to the decimation of the commercial agricultural sector (Bond and Manyanya 2002). The economy has been in free-fall since then. At the time of the research (April 2004 to August 2005) inflation peaked at over 600 per cent, by far the highest in the world (Gono 2005). Compared to 1990, the local currency was worth less than 1 per cent of its value, while incomes were worth only 2 per cent (Clemens and Moss 2005). Unemployment was estimated at an unprecedented 70 per cent while 80 per cent of the population lived below the poverty threshold (IMF 2005). The erosion of incomes and the dwindling prospects of (formal sector) employment contributed to the situation where two in five urbanites were involved in the informal economy (IMF 2005).

The hostile political and economic conditions have directly impacted on the plight and prospects of youth. The economic meltdown has forced youth to turn to informal livelihood practices. The prospect of employment in the shrinking national economy is remote. Further, the few available jobs do not pay enough to stave off poverty, hence the flooding of the informal economy even by gainfully employed people including youth who are supposed to be dependents. The political situation makes generating informal livelihoods risky. Informality is routinely blamed for sabotaging the economy, an accusation that in fact contributed to *Operation Murambatsvina/Restore Order* (Gono 2005; Kamete 2007). The perception of urban centres as centres of dissent (Maroleng 2005: 1) does not help matters.

The study

This chapter is based on a study of (formally) unemployed urban low-income youth (aged between 14 and 25 years) operating in contested urban spaces in Harare. The contestation arose out of the fact that the youth used urban spaces that did not belong to them for purposes that were not in keeping with planning requirements. The research took place between April 2004 and August 2005. The 81 youth were selected from six sites in various contested land parcels: two (Chinhoyi Street and Downtown Cripps Road) in the city centre, two (Bluffhill Industrial Park and Mabelreign Shopping Centre) in medium-income residential suburbs and two (Warren Park Bus Terminus and Kuwadzana Mini Car Park) in low-income residential areas. Only those youth who had been operating on the given land parcel continuously for no less than two years participated in the study.

The youth were all self-employed. The popular lines of business were female-dominated petty trading, and male-dominated petty production (such as sculpture, toy-making, metal fabrication) and semi-skilled services (such as hairdressing and vehicle and household appliance repairs). In terms of education, all the youth had completed at least the seventh grade. More men (>75 per cent) than women (>60 per cent) had completed four years of secondary school education. Of these, only three (all men) had passed their Ordinary Level examinations. None had tertiary or vocational qualifications.

The majority of the youth (64 per cent) had moved to Harare from rural areas or smaller towns. None of them had been in the city for less than two years. All considered themselves 'city-wise'. Only a third (34 per cent women and 37 per cent men) were *born-locations*, that is, they were born and raised in the low-income, high-density residential areas (HDRA) of Harare. Two in three (67 per cent) considered themselves breadwinners. About 31 per cent lived alone or with friends, and were wholly responsible for rent and all household expenses. Even the 33 per cent who were not breadwinners complained that they found themselves increasingly contributing towards their respective household budgets.

A semi-structured interview schedule was used to capture information on the youth's response to official hostility and violence. In the interviews, I sought to determine specific reactions and the efficacy of those reactions to the actions of the various authorities to evict the youth from the contested sites or force them to conform to official conceptions of order. Other issues covered in the interviews included participation in collective action, individual roles in encounters with the system, as well as personal interpretations and explanations of specific encounters and outcomes. In addition, there were 12 focus group discussions. Apart from helping triangulate key information obtained from one-to-one interviews, the discussions captured information on the youth's evaluation of specific tactics of encounter and confrontation, as well as collective action, organization, mobilization and alliances.

The clampdown

Routledge (1997: 70) reminds us how "different social groups endow space with amalgams of different meanings, uses and values." His argument applies to Harare, where it is these differences that "give rise to various tensions and conflicts . . . over the uses of space for individual and social purposes and the domination of space by the state and other forms of social power" (Routledge 1997: 70). Faced with youth who persistently disregarded planning regulations and controls, Harare City Council (HCC), through its planning system, regularly clamped down on planning violations. In two years there were no fewer than 30 overt campaigns on the six sites, this in addition to numerous covert operations (Kamete 2008). Normally, planning enlisted the help of the Harare Municipal Police (HMP). In cases deemed volatile or posing a threat to public order and security, HCC received back-up from the Zimbabwe Republic Police (ZRP), in the form of the regular state police and/or the dreaded Police Support Unit (PSU). Sometimes the state agencies launched their own operations, with or without the knowledge and/or approval of HCC.

The restoration of order – officially termed 'clean-up' operations – normally consisted of lightning raids. A typical raid often involved physical assault, evictions, confiscation of goods, arrests, incarceration, and the exacting of fines. In addition, there were numerous covert operations, in the form of undercover police work by plain-clothes operatives from state agencies responsible for investigating ordinary crimes and state security. Some of these operatives were notorious for demanding bribes in the form of 'protection' or 'shut-up' fees. Youth resistance was therefore in reaction to planning as a 'spatial technology of domination' (Pile 1997: 3) and the unrelenting multipronged assault on their livelihood.

Zimbabwe is not the only African country that has clamped down on illegality and informality in urban areas. For example, Nnkiya (2006: 82) explains that until the mid 1980s the Tanzanian government was "generally hostile towards the informal sector." It made petty trading illegal and banned self-employed people from towns, forcefully repatriating them to the countryside. In Kumasi, Ghana, physical planning schemes make no accommodation for traders "who are continuously harassed and evicted" (King, 2006: 111). In Maseru, Lesotho, "street traders in the CBD West have had to endure periodic raids and eviction by the MCC [Maseru City Council] and the police" (Leduka 2002: 3). What differentiates Zimbabwe from the other countries in sub-Saharan Africa is the ferocity and persistence of the clampdown on planning violations. Whereas in other countries the authorities have in a way mellowed down (Tanzania), become tolerant (Lesotho), or have simply failed to enforce the regulatory framework (Ghana), those in Zimbabwe have been consistent in their bid to ferociously stamp out spatial unruliness.

Resisting the spatial order

The interviews and focus group discussions unearthed a variety of reactions, ranging from complete submissiveness to violent confrontation and retaliation. I noted three modes of resistance: 'docility', 'fighting fire with fire', and 'resistance at the margins'. This is not to imply that there was a neat temporal progression from one form to the other; nor does it mean that there was only one style of resistance at a particular period in a particular place among particular groups of youth.

Docility

For this study, docility is defined as a trait characterized by submissiveness and non-aggression *during* official clean-up campaigns. This acquiescence was common where youth did not agree on a response. At their most docile, the youth were quiet, easy to control, and manageable. Significantly, the majority (74 per cent) in the city centre referred to docility as 'a previous strategy', implying that they had, according to Jonah (18, Cripps Road)[2], "ditched it far, far away." A significant proportion at Mabelreign Shopping Centre (48 per cent) was at the time of the study still leaning towards docility.

Non-aggression was the hallmark of docility. In the event of a raid, the common reaction was flight. According to Madiro (17, Warren Park), at the sight of the police, they would "scatter like birds . . . north, east, south and west," each one of them taking with him or her as much personal property as he or she could. They would passively stand at a safe distance watching as the police destroyed shacks and confiscated merchandise. After the raid and "peace had descended on the area" (Mavis, 22, Warren Park), they would return to the desolated sites and sift through the rubble to salvage what they could. Some, whose goods had been seized, would peacefully go to the HMP headquarters to "plead with the officers and pay the . . . fine for breaking some section of some by-law" (Ellis, 19, Bluff-hill). If they were lucky, they would get some of their goods back.

In cases where the goods had been seized by the ZRP, very few (less than 5 per cent) of the youth bothered or dared to go to the police station to reclaim their merchandise. The process was useless in that the police would "swindle the goods in a bogus auction" (Morris, 16, Chinhoyi Street); it was dangerous because it could lead to visibility and arrests. The ZRP always demanded proof of ownership, mainly in the form of purchase receipts, before releasing the goods. Failure to produce these could lead to detention for possession of stolen property. Needless to say, most youth (on average 97 per cent) did not have the receipts.

The submissiveness extended to dealings with undercover officers. The youth would simply give in to the police officers' incessant demands, pay the 'protection fees', "and just shut the hell up" (Dorika, 14, Cripps Road). Agents suspected (or claiming) to be on undercover missions – "however abusive, unbearable and nasty they were" (Lottie, 17, Mabelreign) – would be left alone to do their work. Where questions were asked, and 'cooperation' was demanded, the youth would willingly play ball, even giving *madhikaz* (slang for 'detectives') helpful hints, and at times, valuable leads.

Docility, described by Janet (12, Cripps Road) as "the safest but most unwise of responses" was, according to Boniface (23, Cripps Road), "a very, very expensive thing." While its virtue lay in avoiding injury, arrest and incarceration, its financial costs were immense. Even youth who favoured docility admitted that there was a link between docility and increasing incidence of harassment, claiming that the police took advantage of the easiness with which "they stole our things" (Dylan, 16, Chinhoyi Street). Melissa (24, Kuwadzana), who had avowedly 'renounced' docility, reflected:

> This was a most stupid thing to do. . . . Whenever these dogs [the police] felt hungry, angry or broke or bored they would remember that there were goldmines to exploit somewhere in town. If they did not have food in their house, they would tell their wives not to worry . . . They would come here, release their anger, lay their hands on free things, have some fun . . . and leave us broke and wounded.

Two in three of the youth felt the strategy was, as Kumbirai (15, Warren Park) put it, "not good for serious business." The majority of them in downtown Harare had no kind words for what Jimalo (17, Chinhoyi Street) dismissed as "this rubbish, unmanly act fit for old, pious churchwomen." According to Jimalo this was "a weapon of hopeless and lazy sissies and cowards"; Ronny (17, Kuwadzana) thought that "this behaving like doves . . . [was] 110 per cent stupid, 200 per cent s***t" as the authorities ended up "treating us like their second wives."

By its very nature, docility did not give the authorities a reason to review their modus operandi. As a result, whatever approach the authorities adopted became normal practice because, in their view, it worked well. For them, every campaign to deal with the 'spatial malcontents' was a success. The 'successful' repressive practices were thus inscribed onto the authorities' operational patterns. The contentious issues, remained the same, unquestioned and therefore unresolved. There was no need to revisit them. Consequently the hide-and-seek games continued, following the same old pattern over and over again: unrelenting raids, unceasing loss of property, sporadic incarcerations, payment of fines. This became the life, the rhythm.

A colleague questioned my labelling docility as resistance. Hers was a typical modernist perspective. On the face of it, the youth were not resisting but "just absorbing punishment . . . like sheep being led to the slaughter" (Ms H, youth worker). If resistance is to be viewed as spectacularly standing in "implacable opposition to power" (Pile 1997: 1), then the docility displayed by the youth pales into insignificance when compared to Kenya's *Mungiki* or South Africa's young anti-apartheid activists (Kagwanja 2005; Seekings 1996). However resistance is not limited to "heroic refusal" (Cresswell 2000: 165). While they did not physically resist the raids, the 'docile' youth did mount implacable opposition to power in denying the authorities the 'clean' and 'orderly' city they desired. Despite ferocious attempts to annihilate their livelihoods they always reoccupied the contested spaces. The desire of the authorities was to eliminate illegality and informality; the desire of the youth was to defend illicit livelihoods. So, simply by returning after the raids and reconstituting their survival practices the youth were acting in opposition to power. Through this tenacity, they continuously resignified and disrupted the official ordering of space through planning. Whereas they did not violently oppose the *event* (raids and assault), they did oppose the *ideals* that the official violence was meant to uphold. This should certainly count as resistance.

'Fighting fire with fire'

Disgruntled with, according to Fadzai (21, Bluffhill), "always getting walloped . . . running away, and paying stupid fines and bribes," some youth resolved that it was in their best interest to resist being "needlessly harassed" by the authorities. As described by Ronny, "the sensible . . . thinking ones" resolved it was in their best interest "to fight fire with fire." The strategy gained currency at about the same time the majority of the urban electorate turned against ZANU-PF at the turn of the century. Most youth began to reinterpret the actions of the authorities as political victimization. Maureen (20, Mabelreign) accused government of "shamelessly avenging its defeat and [political] rejection" on youth. For Marian (18, Warren Park), it was now time to show "the fat cats that they cannot beat us like we are a grade one class." It was then that hostility and aggression began spreading in many sites. Of the six sites, the only ones where the youth did not try 'fighting fire with fire' at least once were the two sites in the middle-income areas of Mabelreign and Bluffhill.

What seemed to incense many of the youth was the fact that government, which they insisted was to blame for 'killing the country', was closing off what limited sources of livelihood were available. In many cases, the youth indignantly asked questions such as: "Do they want us to steal and kill?" "Do they want us to be thieves like them?" "What do they want us to do?" The combined losses and injury from officially sanctioned raids and 'private missions' made those who were advocating resistance win over many of their colleagues. It was at this point that isolated resistance crystallized into full-scale confrontation.

Instead of running away or retrieving what little they could after the 'clean-up' campaigns, the youth began to stand in the way of law enforcement agents, hurling insults, throwing projectiles at the forces, and engaging in what Ronny fondly described as "serious hand-to-hand combat." Typically, during and in the aftermath of raids, some irate youth would vandalize government and council property, especially vehicles and buildings. In some cases they would single out some well-known officers taking part in the raids, trail them when they would be off-duty, and assault them. Significantly, 23 of the 35 women (about 66 per cent), who initially had not favoured such aggression, ended up joining in the pitched battles.

Where head-on confrontation was the dominant style of resistance, under-cover work became dangerous for officers. Apart from refusing to pay bribes, youth began to assault people they suspected of having links to security and law enforcement agencies. In these areas, the presence of an officer on duty was enough ignite a conflagration. Needless to say, some innocent civilians were inevitably caught in the crossfire. Cases were recounted where youth would burst into business premises and go on the rampage simply on the suspicion "that such and such a businessman was responsible for some raid or investigation" (Mrs X, businesswoman), or was a *mupuruvheya* ('sell-out') collaborating with 'the enemy'.

In the end, this response proved to be more costly than docility. Although government forces did lose some battles, they always won the war. As Assistant Inspector M, of Harare Central Police Station boasted, even when the 'law' suffered initial reversals by "being repulsed on the first attempt," reinforcements would invariably be called in and the 'law' would "eventually come out tops and thoroughly crush the thugs." The no-nonsense PSU, specialists in crushing riots, increasingly became a common sight during raids. Notably, among groups favouring this form of resistance there were three times as many arrests as there were among the docile groups. On average the former were subjected to longer incarcerations and higher spot fines than the latter. There were more raids mounted against violently resisting youths than on 'docile' ones. This was partly because of 'rematches', a term used by the youth to describe the 'return' of the now-beefed-up state forces after they had suffered initial setbacks. Increasingly, security forces from outside the city began to be deployed. These were more vicious than their Harare-based counterparts. Because members of these forces were hard to isolate and track, 'revenge missions' became almost impossible. One of the few positive outcomes of 'fighting fire with fire' was a decrease in 'undercover' police missions, and with this a drop in extortion and bribes.

It was the problematic nature of docility that pushed some youth to the other extreme. But under authoritarianism this response was doomed from the start. The best this response could yield were short-lived victories. The worst harm it could inflict on the authorities were temporary setbacks. Each time the authorities were repulsed there was a violent 'rematch' that brought more carnage and repression. Violent confrontation provoked a "rapid and ferocious response" as the state simply chose "to employ more coercion" (Scott 1985: 33, 36). The involvement of the PSU meant that any temporary setbacks were immediately reversed. As a result, the youth did not get any mileage out of aggression. If anything, it made things worse. Raids increased in frequency; the losses mounted; the situation got worse. Economic injury was now compounded by physical injury. Obviously, this kind of resistance was bad for business – and unsustainable.

By all counts, 'fighting fire with fire' is what in the modernist narrative counts as resistance. This is the dramatic form of resistance that Fernandes (1988: 176) describes as 'manifest'. This kind of violent resistance goes beyond the unspectacular oppositional practice in some authoritarian African states (Dorman 2005). It reminds us of how *Mungiki* engage the Kenyan security forces in open confrontation. It also resonates well with how urban youth in apartheid South Africa took on the security forces in a bid to make the country ungovernable (Everatt 2005). From a modernist perspective, the Harare youth, like their Kenyan and South African counterparts, are rational subjects. They have a clear target (the authorities), clear strategies (open confrontation) and a clear objective (transforming a situation of oppression) (cf. Pile 1997; Raby 2005).

Resistance at the margins

Faced with the costly realities of frontal resistance, it was perhaps inevitable that some youth would become 'war weary' and re-evaluate their strategy. As some candidly admitted, confrontation did not improve matters. It was, as Betty (22, Warren Park), pointed out, "bloody and led . . . nowhere." The 'enemy' simply became more stubborn and more resolved to root out what was perceived to be a social vice, an administrative nuisance and a planning nightmare. Resisting violently only fuelled this resolve.

A new strategy began to take hold. Docility was costly; aggression worsened matters. Some youth started inventing novel ways of dealing with the authorities. Having listened to them and the authorities, I decided to label this evolving response *resistance at the margins*. In contrast to simplistic full-frontal confrontation or extreme docility there was a conscious attempt to come up with forms of resistance that, without challenging the authority of an increasingly repressive state, minimized damage, while at the same time bringing some gains, albeit marginal (cf. Scott 1985: 299). In place of confrontation, a proactive approach characterized by some form of negotiation, an alternative discourse, and downright underhand dealings began to catch on.

Youth in all areas invented ways of limiting the exposure of their property to damage or confiscation. They brought out as little as they needed to sell while leaving the bulk of the goods 'in storage'. 'Restocking' would be done as and when necessary. 'Raiders' could only lay their hands on what was available – and there was not much of that on display. With time, the decrease in the amount and value of merchandise at the point of sale translated into a decrease in losses. Inspector P, who admitted to "looting the little criminals on a regular basis," complained that the youth were becoming "cheaper and poorer targets" with the result that raiding them was becoming "less interesting and profitable." The practice soon became popular as more youth began employing this loss-minimizing technique.

Linked to the reduction in on-site merchandise was the adoption of cheap, quick-build collapsible structures. This strategy was invented in the suburban sites, where, up to then, youth had been erecting fairly durable structures. Plastic sheeting, light metal and wood were chosen over cement and bricks as building materials. In the CBD, some extremely mobile structures appeared towards the end of 2003. These 'workshops-on-wheels', originally introduced by phone-shop operators, were a crude version of push carts used by mobile food and ice-cream vendors. The net effect of these contraptions was a curtailment of losses incurred when the authorities razed down illegal structures. Youth would frantically push away their carts as soon as word of an impending raid reached them, which itself was a result of yet another invented loss-minimization strategy: underhand dealings.

Some youth infiltrated the local authority and recruited 'moles' (whom they called *mafesi* [friends]) in council departments, especially the Department of Works, the HMP's 'mother-department'. There was no shortage of council workers who were willing to trade information for material gain. Even in central government, some police officers were willing to trade information for a 'fee'. It did not take long for youth to make some 'business arrangements' with these *mafesi,* whose side of the bargain was to simply warn their recruiters of impending raids. As soon as the 'mole' got wind that a raid was being planned, word was quickly sent to the youth, mostly through coded messages using the mobile phone short message service (SMS). Having been thus warned, the youth would be prepared. To allay suspicions that they had inside information, they would be "manning their posts" (George, 20, Bluffhill) as usual but with reduced merchandise. Another reason for this was that the planned raids could fail to materialize – as was sometimes the case. At any rate, there was a verbal understanding that the youth had to be on site, "acting normally" and even going to the extent of, according to Dylan "faking surprise" when the raids did occur. After the raid had taken place as per the warning, the mole got what George referred to as "a handsome payment." During periods of 'peace' the mole got paid an irregular 'retention fee' in the form of cash payments, free handouts, lunches and access to scarce commodities such as sugar and fuel.

On some sites youth managed to enter into some kind of negotiations with the authorities. In all cases, the negotiations invariably started as petitions, wherein they enlisted the services of an influential personality to intercede in their 'disputes' with the authorities. For example, Bluffhill youth managed to enlist the help of a prominent alderman and former Harare mayor, who ended up being some kind of a patron. From the time he got involved up to his death, there was no raid at the Bluffhill site. In a few notable cases, Warren Park and Kuwadzana youth succeeded in getting their local councillors and at least one member of parliament to, according to Ronny, "reason and plead with these stone walls."

It is partly through these emissaries that the youth managed to break the authorities' monopoly on discourse. This is instructive considering that one reason why informal livelihood practices have been marginalized and rendered invisible in policy is that "the concepts and discourses that could make them 'visible' have themselves been marginalized and suppressed" (Gibson-Graham 1996: x–xi). Prior to this strategy, the authorities had been able to put across their case through an unchallenged discourse centring on, among other things, the principles of town planning, environmental management, crime fighting, and sustainable development, all couched in Harare's image as the 'Sunshine City'. Gradually, the youth, helped by what a planner condemned as "their well-informed, media-loving and smooth-talking sympathizers," developed an alternative discourse which centred on issues such as the lack of employment in the formal sector, the absence of alternative livelihoods, the youth's honesty in choosing to earn a decent living rather than resorting to a life of crime, and their ingenuity in creating employment rather than spending time looking for nonexistent jobs in the formal sector. These discourses were cleverly meant to dovetail with government's emphasis on small businesses, self-reliance and black empowerment. At times helped by their 'sympathizers', youth were able to come up with arguments specifically countering and disputing the authorities' claims in what Ms H, a youth worker, dismissed as "unsubstantiated official propaganda depicting the workplaces of the youth as arenas of crime and vice." In this way, the youth were able to put the official discourse on the defensive as they questioned the veracity of claims and countered technocratic conclusions with moral judgements.

Some of these alternative discourses managed to find their way to parliament, council chambers, and onto community agendas and other forums. The youth also found a means of publicizing their case in the media, both electronic and print. Faced with alternatives to their established discourses, and unable to suppress these other knowledges, the authorities had to find a way to respond. Instead of rushing to raze down structures and confiscate goods, it was not uncommon for the authorities to invest time and effort responding to the counter-hegemonic arguments of the youth as well as protestations and criticism of customers, sympathizers and critics.

It is partially due to this new development that the authorities made some shifts in positions, and some kind of an uneasy informal compromise was reached.

In an attempt to "rein in the rampaging informal sector" (Mr V, planning officer), the authorities started putting in place some conditions that should be met in order for the youth to be allowed to operate. Many youth in the contested spaces became beneficiaries of what Mr V described as the "new paradigm," variously referred to as being 'enabling', 'sympathetic', 'facilitative', and 'empowering', among other descriptive tags.

'Resistance at the margins' was by far the most effective resistant practice among the youth. Docility was bad for business; so was open confrontation. Maybe the solution lay somewhere in between. Storage and building tactics coupled with the infiltration minimize losses and injury; inventing an alternative discourse disrupted the state's dominant discourse; and enlisting the protective influence of patrons helped youth engage the authorities and secure some (precarious) concessions. Crucially, 'resistance at the margins' bought them some time, bringing periods of (tenuous) relief and "a whole lot of sympathy" (Mr V). Significantly, it included an effective response to epistemic violence (see Spivak 1988; Castro-Gómez 2002) and forced the authorities to rebrand their discourse and rethink their spatial practices in the face of an unsettling alternative discourse that was beginning to punch holes into the age-old rationalistic and 'technocentric' discourse (cf. O'Riordan 1986).

'Resistance at the margins' is a hybrid strategy which can best be understood and explained in the postmodernist conceptualization and theorization of resistance. The multiplicity of resistance tactics makes sense in the light of Foucault's assertion that power is everywhere and that resistance coexists with power (Foucault 1998). It is through postmodernist lenses that the "complex flows of power relations, fragmented and constricted subjectivities and local and individualized activities" (Raby 2005: 161) can usefully be analysed. In utilizing urban land for informal livelihood practices that go against official planning stipulations, Harare's youth created alternative economic geographies (Leyshon and Lee 2003). These illegal economic spaces could not be defended from official repression by simplistic docility or impetuous confrontation. In resisting at the margins, the youth adopted a combination of among other tactics, negotiation, patronage, subterfuge and counter-discourse. Biaya (2005: 222) paints a similar picture of Kinshasa's *shege* stating that "the city ceases to be a simple space of administrative . . . control, but rather a space for creative and recreative activity." Hence,

> In this space young people can simultaneously develop a mode of territorial control, a culture of illegality and a political base from which their action takes on a signification of opposition to the state, its project of dominance and its practices.
>
> (Biaya 2005: 222)

This amalgam of strategies has also been used by vendors in Maseru, Lesotho. Setšabi's modernist narrative of the response of vendors to evictions reveals a combination of strategies: negotiating with the authorities, refusing to relocate to officially designated marketplaces, increasing mobility, going to court, and joining vendors' organizations (Setšabi 2006). Interestingly, betraying how restrictive the modernist analysis of resistance is, Setšabi refers to these acts of resistance as 'coping strategies' (2006: 147).

Enter Operation Murambatsvina/Restore Order

In May 2005, the government launched *Operation Murambatsvina/Restore Order* (OM/RO). Officially, OM/RO was the embodiment of the authorities' rejection *(kuramba)* of filth *(tsvina)*. The operation was reportedly intended to restore order and sanity to urban space (Government of Zimbabwe 2005) by eradicating the 'filth': illegal structures, informal livelihoods (Potts 2006) and people benefiting from these. Though according to Bratton and Masunungure (2006: 33) youth were victimized more than any other group, they were not the specific target. OM/RO was a nationwide urban campaign of demolitions, evictions, arrests and detentions. According to the state's rationalization, the filth had reached such intolerable proportions that "the risk to public health and morality, national security and the economy" could no longer be tolerated (Government of Zimbabwe 2005: 30). In Harare, the crackdown almost annihilated informality. In just over two months it accounted for some 38,000 informal housing units and 9,000 informal business structures (Tibaijuka 2005: 85).

It seemed OM/RO was the decisive blow against informality. There was some initial resistance, mostly by youth in parts of Harare and Chitungwiza; but it was short-lived. For the next two months the youth did not "try even the feeblest attempt" (Betty). Kumbirai reflected, "The *Tsunami* [the popular name for OM/RO] was too fast, too big . . . too risky. We were mesmerized; no movement, no reaction except to protect oneself. We couldn't do anything *while it was going on*" (emphasis mine). Lena (17, Mabelreign) pointedly said they did not resist because *"tanga tisanganzwe kuda kufa* [We did not feel like dying]." For two months, the contested spaces were virtually free of filth; and for months thereafter, the youth were conspicuous by their near-absence. To be sure, some did try to stage a comeback. But each time they attempted a "resurrection from the coma" (Lloyd, 22, Warren Park), they were quickly and ferociously subdued.

The 'anti-filth operation' almost succeeded. It is instructive to find out why. When the operation came, there was no way of stopping it; nor was there a way of avoiding it. It was unprecedented, large scale and swift. Existing forms of resistance could not mutate fast enough to counter it. Tellingly, Ronny said, "*Zvaive zvinyowani izvi*" [This was new]. He was right. Instead of having to deal with the 'usual suspects' – the planning system, the municipal police and the ZRP – the youth found themselves facing the full might of the state. Everything that symbolized the force and brutality of the authoritarian state was ranged against them.

The nature of the blitz and its scale, as well as its scope, speed and sheer destructiveness did not give youth time to contrive a resistance strategy. For one, they had neither the opportunity nor clue to infiltrate the system. They were dealing with central government, reportedly with the full blessings of the Politburo, the highest decision-making organ in the governing political party. The youths' crude tactics could not work this time round. On the ground, the operation was faceless and impersonal and there was no way of negotiating with and around it. They could not identify weak links. The operation could not be threatened, bribed, or petitioned. It could not be contested either. Nathan (25, Chinhoyi Street), a self-confessed former 'militia-boy', mused, "The owners of the country were in a hurry. Stones, reason, tears or prayers could not stall them. They wanted to get the devil's job done."

So it was that by the beginning of July, the youth were no longer the masters of places they had illegally colonized. They could not make a living by resisting the official spatial order. Although modernist planning might not have been the architect of the operation – which incidentally was executed largely in the name of planning – its vision of the city had triumphed. Like others who depended on illicit livelihood practices throughout urban Zimbabwe, the youth were swept away into the background; most were driven underground. To be sure they did come out, occasionally, but the authorities always swept them away, again and again. It took them months to adapt 'resistance at the margins' to suit the new perilous conditions (see Kamete 2012). By the time this chapter was completed, the newer versions of resistance at the margins were still in the 'trial' phase. Notably, though still subject to being forcibly removed and punished, the youth started to survive longer with fewer losses on the contested spaces. But it cannot be claimed that they had fully recovered. It was no longer 'their' territory.

OM/RO almost eradicated visible spatial disorder. Why was the operation able to paralyse the youth? As noted above, in its scale, speed and ferocity, OM/RO was unprecedented. It was something new, something strange. It was a monster against which existing modes of resistance could not work. For a time, this was an emergency that could not be contained because there was nothing in the existing toolkit to tackle it. At this point, one can claim with a measure of certainty that whatever the successful resistance strategy of the future entails, it is sure to be a version of resistance at the margins (cf. Kamete 2012).

Conclusion

This chapter has made three key observations. First, confirming that planning is a spatial technology of domination, the chapter shows that 'subaltern actors' (Moore 1997: 87) have an array of resistant activities they can choose from to protect alternative spaces that they continuously create and recreate and which the state, through planning, continuously attempts to erase. Second, the resistant activities of the youth in Harare have important similarities and differences with other forms of resistance in urban Africa. Third, the postmodernist conceptualization and theorization of resistance yields deeper and more useful insights than modernist narratives, as it exposes the messy and complex realities in resistances that are both open and hidden, simple and subtle, 'pure' and hybrid.

In Harare, the authorities' heavy-handed enforcement of planning regulations and controls amounted to an all-out attempt to wipe out illicit livelihood practices. Embattled youth 'resisted' this bid to eradicate their way of life by defending their livelihoods against the onslaught that threatened their very existence. In its most simplistic form, the characteristic 'neatness' and certainty in modernist perspectives comes out in the study. There are two clear sides to the contestation. On one side there is the dominating state controlled by the elite. They 'possess' power, which they use through planning and the state's repressive apparatus to impose and maintain their domination of urban space (see Pile 1997: 1). State power is used to repress 'alternative geographies' (Leyshon and Lee 2003). On the other side is the subordinate class, the 'subaltern actors' consisting of 'powerless', dominated and marginalized youth. As rational beings they are aware of the source of their problems; they have clear choices for opposing the state's dominating and repressive power (see Raby 2005). Hence, docility, 'fighting fire with fire', and 'resistance at the margins' are no more than manifestations of particular forms of rationality.

There are some problems with this modernist narrative. For one, the youth did not stage anything remotely resembling a frontal assault on the system of dominance; nor did they fully submit. Accordingly, as Routledge (1997: 68–9) complains, this theorization is "fraught with an intellectual taming that transforms the poetry and intensity of resistance into the dull prose of rationality." As shown in this chapter, even when youth resorted to violent confrontation, resistance consisted of them *not* trying to stage a revolution by 'overthrowing' the planning system or radically changing the rules. They never contested "the formal definitions of hierarchy and power" (Scott 1985: 33). They merely protected or defended their livelihoods. They did not 'rise up', rationally citing as their demand the changing or abolition of the instruments that regulated space and criminalized their activities. Significantly, they did not openly question or contest the authority of the institutions that crafted and wielded the repression. Rather, theirs was a defensive and reactive strategy encapsulated in simple disobedience, violence and tenacity. The resistance lay in them doing and persisting in doing what, according to the dominating power of planning, was not supposed to be done.

In the end, the youths' actions did prevent the planning system from delivering the city desired by its sponsors. Defiance permeated even the passive and pacifistic modes of resistance. After every act of state violence, the dominated ones always reoccupied the contested spaces, wounded but defiant and determined to continue. Thus even at their most docile, the practices of the youth should count as "oppositional practice" (Aggleton and Whitty 1985: 62). In a way, they did undermine "the reproduction of oppressive social structures and social relations" (Walker 1985: 65).

A key shortcoming of modernist perspectives is their insistence on power being an exclusive possession of the dominant group (the authorities) which is deployed against the subordinate group (youth). Cresswell (2000: 259) is understandably concerned about resistance becoming everything, and thereby becoming meaningless and theoretically unhelpful. However, Raby (2005: 151) rightly cautions that limiting "configurations of resistance to collective, directed actions taken by a subordinate group towards a dominant one (i.e. a strike) can neglect other potentially subversive activities." The empirical material in this chapter suggests that it is more fruitful to conceptualize resistance as comprising all "engagements that in many different ways challenge specific social, moral and economic orders" (Keith 1997: 277). The cases presented in this chapter endorse Pile's observation that "people are positioned differently in unequal and multiple power relationships" which makes acts of resistance "contingent, ambiguous and awkwardly situated" (Pile 2997: 2–3).

As argued by Foucault (1998: 95) "where there is power there is resistance." The present discussion suggests this is the case also where that power is manifested through spatial technologies of domination backed by an authoritarian state's infrastructure of repression (see Sharp et al. 2000). Moreover, "there is a plurality of resistances, each of them a special case: resistances that are possible, necessary, improbable" (Foucault 1998: 96). It is therefore not fruitful to think of relations of power that have no resistances (see Gordon 1980). Hillier (2002: 51) explains that the very form of power that subjugates also produces the possibility of refusal. This is born out in Harare's contested spaces. Faced with a relentless onslaught on their way of life, the youth experimented with a 'plurality of resistances'. All these acts were driven by the intent to prolong and protect other spaces in a hostile urban environment. Each of them was an experiment and a localized strategy. Indeed, as shown in the comparative examples mentioned in this chapter, "resistance is understood where it takes place" (Pile 1997: 3).

Notes

1   According to Biaya (2005: 220) *shege* "denotes the condition of the clandestine migrant in the West or in the Congo."
2   The figure and the name in parenthesis refer, respectively, to the age and location of the respondent.

References

Abowitz, K. K. (2000) "A pragmatic revisioning of resistance theory," *American Educational Research Journal*, 37.4: 877–907.

Aggleton, P. J. and Whitty, G. (1985) "Rebels without a cause? Socialization and subcultural style among the children of the new middle classes," *Sociology of Education*, 58.1: 60–72.

Allen, J. (2003) *Lost Geographies of Power*, Oxford: Blackwell.

Amin, A. and N. Thrift (2002) *Cities: Reimagining the Urban*, Cambridge: Polity.

Atkinson, E. (2002) "The responsible anarchist: postmodernism and social change," *British Journal of Sociology of Education*, 23.1: 73–87.

Biaya, T. K. (2005) "Youth and street culture in urban Africa: Addis Ababa, Dakar and Kinshasa," in Honwana, A. M. and de Boeck, F. (eds), *Makers and Breakers: Children & Youth in Postcolonial Africa*, Oxford: James Currey.

Bond, P. and Manyanya, M. (2002) *Zimbabwe's Plunge: Exhausted Nationalism, Neoliberalism and the Search for Social Justice*, London: Merlin.

Bratton, M. and Masunungure, E. (2006) "Popular reactions to state repression: Operation Murambatsvina in Zimbabwe," *African Affairs*, 106.422: 21–45.

Butler, J. (1997a) *Excitable Speech: A Politics of the Performative*, New York: Routledge.

Butler, J. (1997b) *The Psychic Life of Power: Theories in Subjection*, Stanford, CA. Stanford University Press.

Castro-Gómez, S. (2002) "The social sciences, epistemic violence and the problem of the invention of the other," *Napantla: Views from the South*, 3.2: 269–85.

Cato, J. W. (2002) "Theory versus analytics of power: the Habermas–Foucault debate," *Agora*, 3.1: 1–13.

Chikuhwa, J. (2004) *A Crisis of Governance: Zimbabwe*, New York: Algora.

Clark, G. (1994) *Onions Are My Husband: Survival and Accumulation by West African Market Women*, Chicago: University of Chicago Press.

Clemens, M. and Moss, T. (2005) *Costs and Causes of Zimbabwe's Crisis*, Washington, DC: Center for Global Development.

Cresswell, T. (2000) "Falling down: resistance as antagonistic," in Sharp, J. P., Routledge, P., Philo, C. and Paddison, R. (eds), *Entanglements of Power: Geographies of Domination/Resistance*. London: Routledge.

De Certeau, M. (1984) *The Practice of Everyday Life* (Trans. S. Rendall), Berkeley: University of California Press.

Deleuze, G. (1988) *Foucault*, London: Athlone.

Dorman, S. R. (2005) "Past the Kalashnikov: youth, politics and the state in Eritrea," in Abbink, J. and van Kessel, I. (eds), *Vanguard or Vandals? Youth, Politics and Conflict in Africa*, Leiden: Brill.

EIU (Economist Intelligence Unit) (2006) "Country profile: Zimbabwe 2006," London: Economist Intelligence Unit.

Everatt, D. (2005) "From urban warrior to market segment? Youth in South Africa, 1990–2000." Available at www.academia.edu/522667/From_urban_warrior_to_market_segment_Youth_in_South_Africa (accessed 9 December 2012).

Ferguson, J. (1999) *Expectations of Modernity: Myths and Meanings of Urban Life on the Zambian Copperbelt*, Berkeley: University of California Press.

Fernandes, J. V. (1988) "From the theories of social and cultural reproduction to the theory of resistance," *British Journal of Sociology of Education*, 9.1: 169–80.

Foucault, M. (1998) *The Will to Knowledge: The History of Sexuality 1* (Trans. R. Hurley), London: Penguin.

Friedmann, J. (1993) "Toward a non-Euclidean mode of planning," *American Planners Association Journal*, 59.4: 482–485.

Gibson-Graham, J. K. (1996) *The End of Capitalism (As We Knew It): A Feminist Critique of Political Economy*, Oxford: Blackwell.

Gono, G. (2005) "The 2005 post-election and drought mitigation monetary policy framework," Harare: Reserve Bank of Zimbabwe.

Gordon, C. (1980) *Michel Foucault: Power/Knowledge: Selected Interviews and Other Writings, 1972–1977*, Brighton: Harvester.

Government of Zimbabwe (2005) "Response by Government of Zimbabwe to the report by the UN Special Envoy on Operation Murambatsvina/Restore Order," Harare: Government Printer.

Hillier, J. (2002) *Shadows of Power: An Allegory of Prudence in Land-Use Planning*, London: Routledge.

HRW (Human Rights Watch) (2006) *World Report, 2006*, New York: Human Rights Watch.

ICG (International Crisis Group) (2006) "Zimbabwe's continuing self-destruction," Brussels: ICG.

IMF (2005) "Staff report for the 2005 Article IV consultation," Washington, DC: IMF.

Jimu, I. M. (2005) "Negotiated economic opportunity and power: perspectives and perceptions of street vending in urban Malawi," *Africa Development*, 30.4: 35–51.

Kagwanja, P. M. (2005) "Clash of generations? Youth identity, violence and the politics of transition in Kenya, 1997–2002," in Abbink, J. and van Kessel, I. (eds), *Vanguard or Vandals? Youth, Politics and Conflict in Africa*, Leiden: Brill.

Kamete, A. Y. (2007) "Cold-hearted, negligent and spineless? Planning, planners and the (r)ejection of 'filth' in urban Zimbabwe," *International Planning Studies*, 12.2: 153–71.

Kamete, A. Y. (2008) "Planning versus youth: stamping out spatial unruliness in Harare," *Geoforum*, 39.5: 1721–33.

Kamete, A. Y. (2012) "Not exactly like the phoenix – but rising all the same: reconstructing displaced livelihoods in post-clean-up Harare," *Environment and Planning D: Society and Space*, 30(2): 243–261.

Keith, M. (1997) "A changing space and a time for change," in Pile, S. and Keith, M. (eds), *Geographies of Resistance*, London: Routledge.

King, R. (2006) "Fulcrum of the urban economy: governance and street livelihoods in Kumasi, Ghana," in Brown, A. (ed.), *Contested Spaces: Street Trading, Public Space and Livelihoods in Developing Countries*, Rugby: ITDG.

Leduka, R. C. (2002) "Contested urban space: the local state versus street traders in Maseru's CBD West." Paper presented at the conference on "Gender, Markets and Governance," Bamako, 7–12 September.

Lefebvre, H. (1991) *The Production of Space* (Trans. D. Nicholson-Smith), Oxford: Blackwell.

Leyshon, A. and Lee, R. (2003) "Introduction: alternative economic geographies," in Leyshon, A., Lee, R. and Williams, C. (eds), *Alternative Economic Spaces: Rethinking the Economic in Economic Geography*, London: Sage.

Lovell, N. (2006) "The serendipity of rebellious politics: inclusion and exclusion in a Togolese town," in Christiansen, I., Utas, M. and Vigh, H. (eds), *Navigating Youth, Generating Adulthood: Social Becoming in an African Context*, Uppsala: Nordiska Afrikainstitutet.

Mann, M. (1986) *The Sources of Social Power, I: A History of Power from the Beginning to A.D. 1760*, Cambridge: Cambridge University Press.

Maroleng, C. (2005) "Zimbabwe: increased securitization of the state?," ISS Situation Report, Pretoria: Institute for Security Studies.

Martin, B. (1988) "Feminism, criticism and Foucault," in Diamond, I. and Quinby, L. (eds), *Feminism and Foucault: Reflections on Resistance*, Boston, MA: Northeastern University Press.

Mills, C. (2000) "Efficacy and vulnerability: Judith Butler on reiteration and resistance," *Australian Feminist Studies*, 15.32: 265–79.

Mooney, G. (1999) "Urban 'disorders'," in Pile, S., Mooney, G. and Brook, C. (eds), *Unruly Cities? Order/Disorder*, Routledge: London.

Moore, D. S. (1997) "Remapping resistance: grounds for struggle and the politics of place," in Pile, S. and Keith, M. (eds), *Geographies of Resistance*: London: Routledge.

Mugabe, M. (2005) "Militarization of civic affairs in Zimbabwe," Politicalaffairs.net. Available at www.politicalaffairs.net/militarization-of-civic-affairs-in-zimbabwe/ (accessed 9 December 2012).

Nnkiya, T. (2006) "An enabling framework? Governance and street trading in Dar es Salaam, Tanzania," in Brown, A. (ed.), *Contested Spaces: Street Trading, Public Space and Livelihoods in Developing Countries*, Rugby: ITDG.

O'Riordan, T. (1989) "The challenge of environmentalism," in Peet, R. and Thrift, N. (eds), *New Models in Geography*, London: Unwin Hyman.

Pile, S. (1997) "Introduction: Opposition, political identities and spaces of resistance," in Pile, S. and Keith, M. (eds), *Geographies of Resistance*, London: Routledge.

Potts, D. (2006) "'Restoring order?' Operation Murambatsvina and the urban crisis in Zimbabwe," *Journal of Southern African Studies*, 32.2: 273–91.

Raby, R. (2005) "What is resistance?" *Journal of Youth Studies*, 8.2: 151–71.

Robertson, C. C. (1997) *Trouble Showed the Way: Women, Men, and Trade in the Nairobi Area, 1890–1990*, Bloomington: Indiana University Press.

Rose, M. (2002) "The seductions of resistance: power, politics, and a performative style of systems," *Environment and Planning D: Society and Space*, 20.4: 383–400.

Routledge, P. (1997) "A spatiality of resistances: theory and practice in Nepal's revolution of 1990," in Pile, S. and Keith, M. (eds), *Geographies of Resistance*, London: Routledge.

Sandercock, L. (1998) *Towards Cosmopolis: Planning for Multicultural Cities*, Chichester: John Wiley.

Schutz, A. (2004) "Rethinking domination and resistance: challenging postmodernism," *Educational Researcher*, 33.1: 15–23.

Scott, J. C. (1985) *Weapons of the Weak: Everyday Forms of Peasant Resistance*, New Haven: Yale University Press.

Scott, J. C. (1990) *Domination and the Arts of Resistance: Hidden Transcripts*, New Haven: Yale University Press.

Seekings, J. (1996) "The 'lost generation': South Africa's 'youth problem' in the early-1990s," *Transformation*, 29: 103–25.

Setšabi, S. (2006) "Contest and conflict: governance and street livelihoods in Maseru, Lesotho," in Brown A. (ed.), *Contested Spaces: Street Trading, Public Space and Livelihoods in Developing Countries*, Rugby: ITDG.

Sharp, J. P., Routledge, P., Philo, C. and Paddison, R. (2000) "Entanglements of power: geographies of domination/resistance," in Sharp, J. P., Routledge, P., Philo, C. and Paddison R. (eds), *Entanglements of Power: Geographies of Domination/Resistance*, London: Routledge.

Short, J. R. (1996) *The Urban Order: An Introduction to Cities, Culture, and Power*, Oxford: Blackwell.

Simone, A. M. (2004) *For the City Yet to Come: Changing African Life in Four Cities*, Durham, NC: Duke University Press.

Spivak, G. C. (1988) "Can the subaltern speak?," in Nelson, C. and Grossberg, L. (eds), *Marxism and the Interpretation of Culture*, Urbana: University of Illinois Press.

Swilling, M., Simone, A. and Khan, F. (2002) "'My soul I can see': the limits of governing African cities in a context of globalisation and complexity," in Parnell, S., Pieterse, E. Swilling, M. and Wooldridge, D. (eds), *Democratising Local Government: The South African Experiment*, Cape Town: University of Cape Town Press.

Tibaijuka, A.K. (2005) "Report of the fact-finding mission to Zimbabwe to assess the scope and impact of Operation Murambatsvina by the UN Special Envoy on Human Settlements Issues in Zimbabwe," New York: United Nations.

Tilly, C. (1991) "Domination, resistance, compliance . . . discourse," *Sociological Forum*, 6.3: 593–602.

Tranberg-Hansen, K. (2004) "Who rules the streets? The politics of vending space in Zambia," in Tranberg-Hansen, K. and Vaa, M. (eds), *Reconsidering Informality: Perspectives from Urban Africa*, Uppsala: Nordiska Afrikainstitutet.

Tripp, A. M. (1997) *Changing the Rules: The Politics of Liberalization and the Urban Informal Economy in Tanzania*, Berkeley: University of California Press.

Walker, J. C. (1985) "Rebels with our applause? A critique of resistance theory in Paul Willis's ethnography of schooling," *Journal of Education* 167.2: 63–83.

Yiftachel, O. and Huxley, M. (2000) "Debating dominance and relevance: notes on the 'Communicative Turn' in planning theory," *International Journal of Urban and Regional Research*, 24.4: 907–13.

Chapter 4

Unsettling insurgency

Reflections on women's insurgent practices in South Africa

Paula Meth

This chapter builds on work which celebrates insurgent planning practices, and which recognises the possibilities for repression inherent within these. Calling for more attention to the practice of so-called repressive insurgencies, it uses two case studies from Durban, South Africa to unsettle some assumptions arguably embedded in notions of "anti-democratic" or repressive insurgency. The cases tell the stories of marginalised women who participate through insurgency in shaping their city. Their contributions to resolving unmet housing and employment needs represent acts of insurgency against a state which has, in part, retreated from the provision of shelter and employment through its commitment to a neoliberal agenda. These insurgent practices parallel other celebrated insurgent contributions to cities. The women, however, also manage crime and violence in their local areas, using a range of strategies, some of which can be considered insurgent, as they directly challenge the authority and competence of the state. These crime management practices are, however, at times very violent, as the women's insurgent practices involve forms of vigilantism to achieve their purposes. Yet given the marginalised status of the women, and the reality of an absent state, trying to make sense of these practices (from the perspective of planning theory) proves challenging. Labelling them anti-democratic and repressive is arguably inadequate. This chapter makes use of this contradiction to unsettle the concept of insurgency and develop further ideas about the difficulties of celebrating or condemning the contributions of the marginalised to diverse and unequal cities.

Submitted by the Association of European Schools of Planning (AESOP).

Originally published in *Planning Theory & Practice* 11:2: 241–263 (2010). © 2010 By kind permission of Routledge and the Royal Town Planning Institute.

Introduction

The concept of insurgent planning practices has offered planning theorists an exciting lens through which to explore the alternative actions of people operating outside of, or alongside, the "formal" planning framework. Research in this area has focused on the vital contributions that citizens have made to shaping their cities and both the intellectual task of unearthing insurgent practices, and the practices in and of themselves, are celebrated by theorists as transformative and emancipatory (Sandercock 1998: 2, 6).

However, planning theorists (and others) have also acknowledged the potential for repressive outcomes from insurgent practices (Holston 1998: 54), citing occasional examples. This chapter argues that this tendency has received inadequate empirical and analytical attention in its own right,[1] with the result that the multiplicity and gendered politics of repressive insurgency have been particularly overlooked. In order to contribute to this debate, the chapter draws on the everyday experiences of marginalised African women (and some eyewitness accounts from men) living in the areas of Cato Manor and Warwick Junction in the city of Durban, South Africa. They all live either informally or in precarious conditions, struggling with poverty, unemployment, or informal employment. Their lives are framed by the neoliberal patriarchal democracy of post-apartheid South Africa, where poorer black residents have ironically gained rights (voting and constitutional rights, for example) but have also suffered the consequences of rising inequality (Miraftab 2009: 40). Yet this is not a simple story of state retreat from service provision (see Roy 2009b; Meagher 2007 for comparison). Post-apartheid, the South African state actually invested quite heavily in, and improved their approaches to, the delivery of particular services, such as housing provision and policing, but less attention was paid to areas such as employment creation, for example. Investment in services does not necessarily result in great amelioration of inequality, and such improvements can sometimes occur alongside other state-directed programmes which work to undermine progressive advances. Furthermore, the consequences of apartheid's violent methods of suppression and state-driven inequality means that simply increasing police resources cannot solve soaring crime rates, allegations of police corruption, and rising drug-related violence. Durban presents a particular challenge to the changing politics of national policing as a result of wider political violence in the region of KwaZulu-Natal during the dying years of apartheid, and the legacy of this brutality for residents today. The role of the police in this violence was prominent, and this history shapes the ways in which residents currently relate to the police.

Seen within this wider context, the women involved in the two projects all contribute in different ways to the shaping of their urban spaces, through self-help housing, entrepreneurial practices, collective action, and resistance to particular state interventions (particularly the threat of eviction). Their contribution to the planning and production of their spaces is evidence of a productive insurgent counterplanning against a state which has failed to deliver, or to deliver effectively, on various fronts (Miraftab and Wills 2005). Yet their insurgent practices are broader than this, and also encompassed collective efforts to manage and resist very high levels of crime and violence. Though this is arguably an essential component of urban management, in this case, it incorporated practices of vigilantism against suspected criminals. This chapter illustrates how vigilantism is a form of insurgency, as residents directly mobilise against the state by challenging its authority to maintain law and order. By focusing attention ultimately on the practice of vigilantism, the chapter aims to "unsettle" some of the assumptions implicit in the celebration, on the one hand, of the transformative possibilities of insurgency, and on the other, the denouncement of repressive practices, by exploring the complexities of the less palatable actions of these women in their daily endeavours to manage high levels of crime and injustice.

The chapter unsettles ideas of insurgency through four key points illustrated by empirical material. First, anti-democratic or repressive insurgency is often represented in planning theory literature as dichotomous to transformative practices. This binary logic obscures the complex interconnections between different types of insurgent practice. Transformative and repressive practices can be mutually constitutive, and can also be enacted by one and the same "community." Second, the interconnections between gender, politics and the possibilities of "repressive" insurgency have not been adequately explored. Marginalised women's actions are often assumed to be transformative, simply by virtue of their more oppressed and excluded position; an assumption that needs to be questioned. Third, the rationalities, moralities, rights, and values underpinning repressive practices deserve fuller attention, given the ways in which such practices work to justify the actions of mobilised communities. Finally, this chapter adds its voice to a chorus of calls from theorists for an approach to researching insurgency that is both context-driven and informed by empirical analysis.

The overall aim of this chapter is thus to unsettle the concept of insurgency and to contribute to debates within planning theory about insurgency. In particular, the acting of multiple insurgent practices (some transformative, some repressive) by poor women presents difficulties for planning theory, which in its more progressive form celebrates the empowerment, participation and transformative possibilities of the oppressed – often women.

Planning, insurgency and vigilantism

Sandercock defines insurgency as follows:

> The very word "insurgent" implies something oppositional, a mobilizing against one of the many faces of the state, the market or both. Insurgent

planning is insurgent by virtue of challenging existing relations of power in some form. Thus it goes beyond "participation" in a project defined by the state. It operates in some configuration of political power, and must formulate strategies of action. Insurgent planning practices may be stories of resistances, and not always successful . . . of resilience . . . or of reconstruction.

(Sandercock 1999: 41)

Sandercock's definition draws on her original use of the term (Sandercock 1995), following Holston's (1998, 1999) work on spaces of insurgent citizenship. Holston calls for a new imaginary of the city, one that focuses on alternative sources of "citizenship rights, meanings and practices" in opposition to the assumed legitimacy of the modernist state (1998: 39). Although his notion of (absolute) opposition to the state is challenged below, Holston's insistence that planning theorists should look beyond the state in their understanding of city practices is very powerful. Theorising insurgency calls for a broader definition of "planning" as Sandercock clarifies: "insurgent planning practices are instigated by mobilized communities, acting as planners for themselves" (1999: 42). The tendency to view planning as a regulatory practice has meant that it is possible to "miss its transformative possibilities, which in turn may be connected to histories of resistance" (Sandercock 2003: 40).

Ideas of insurgency draw strongly on notions of resistance, defined as "any action, imbued with intent, that attempts to challenge, change, or retain particular circumstances relating to societal relations, processes, and/or institutions" (Routledge 1997: 69) and Scott distinguishes between "the open, declared forms of resistance . . . and the disguised, low-profile, undeclared resistance" including foot-dragging, threats, aggression, gossip, and rumour (Scott 1990: 198). However, though a breadth of resistive practices may be identified, it is Sandercock's insistence that insurgency is specifically mobilising against the state, the market, or both, which underscores the significance of such practices for planning.

Sandercock's challenge to make the "invisible visible" (Sandercock 1995) and to embrace the view from below (Sandercock 2003: 43) has been pursued within radical planning. Its focus is twofold: first, rethinking planning histories in terms of who has been excluded (women, gay people, ethnic minorities, etc.); second, on analysing practices and sites of insurgency, often in terms of its initiation and enactment by the poor. However, the political agendas and status of insurgent groups can vary widely, with Holston (1998) identifying examples that include: "the homeless, networks of migration, neighbourhoods of Queer Nation, constructed peripheries [built by the poor] . . . ganglands, fortified migrant labour camps, sweatshops." He cites the squatting of Brasilia's periphery as an example of this diversity (Holston 1998: 44–45, 48). Rethinking insurgency has drawn attention to the varied viewpoints of those inhabiting a particular space: in the context of South Africa, Miraftab and Wills argue that the oppositional insurgent practices of residents involved in anti-eviction campaigns must be taken as seriously as sanctioned actions (2005: 212). Finally, planners have used ideas of insurgency to develop theoretical frameworks through which they can understand the future differently. As Sandercock implores: "[if we can] demonstrate its [planning's] multiple and insurgent histories, we may be able to link it to a new set of public issues" (2003: 47).

It is evident that certain practices of the urban poor, such as informal house-building and informal employment, are commonly used within broad definitions of insurgency (Holston 1998; Nesvåg 2000). Across the world, poorer residents are mobilising against the state (and against the formal market and even fellow citizens in particular cases) in efforts to counter a failure to provide homes and jobs. Through these practices they are contributing to the planning and making of their cities. However, the same residents are also constantly managing very high levels of crime and everyday violence, using a diverse set of tactics including the employment or membership of vigilante groups. Following Holston's call for engagement with "new territorializations of power and violence and the new paradigms of citizenship in urban peripheries . . . throughout the global South" (Holston 2009: 16), and Roy's rhetorical questioning of whether urban violence is "not the face of planning in poor neighbourhoods all over the world?" (2009a: 10), this chapter not only recognises vigilantism as a constituent part of urban reality, but argues that it should viewed as an insurgent practice.

Harris provides a working definition of vigilantism:

> Vigilantism is a blanket term for activities that occur beyond the parameters of the legal system, purportedly to achieve justice. It covers a wide range of actions and involves an eclectic assortment of perpetrators and victims. Vigilantism can exist as an isolated, spontaneous incident or as an organised, planned action. . . .
>
> (Harris 2001: 6)

Vigilantism is eclectic and ambiguous (Buur and Jensen 2004; Meagher 2007) and changes over time and space. It is not a bounded analytical construct, but rather a practice (Buur and Jensen 2004: 148), prone to change and contradiction. Importantly, the way that vigilante practices work, and the arguments by which they are justified, relate closely to the practices of insurgency outlined above. Through vigilantism, citizens and communities mobilise against the state in an explicit challenge to its authority and competence, and in explicit response to its failures (Sandercock 1999). In particular, vigilantism challenges the role of the police, as well as the courts and judicial system, in the management of crime. Its moral justification is that it purports to achieve particular forms of justice (often for the alleged benefit of poorer residents) and to reduce the incidence of random criminal acts and violence.

Vigilantism, like other insurgent practices, is often understood as occurring in opposition to the state (Sandercock 1999), in response to perceived failures of the state. Various theorists (Buur and Jensen 2004; Meagher 2007; Oomen 2004) have illustrated how the practice of vigilantism is linked to, and even constitutive of, the state. In complex ways, vigilante activities can often become state-like performances (Buur and Jensen 2004: 144). This oppositional, but also mimicking, characteristic of vigilantism parallels the ways in which insurgent practices are variously situated and embedded within governance structures, and arguably can never be entirely separated from state practices by their very definition: they "empower, parody, derail, or subvert state agendas" (Holston 1998: 47). Relations with the state are central to understanding vigilantism, and Meagher argues that analyses of vigilantism need to understand its relationships with processes such as state withdrawal and political liberalisation (2007: 92), and political processes have been identified as fundamental to insurgent practices more generally (Roy 2009b).

In complex ways, the blurring of the boundaries between vigilantism and the state maps onto notions of justice and rights that underpin much vigilante activity. Holston (2009) has examined this in relation to the changing discourse of rights within a Brazilian context, while Oomen identifies the ways in which South African vigilante leaders and members of an organised vigilante movement known as Mapogo express their claims for justice in terms of a morality that opposes the morals, values, and rights embodied within the post-1994 neoliberalist constitutionalist discourse of the post-apartheid state (Oomen 2004: 154). Mapogo members criticise the rights now seemingly available to criminals, arguing that they represent an extension of "European" rights that run counter to their African traditional values. Posel describes this claim as "the burden of rights" (Posel 2004: 233), arguing that it appears to be informing insurgent practices elsewhere, beyond South Africa (see Holston 2009). The assertion of traditional values is enacted through the formation of "moral communities" through which vigilante practices are justified (Oomen 2004; Buur and Jensen 2004), and the significance of such community mobilisation underscores the ways in which vigilantism is a form of insurgent practice. One example of a moral community, drawn upon within this chapter, is that it is formed by women in opposition to practices of sexual violence, performed usually by men. As Posel notes, the reinvigoration of vigilantism in post-apartheid South Africa is linked to the politicisation of gender and sexual violence in this context (2004: 235). In other words, the mobilisation of women against the state's perceived failure to manage sexual violence is indicative of wider, gender-based insurgent practices, and the persistence of moral complexity underpinning insurgent practices is recognised by Sandercock (1999: 43).

The formation and operationalisation of these moral communities, from within which vigilantism is justified, is a key form of community building through participation, embedded at times in formalised membership structures. Meagher has described the roles performed by one Nigerian vigilante organisation, the Bakassi Boys, pointing out that though the actions themselves were problematic, they were also illustrative of a positive social impulse, namely a demonstration of popular initiative and a "flourishing of social capital" (Meagher 2007: 96). The social formation of vigilante groups also parallels the organisation of wider community structures, drawing on a range of membership structures. This has led Buur and Jensen (2004: 139) to distinguish between vigilantism as an "organisation" and as a "formation", and to compare more formal to more ad hoc forms of association. These patterns of constitution shape power relations within and between vigilante groups as well as between vigilantes, citizens and the state in its various forms.

Vigilante actions, like other insurgent practices, are tied to place, as justifications for violence often centre on the protection of property rights (Meagher 2007: 96) or the promotion of personal safety within a community. The materiality of place, and the politics of this materiality thus lie at the heart of urban insurgency (Roy 2009a). Urban security and the management thereof through vigilantism is an insurgent strategy with much legitimacy among urban dwellers. People's presence within, and use of, place is controlled through vigilantism, and those criminal elements that are considered "out-of-place" and "evil" by moral communities face "exorcism" from the locale (Buur and Jensen 2004: 147). Further, this imperative to control particular bodies in space is highly reminiscent of the regulatory elements of planning, affecting physicality, sociality, and spatiality (Sandercock 1998: 6) whether formalised, insurgent or otherwise. Finally, Sandercock and Holston both emphasise the significance of, and possibilities for, transformation, future change, and empowerment through insurgent practices (Sandercock 1998: 20; Holston 1998: 39). These mirror the alternative imaginings of justice, the distinctive forms of governance, and the different types of social and political organisation, based on different values, held by moral communities reliant on vigilantism (Oomen 2004: 163).

It can thus be argued that vigilantism, as an urban practice mobilising against the state, should be seen as a form of insurgency. This chapter now turns to the wider concept of insurgency and raises some concerns, alongside other theorists, about the ways in which it is framed and used. The argument above, which positions vigilantism as a form of insurgency, informs these wider and more generic concerns.

Unsettling insurgency

Within studies of resistance, theorists have voiced concerns over the romanticisation of certain (often banal) acts and the subsequent labelling of these as forms of "resistance" (Abu-Lughod 1990; Cresswell 2000). As Cresswell points out, defining everyday acts, such as watching television, walking, telling jokes, and purchasing westernised consumer items as "resistance" can also render the concept "meaningless and theoretically unhelpful" (2000: 259). There is also an increasing recognition of the fact that acts of resistance can be both liberating and subordinating at the same time (Routledge 1997: 90), a theme echoed within analyses of insurgency. Holston, for example, draws attention to the possible paradoxical outcomes that can result from residents' participation in the planning of their communities, particularly in relation to the treatment of their fellow citizens, and uses the term "anti-democratic" to describe these (1998: 53–4). Vigilantism itself is an obvious, although extreme, example of paradoxical insurgency, as gains are achieved by some residents while others suffer immeasurably. Yet, the declaration of particular practices as inherently anti-democratic can be inadequate and lead to an analytical dead-end. Fuller interrogation of these paradoxical and contradictory practices is absolutely central to a meaningful understanding of insurgent planning.

First, although planning theorists have emphasised the paradoxical outcomes of insurgent practices, stories of insurgency have tended to present transformative practices in dichotomous opposition to anti-democratic practices (although see Holston 2009), or have referred to anti-democratic practices only as a cautionary after-thought to stories celebrating insurgency. This has had a range of analytical impacts. By whom and by what means have practices been defined and declared "anti-democratic"? The politics of this question have gone largely unaddressed, with the result that the power implications of such value judgements have been ignored. Furthermore, the dichotomy between insurgency and anti-democratic practices is arguably false (as others have recognised: see Abu-Lughod 1998). Defining particular practices as anti-democratic runs the risk of oversimplifying both the practices themselves and the citizens or communities enacting them. Practices of insurgency which are transformative can also be repressive, and vice versa; indeed the practices can actually be mutually constitutive. In addition, the actions of citizens and communities cannot always be labelled as "democratic or anti-democratic" because a single person or community may be both at the very same time. Thus a group of residents may practise some acts which are empowering (challenges to the state over housing, for example) but also acts which are repressive (vigilantism, for example). Furthermore, these variable practices can be interpreted as evidence of undemocratic settlement invasions, or legitimate forms of crime management, depending on the perspective of the observer.

Second, references to repressive or anti-democratic insurgent practices would benefit from a fuller engagement with the politics of gender to deepen the relevance of such analyses. Women's innocence and presence as "planning visionaries" (Sandercock 1998: 6) is often assumed within accounts of insurgency, rather than demonstrated or contested. As Haraway (1988) points out:

> There is a premium on establishing the capacity to see from the peripheries and depths [i.e. the subjugated]. But here there also lies the serious danger of romanticizing and/or appropriating the vision of the less powerful while claiming to see from their positions . . . The positionings of the subjugated are not exempt from critical reexamination, decoding, deconstruction, and interpretation . . . The standpoints of the subjugated are not "innocent" positions.
>
> (Haraway 1988: 59–60)

This recognition of the gendered politics of subjection contributes to arguments put forward by Watson (2003) and Wirka (1998), among others, who highlight the problematic roles played by both poor and powerful women in generating alternative planning histories and contributions.

Gender and the politics of gender relations are not static, but always changing and deeply interconnected with multiple other social relations (class, race, sexuality, etc.). This interconnectedness must inform analyses of insurgency. A key example is that the gendered politics of women are forged in relation to male performances of masculinity and vice versa, many of which are fragile, marginalised, and changing. Some marginalised men may be sexist, but they are still marginalised and an appreciation of masculine vulnerabilities is also essential to understanding gender relations. Similarly, the actions of poorer women are central to ideas of insurgent planning, yet as often as they are overlooked, they are also frequently celebrated in an uncritical manner. Arguably, women's agency is shaped by complex social relations, and marginalised women often suffer greater powerlessness because of poverty and their material living circumstances. However, although they are operating within frameworks of patriarchal relations, these women are not powerless, but express their agency in various, although often constrained, ways (see Watson 2003). A critical politics of gender underscores an appreciation of the ways in which insurgent practices are multiple and complex, and this chapter supports other work which aims to move beyond simplistic gendered conceptualisations of women as weak victims and men as strong perpetrators. Rather, the gendered politics of insurgency are far more complicated, a function of diverse gendered histories, changing discourses of sexuality and rights, and the simultaneous rejection and embracement of traditional patriarchal values for different purposes by both men and women (Oomen 2004; Robins 2008).

Third, complex questions of rights, moralities, and rationalities underpin insurgent practices (Holston 2009; Watson 2003, for example). These issues are central to the justifications underpinning insurgent practices, but they are arguably not singular or universally applicable. For instance, Watson (2003), writing mainly about South Africa, argues forcefully that ideas of rights, values and rationality are deeply contested and points to "deep differences" often overlooked within planning theory (Holston 2009). Watson does, however, state that particular values are universal and she argues that "murder . . . and criminality . . . are no more acceptable to ordinary men and women in Africa than they are in any other part of the world" (2003: 405). This chapter illustrates how even these apparently universal moral frames are unstable and ambiguous, since vigilantism, an insurgent practice, often adopts violent strategies to administer justice. The normalisation of this by ordinary South Africans is a function of unstable relations between citizens and the state (Meagher 2007), one consequence of the weak institutionalisation of the latter under colonialism, which resulted in an "informalization of politics" and the growth of "patrimonial forms of power" (Watson 2003: 401). The effects of this have been exacerbated by the withdrawal of the state from particular roles as a function of neoliberalist policies (Meagher 2007: 92).

Furthermore, when interrogating insurgency, it is important to recognise that a focus on rights should not simply be framed in terms of their alleged universality (or lack thereof). This chapter illustrates that when insurgent citizens appeal to rights, this rhetoric is used less to draw attention to inequities created by a deficiency of rights than to highlight the injustices created by a perceived excess of rights. For instance, in South Africa, particular citizens believe that criminals (among others) have gained undeserved rights as a function of the country's changed constitution (see Holston's similar account in Brazil (2009: 23)). Rights discourse is employed to justify action in such cases, and to counter a perceived "burden" of rights. Finally, residents draw on particular and local ideas of morality to inform and rationalise their practices. This chapter illustrates how they construct "moral communities" (Buur and Jensen 2004) which function to justify particular, if not all, vigilante practices, and to mobilise participants in their cause.

Fourth, acts of insurgency are often defined through the particularities of time and space (Holloway and Hubbard 2001: 220). Therefore, analyses of insurgency must be context-driven, alive to the specificities of everyday lived realities, and shaped by empirical analysis. This means that a critical engagement with insurgent practices must have a very strong appreciation of the local context, an awareness of the politics at varying scales, and a sense of the historical influences at play. For instance, across the Global South, insurgency is intimately bound up with informality, as Roy explains: "insurgence often unfolds in a context of informalization where the relationship between legality and illegality, the recognized and the criminalized, the included and the marginalized, is precisely the cause of counter-politics" (Roy 2009a: 9).

This call for a strong appreciation of the specificities of a particular case has methodological implications. It requires the analyst to move beyond the theoretical and instead to draw extensively on everyday empirical realities. Watson calls for planners to "return to the concrete" (2003: 403) as a way of thinking about difference and variability across contexts, and this emphasis on conducting empirical research (396) as well as exploring the situated nature of such research, is vitally important. Similarly, academics working on vigilantism point to the essential need for grounded ethnographic work that is also historical in nature (Buur and Jensen 2004: 143) revealing the nature of change over time, as well as the persistence of place in shaping particular events. Particular forms of qualitative research, arguably a mixed-methods approach, are able to develop the depth and sensitivity required to understand highly complex and changeable phenomena. Importantly, the perspective of "the everyday" (Buur and Jensen 2004: 141) is central to appreciating the lived experiences and rationalities of phenomena, and also in viewing and unearthing insurgent practices from the bottom up. This chapter thus draws on qualitative research using a case studies approach with its primary focus on the everyday experiences of men and women. It is to the practicalities of such methodology that this chapter now turns.

Methodology and case studies

Two related projects were conducted in South Africa between 2002 and 2007, both examining the gendered experiences of violence and place. The first focused on women, the second on men. The research aimed at unpacking the problematic dualism between domestic and public space, and to complicate the concept that women are victims, and men perpetrators, ideas that underscore much work on gender and violence. A complex and differentiated mix of qualitative methods were used in both projects to examine the spatial, historical, and emotional experiences of violence (Meth 2003; Meth and McClymont 2009). The range of methods (focus groups, interviews, diary writing, photography, drawing, life-history, and evaluation interviews) afforded different spaces of disclosure for the participants, allowing them to explore difficult memories in a manner over which they had relative control. The projects were facilitated by two Zulu-speaking female African research coordinators who worked as interpreters and translators. Questions of researcher subjectivities and the ethics and emotions of working on violence are central to the methodologies of both projects and have been explored in detail elsewhere (Meth and Malaza 2003; Meth and McClymont 2009). Suffice to say that both studies were fraught with tensions, but that these served to productively inform the outcomes of both projects and to shift the relations between participants and researchers in useful ways.

The two projects drew on distinct but overlapping cases in their examination of space and place. The project that focused on women concentrated on three different case study locations: an informal settlement (Cato Manor), an inner city market (Warwick Junction), and a more formal township (Kwa Mashu). The idea was to compare and contrast the geographies of violence with the nature of housing and place in each location. The project that focused on men explored in more detail the particularities of informal space, concentrating solely on Cato Manor.

This chapter draws on material from the studies of both Cato Manor and Warwick Junction. Both areas are centrally located within the city boundary: Warwick Junction forms part of the core inner city, while Cato Manor lies within 7 km of the city centre. Warwick Junction is a bustling entrepreneurial hub and a key transport node. Informal trading of a wide range of goods and services (food, clothing, muthi,[2] hair dressing, etc.) occurs alongside formal businesses. Some informal traders operate within managed market areas (with services and some shelter), while others are located on the pavements. Cato Manor, by contrast, is a large residential area and the site of intensive, state-led regeneration with an emphasis on the provision of small formal housing units and related community and business facilities. Housing is a mix of informal and formal: some sections await redevelopment, while others have experienced invasions by squatters desperate for land. Further characteristics of the two areas will be presented below, through the analysis of women's insurgent practices in both locations.

Stories of insurgency

This chapter focuses on three areas of insurgency: claims to place and housing, insurgent entrepreneurialism, and vigilantism. Though there is not space to examine the roles played by women in broader practices of South African resistance, particularly in relation to the apartheid state and its repressive segregationist policies (Bonnin 2000: 302), it is important to note that women's historical politicisation through opposition to the state shapes their current insurgent practices (see Miraftab 2009 for a similar argument). Apartheid was explicitly gendered and the spatialities of segregation were defined by gender relations: women's access and rights to cities were highly circumscribed and the generic relegation to homeland regions had a huge impact on poor urban women. Yet despite this political regulation, cities became sites of relative freedom and greater equality for women, spaces where they could challenge patriarchy and the state. Arguably, this still remains the case today.

Women in Cato Manor were highly active in the anti-pass[3] movement and demonstrated considerable resistance to forced removal. Their actions in the 1950s and 1960s were described by the resident African National Congress

Women's League (ANCWL) as "heroic, their violent actions justified" (Edwards 1996: 102 cited in Beall et al. 2004: 320). This history of insurgency has framed current relations between citizens and the post-apartheid state in complex ways. The majority of women support the ruling ANC, but are angry about its failure to deliver on its election promises. Their memories of fighting for freedom are still fresh, and they express a growing resentment about the policies and interventions of the constitutionalist neoliberal democracy, particularly the failure to provide employment and changes to criminal legislation.

Claims to place: insurgency and housing

Both Cato Manor and Warwick Junction have histories of insurgent claims to place and are the sites of ongoing conflicts over access to housing. Cato Manor was settled in the early 1900s, the land leased out to African residents by Indian landowners. The area expanded rapidly, housing around 45,000–50,000 people by 1950 (Maharaj and Makhathini 2004: 28, 30). In 1959, a violent period of unrest broke out as residents were threatened with eviction by the government. The unplanned squatter settlement was perceived to be "a hotbed of crime" and the authorities were desperate to eradicate it (Walker 1991: 230). Women "formed a large percentage of the 'illegal' population who were threatened with removals" (Walker 1991: 231). Between 1958 and 1963, as the policies of grand apartheid were implemented, evictions from Cato Manor of both African and Indian residents occurred (Maharaj and Makhathini 2004: 31). The area was effectively cleared of almost all its residents, remaining "empty" right up until the late 1980s.

From this time onwards, the rising urbanisation of the area by Africans was tolerated, while the apartheid state struggled with crises and talks of political negotiation grew. The authorities controlling Cato Manor increasingly ignored the illegal occupation of land through invasions. Although Beall et al. (2004) caution that little is known about the gender dynamics underpinning this process in this particular case, they acknowledge that "the widely held perception is that women settlers were in the majority and were proactive in occupying land, building shacks and extending settlement" (2004: 320). Large-scale invasions occurred in July 1993 and 1995 (Maharaj and Makhathini 2004: 34), a time of great political and spatial flux, and it was during this period that the majority of the 15 female project participants in this study arrived in Cato Manor. Their movement formed a particular strategy, designed either to escape escalating violence in other parts of Durban, or to access the possibility of new housing and employment in the area. Their actions ran counter to the formal emphasis of National Housing Policy (Watson 2003) and to the plans of the local state, which, along with the recently formed Cato Manor Development Association (CMDA) (CMDA 2009) had begun to plan intensive formal housing development in the area. Despite this, the actions of squatters rapidly changed the urban fabric of Cato Manor and revealed a certain strength of purpose amongst particular urban citizens. Attempts by the state to control and evict squatters were of variable success and women played a central role in resisting state demolitions, as this eyewitness account from Nhlahla (a male resident of Cato Manor) illustrates:

> The police . . . for two years . . . [came] . . . to destroy our houses. We were always rebuild[ing] houses after the police [were] gone because we didn't have [a] place to sleep. The women who were with us were upset about that. They form a group of women only and told us they want to solve that problem because [they] were tired. The women were combining in that mass action in the road; it was like a toyi toyi[4] of the women only. The municipal police were come again as usual to destroy the houses [that] were built. The women were standing [in] a line in the road while the police come. They talk first with the police and those police didn't listen [to] them, then the women made a plan to remove their underwear clothes and they were naked and face the municipal police with their bums and were opening the bums. The police were end there to come and destroy the houses. They [felt] embarrassed and fear[ful] to face the bums of the women, other bums were big. The men were staying at home [during that] . . . incident because of the women who were done that. They were helping us to come out with the good solution about the police.
>
> (Nhlahla,[5] Diary, Cato Manor)

The women within this story of nudity resemble Sandercock's "planning visionaries" (1998: 6), employing subversive strategies to counter the regulations of planning (and see Rangan and Gilmartin (2002) for a similar account of the politics of nudity and the use of transgression as a form of opposition). Nevertheless, their insurgency is tempered by their poor living conditions and their simultaneous desperation for state intervention. The men and women of Cato Manor felt abandoned by the state (in terms of service and infrastructure provision), pointing to a contradictory and multiple set of relations with the state viewed both as a strong regulator and as a failing provider.

In Warwick Junction, historical attempts to resist state policy, particularly over claims to place, were relatively effective. During apartheid, the racial categorisation of the area was a constant site of contest: despite being declared as white in 1963, it retained a mixed character (Maharaj: 1999: 252). In the 1970s, many Indian residents were forcibly relocated to outlying townships, promoting urban blight. Following this, much of the resistance to removals occurred in the post-1976 period (following an uprising in Soweto) and a residents' association was formed to ensure that the area's "grey" (mixed race) character remained (Maharaj 1999: 256). A gendered analysis of this history is however largely absent, and

Maylam comments that in the early 1900s, Indian customs in Durban "kept most women in the household and out of public life" (1996: 114), although they were engaged in welfare and charity work.

In more recent years, however, Warwick Junction has witnessed a dramatic rise in informal traders, and an accompanying escalation of claims to place, centring on the right to live informally in the area. Project participants explained that, to maintain their income, they were forced to break the law and sleep at their trading site in the market area, using plastic to shelter under.

Other female traders slept illegally in nearby buildings, squatting on a nightly basis, or in the underground garages of formal high rise blocks with their children. In doing so, they were resisting the expectations of the local authorities that, as traders, they would return nightly to their "homes" in adjacent townships. They were also subverting the notion of residence-free zones, or zones of discreet formal housing within the city centre, and thus actively shaping the spaces of the inner city. While optimism could be gleaned from their efforts to reimagine the city in different ways (Holston 1998; Sandercock 1998), their lived realities (unsafe and with no privacy) challenge a celebration of counterplanning.

Insurgent entrepreneurialism

The project participants from Cato Manor relied extensively on informal employment to earn their livelihoods, actively shaping their local urban economies. This practice has an extensive history (relating primarily to beer brewing) (Walker 1991: 231). More recently, many of the women had chosen to live without men, and this resistance to patriarchal norms had both negative and positive financial implications for them. Selling was popular with more than half of the women who grew, collected, or purchased goods for sale, namely vegetables, scrap metals, candles, ironing board covers, second-hand clothes, food, cigarettes, and underwear. The authorities have largely turned a blind eye to informal hawkers, but traders do suffer from police raids on their property. Some of the women had set up informal sewing and gardening associations to support their entrepreneurial efforts but many of these suffered from a lack of capital. This "survivalist" precariousness is both gendered and fairly typical (Rogerson 1996).

Warwick Junction also has a politicised past, but given its central location in the inner city, entrepreneurialism has played a more defining role in the shaping of urban space, compared with Cato Manor. The area was an important business and residential district, containing both Indian and white residents (Maharaj 1999: 252) with African traders (selling muthi – traditional African medicine, and beer) able to enter on five-day passes (La Hausse, 1984 cited in Nesvåg, 2000: 36). However, following severe restrictions by the government, from the 1940s to the 1970s, very few traders remained in this area, although a few female African traders "braved the law" (Nesvåg 2000: 37–39). By the early 1980s attitudes towards trading softened and the numbers grew dramatically: by 1990, 900 traders were identified in the area (Nesvåg 2000: 45).

Clashes between mainly African traders and the local government continued into the post-apartheid era: in 1995, unregulated street trading was highlighted by the Durban Metro Council as "undesirable" (Maharaj 1999: 263). Since the fall of apartheid, Warwick Junction has become a fully entrenched transport and economic hub of Durban's inner city. Estimates reveal that around 300,000 commuters use the area on a daily basis and that around 5,000 informal traders work in the greater Warwick Junction and Grey Street area (Khosa 2003). The majority of these traders are African women seeking economic independence (Khosa 2003 after Naidoo 1993). The area has been the focus of much local government planning and renewal (Khosa 2003), yet despite this, female traders suffer from inadequate facilities and strained relations with the local authorities. Nesvåg argues that, in view of the anti-street trading ideology still prevalent in post-apartheid South Africa, this practice can be viewed as a critical form of resistance (2000: 341) and thus gendered insurgency.

All 13 of the women interviewed at Warwick Junction were muthi traders. The majority arrived in the mid 1990s, possibly capitalising upon the change in government, presuming new trading freedoms and new rights to reside in urban areas. Most of the women are members of the Self Employed Women's Union (SEWU), a vocal and active organisation formed in 1993 (Lund et al. 2000) that concentrates on skills development as well as liaison work with city authorities. During 1996, SEWU formed a powerful voice of resistance when the Durban Metropolitan Council introduced new street trading by-laws, forcefully resisting these (along with other organisations), condemning raids and securing the provision of market stall facilities (toilets, wash stands, and overhead cover), influencing the planning of inner city space. As this illustrates, in Warwick Junction, organised forms of gendered insurgent practice have proved essential to the success of entrepreneurialism. Informal trading continues to be a defining feature of this location, and the markets are a substantial draw for customers from the wider region.[6]

Vigilantism as insurgent crime management

Levels of crime, violence, rape, and murder in South Africa are disturbingly high (Jewkes et al. 2006). Crime is particularly acute in areas of informality, and plays out in complex gendered ways: women and children are highly vulnerable to rape, and young men to murder, although neither exclusively. Crime statistics for the two areas under study are highly problematic for two reasons. First, allegations of the misreporting of crime by police in South Africa are widespread. Second, area-specific analysis is not possible, because Cato Manor and Warwick Junction fall into two and three different areas respectively for the collection of crime statistics (Cato Manor and Mayville; Durban Central, Umbilo and Berea) (South African Policing Services (SAPS) 2009, personal communication). Despite these reservations, the statistics suggest that between 2001 and 2008 these areas saw a substantial increase in drug-related crimes and assault with intent to inflict grievous bodily harm, while rates for murder and rape show unclear trends (SAPS 2009).

In both Cato Manor and Warwick Junction, women relied on a variety of practices to deal with high crime and violence rates, including reporting incidents to their local committees, involving the police and non-governmental organisation (NGO) representatives, self-defence, employing witch doctors, and vigilantism. These strategies intersect in ambiguous ways, and in different combinations, at varying times. Inevitably the boundaries between formal or legal and informal or illegal strategies were regularly blurred, with insurgent practices often evolving alongside more regular interventions. Within Warwick Junction, the presence of an active NGO (SEWU) offered a more conventional method for counter-managing crime. The women's allegiance to SEWU structured their strategies, as traders explained: "we report to our leaders" and "our leaders are there for us, they seek to ensure that we are treated well" (Warwick Junction, Focus Group 2). The women also made use of other formal partnership structures to manage crime including community policing forums, such as Warwick Junction's Traders Against Crime (TAC).

Local committees are also often formed within residential areas with the sanction of local governance structures and the police, such as the "peace committees" formed in Cato Manor (Goodenough and Forster 2004). Yet the practice of reporting to a local-area committee could sometimes extend to organised vigilantism, as it conducted mock court-hearings and, at times, meted out punishment before escorting suspected criminals to the police. Several women spoke of using this strategy, and of the variable tolerance of the police towards this. But they also distinguished neighbourhood action from the formation of more ad hoc vigilante groups (Buur and Jensen 2004), which several women had witnessed, and in which some had participated.

The involvement of women in vigilante membership is not unusual. Oomen (2004) found that 68 per cent of the supporters of the Mapogo vigilante organisation in a different part of South Africa were women. However, the participation of women, historically and currently, in actual acts of vigilante violence is less reported and, seemingly, less common. Vigilante practices are, however, gendered in complex ways, beyond membership and perpetration. Mapogo strongly advances the traditional authority of men as one of its key values, with the support of its female members (Oomen 2004), while other vigilante practices have arisen precisely in response to sexual violence against women and children.

Indeed Posel argues that there is a rising politicisation of sexual violence, particularly through changes in legal sanctions because of democratisation. She describes a "moral panic about the moral frailty and sexual menace of manhood" as male sexuality increasingly becomes the focus of critical attention (Posel 2004: 235).

Supporting Posel's "burden of rights" argument, female traders emphasised the rights gained by alleged criminals through changes to sentencing legislation in particular, and the broader implementation of human rights in general:

> I think in the olden days killing a person was not easy because people knew that if you do it you are going to be given a death sentence. Now that the death sentence is not applicable any more, killers know that if they kill someone they will spend a couple of days in prison and thereafter move on with their lives . . . the punishment that killers and rapists get for their deeds is not enough to change their attitudes.
>
> (Nonkululeko, Diary, Warwick Junction)

Through the formation of a moral community, the majority of women felt that vigilante practices were justified and were a way of teaching criminals appropriate moral codes and "correcting immoral . . . behaviour" (Buur and Jensen 2004: 147). They described the ad hoc formation of vigilante groups as "using the strategy of telling neighbours", which would then lead to the formation of a group for the explicit purpose of searching out, capturing, and punishing a suspected criminal. Amanda from Warwick Junction explained that "thugs do get a good lesson that crime does not pay at all" (Diary), deeming vigilante formations effective: "I've seen it myself, it works." The women explained that certain types of criminal were the objects of their attention, namely rapists, burglars, and repeat offenders (Cato Manor Focus Group 2).

> In November 2001 a thief was caught. He was severely beaten by the residents . . . When he recovered he stole again. He broke into a house and was caught. People were so angry that they tried to take out his eyes by knives saying it is the eyes that see the things that he desires to steal.
>
> (Zandile, Diary, Cato Manor)

These moral communities are particularly entrenched by the perceived failings of the state to administer justice and manage crime, a claim echoed nationally (Harris 2001; Oomen 2004). The police were specifically identified as failing in their duties, and the policing gap was increasingly filled by informal and privatised policing practices, now very common across South Africa (Hornberger 2004; Buur and Jensen 2004). These include the hire of private security firms by wealthy citizens (Badboyz 2009), with deep consequences for the moral landscape in which vigilantism unfolds.

Following their lack of faith in the police and the criminal justice system, the women preferred that suspects were punished prior to reporting them to the police: "Yes, it's the best way because they [the criminals] are not jailed any more." When asked if they were personally involved in the violent beating of a suspected child rapist, they explained: "Yes, we were beating up someone who has raped my neighbour's child" (Cato Manor Focus Group 2). Their assessment of vigilante responses were largely positive; one woman even described herself as happy because the criminals who were suspected of breaking into her house were "severely beaten" before being handed over to the police (Thandekile, Diary, Cato Manor). An additional benefit of such vigilante action can be the retrieval of stolen goods.

The tragedy of vigilantism is self-evidently the prospect of the misidentification of criminals. Thandekile explained how her brother was wrongly targeted by a vigilante group in search of stolen goods:

> They went to his place . . . [they] decided to beat him up. When he tried to get the reason why they were doing that to him or explain his case they never listened to him. On the night of the burglary he was at Umlazi [a different township]. The problem was that he was a friend to one of the criminals . . . They hang him in a tree and beat him severely . . . They beat him until they were satisfied . . . He was admitted for a week in hospital.
>
> (Thandekile, Diary, Cato Manor)

Violent vigilantism points more widely to the cracks in a moral community, and to the unequal gendered consequences of such actions for marginalised men.

The power of vigilante formations extends to controlling claims over place. Lihle described the treatment of a suspected child rapist following his release on bail: "The community decided that it was unsafe to stay with such a person and thus destroyed his shack" (Lihle, Diary, Cato Manor). The regulation of space through eviction parallels regulatory practices of planning more commonly associated with the state (Sandercock 1998: 6), forming a kind of parody of official action (Holston 1998: 47). However, it also echoes Buur and Jensen's (2004) notion of "exorcism", in which criminals are deemed out of place and hence require removal. Exorcism is more likely with suspected acts of sexual violence, particularly those involving children as victims, following the growing politicisation of sexuality (Posel 2004) and in these cases it is often achieved through murder. Suspects have also been killed for abducting young girls for purposes of prostitution:

> Unfortunately, for him, the community saw him. They called on everyone in the area to catch the thug. He was caught and beaten . . . He asked for mercy, but nothing of the sort could be obtained from the angry community . . . He was crying so loud that you could hear the echo from the dongas [gullies] around. He cried until he died. They used all sorts of weapons. For instance

he was stoned until he died. Having satisfied themselves that he was dead, they stamped him with a huge stone on his face.

(Phiwe, Diary, Cato Manor)

Similarly, Nonkululeko told a painful story that unsettles insurgency, questioning the legitimacy of the moral community formed in response to a brutal gang rape and murder:

> I witnessed what I'm talking about. I saw someone being killed brutally . . . They raped her . . . and stabbed her to death. They were eight and they all took turns during the rape. It is said that what goes around comes around . . . These boys were caught before her funeral. They were brutally killed. The committee took them to the sports ground. Everyone did what s/he thought they deserved. Some beat them whilst others stoned them. Some people were using bush knives to cut wherever they like in their bodies. They all died. This created another picture in my heart. I understand they were wrong by killing her. I cannot say what was done to them was even worse or it was not supposed to happen because if they were reported to the police they would be out here harassing more people. However, at that very area, crime still exists. They never learnt from what happened to their friends.
>
> (Nonkululeko, Diary, Warwick Junction)

Nonkululeko identifies the futility of vigilantism in her story, yet much research points to the success of vigilante practices for reducing crime for particular periods of time (Meagher 2007; Oomen 2004). As a form of insurgency, however, it rests quite regularly on the use of unpalatable violence. In various cases, however, references to the killing of suspects are made quite casually, pointing to a situated set of values about murder that regard it as justified in some cases, rather than a crime necessitating universal condemnation (see Watson 2003: 405 for comparison). In part, these moral justifications rest on a set of beliefs about the humanity of the criminal, who, as the "embodiment of evil" becomes "subhuman" – a description given by a government minister (Buur and Jensen 2004: 147). In both projects, men and women repeatedly compared criminals to animals, and dogs in particular, to describe their inhumanity and to dehumanize them, underscoring their lack of rights. Following this, several of the women criticised the absence of the death penalty and called for its reinstatement, along with castration arguing that "if the death sentence was in place such people would be dead and the number of criminals would drop" (Nobuhle, Diary, Cato Manor). This support for the death penalty represents a clear challenge to the views currently entrenched in South Africa's democratic constitution which is founded on principles of justice and human rights.

Concluding comment

This chapter has aimed to open up discussions about insurgency by illustrating the complex and variable ways in which insurgent practices are performed, and by critically exploring the gendered politics underpinning these. These conclusions draw out a number of key themes from the chapter, but also aim to highlight the relevance of these arguments for planning theory more broadly.

Building on work which recognises that insurgency can be transformative yet also repressive (Holston 1998; Sandercock 1998), the chapter has focused on practical aspects of this contradiction in relation to gendered politics. The emphasis on multiple forms of insurgency, responding to the management of housing, work, and crime, has illustrated that insurgency is shot through with ambiguities which cross simple dichotomies. Assumed oppositional relations with the state, assumed patterns of power (marginalised versus dominating), and the very binary notion of citizens as transformative or repressive, are revealed to be inadequate for analysing insurgency. Furthermore, through a critical examination of gendered vigilantism, the acute suffering caused by repressive insurgency is brought to light, raising questions about the moral and ethical frameworks informing insurgency, while also recognising that for many members of vigilante groups, this so-called repression is liberating and largely rational.

The empirical evidence here points to the influence on insurgency of multiple and changing relations between the state and its citizens. These can be contradictory, or oppositional at times (witnessed through resistance to housing demolitions) but also mutually constitutive, evidenced by vigilante practices which often go unchallenged by the police and have a history, at a national level, of garnering state support (Meagher 2007; Oomen 2004). The state is not a monolithic dominating force, but rather a complex network of complementary and contradictory power relations, which play out in different ways at different scales. Furthermore, the insurgent practices of informal house-building and protection, and informal entrepreneurialism and vigilantism, work to parody and perform state-like functions through their material provision of infrastructure and services. Acts of vigilantism can work to regulate space and manage the gendered politics of place, by evicting unwanted criminal citizens and protecting private property interests. In these respects, they differ very little from the more developmental and regulatory elements of formal planning.

This recognition of the regulatory characteristics of insurgency is significant for planning theory more broadly. Vigilantism can perform a key role in shaping place, and social relations within place. Thus, although it can be defined as repressive, it cannot be written off as an insurgent aberration within more official, formal planning structures. Instead, it is a practice that must be accounted for. This chapter calls then on planning researchers to engage more fully in the empirical investigation of insurgency, focusing where possible on the paradoxical nature of insurgent practices.

Both case studies revealed that practices of insurgency can and do point towards alternative futures (following Sandercock's 1998 celebration of this characteristic), such as the desire for productive employment witnessed through entrepreneurial practices. However, these alternative future visions also include the death sentence and a future where constitutional rights are curbed for particular citizens (often male), because of their actions as suspected criminals. This raises crucial and difficult questions about the values that underpin these insurgent-inspired alternative futures, and illustrates that some of these values may not be universal or tied to the principles of democracy as conceptualised in the west (Holston 2009).

This fact raises complex but pertinent issues for planning theorists. Do we only celebrate insurgency if it embodies (particular) democratic values, and on what basis can we discriminate between transformative versus repressive practices? Is this even a useful question? Watson's (2003) work is very helpful here, as she calls for recognition of deep differences between different cultures and citizens, with serious implications for theorising participation in planning. Deep differences are not, as Watson explains, unique to South Africa, or indeed even to Africa. Holston supports this, explaining that divergent claims to democracy are not simply "aberrations" which can be dismissed (Holston 2009: 13) by theorists in the global north. Within South Africa, an appreciation of the resurgence of traditional African values, for example, must be acknowledged by planners if their work is to have relevance to the diverse citizenry. By extension, at a global level, differences can be identified in all cities. However, respecting difference may also entail problematic consequences. For instance, appreciation of traditional African values may have problematic gendered implications. Women's equality gains can be reversed through claims to tradition (Robins 2008), and feminist planners must think through the potential antagonisms between gender equality and respect for, and of, cultural difference. A more complex gender analysis is required; one that demonstrates, rather than simply assumes, women's roles as "planning visionaries". Similarly the treatment of men deserves substantial unpacking if we are to understand their roles in insurgency.

By calling for analyses of insurgency that are "situated" (Watson 2003), this chapter aims to contribute to wider claims within planning regarding the politics of methodology. It argues that the political and gendered context of insurgent practices, as well as the political and practical outcomes it seeks in particular cases, are fundamental to the capacity of planners to make sense of insurgency. Planners need to ground themselves fully in the politics of insurgency and the concrete practices of it. This can only be achieved on a case-by-case basis, using empirical evidence and analysing this evidence through a political lens. In the case of women in Cato Manor and Warwick Junction, their practices are a function of complex political realities: the retreat of the state following neoliberal principles and subsequent delivery failures; the underfunding and illegitimacy of the police (related in part to their history of apartheid brutality); the cultural conflicts over ideas of citizenship and rights; the rise of gendered politics through the partial advancement of gender equality (Bonthuys and Albertyn 2007); and the politics of a national economic strategy which has largely bypassed the poor. Only within this messy framework can insurgency be evaluated.

Finally, planning theory would benefit further from exploring the interconnections and spaces between transformative and repressive insurgencies, given their inextricable links. These interconnections embody the everyday, in all its ambiguities. The paradoxical nature of insurgency is a process in need of critical examination. Analyses must avoid narrow binary accounts of good and bad, instead exploring the interconnections between multiple insurgent practices and their mutually constituting nature at the heart of urban politics. Further work on moralities, values, rationalities, and rights is therefore required in order to appreciate the situated nature of paradoxical insurgencies.

Acknowledgements

This chapter has benefited substantially from the valuable comments of the editors, four reviewers and subeditor, and I am very grateful for their input. I would also like to thank Tanja Winkler, Glyn Williams, Sarah Charlton, Linda McDowell, Natasha Erlank and the late Graham Drake for comments on various drafts of this chapter; and Khethiwe Malaza, Sibongile Buthelezi and Kerry Philp for their invaluable research work. The research was funded by the Nuffield Foundation UK and I would like to thank the foundation for this financial support.

Notes

1   See, however, Holston's 2009 paper on gang violence in Brazil, which he uses to critically rethink ideas of democracy and a discourse of rights. Holston's case is extreme (although globally prevalent), as the practices are described as "exceptionally brutal" (2009: 15) and also tied to criminal and drug elements. The example thus does not necessarily blur the binaries of emancipatory/repressive insurgency as easily as those considered in this chapter.
2   Muthi is traditional medicine administered by healers and witch doctors.
3   Passes were documents required by all black South Africans permitting entry into "white urban areas" during the apartheid era.
4   "Toyi toyi" is a form of protest dance-march popularised during the anti-apartheid struggles.
5   Note, all participants' names are pseudonyms.

6   Recently, the area of Warwick Junction, and the informal traders in particular, have come under sustained threat of removal because of municipal plans to demolish the "market" and to invest in a central mall (Skinner 2009).

References

Abu-Lughod, J. (1998) "Civil/uncivil society: confusing form with content," in Douglass, M. and Friedmann, J. (eds), *Cities for Citizens*, pp. 227–237, Chichester: John Wiley.

Abu-Lughod, L. (1990) "The romance of resistance: tracing transformations of power through Bedouin women," *American Ethnologist*, 17(1): 41–55.

Badboyz (2009) Available at www.badboyzsecurity.co.za (accessed 10 August 2009).

Beall, J., Todes, A. and Maxwell, H. (2004) "Gender and urban development in Cato Manor," in Robinson, P., McCarthy, J. and Forster C. (eds), *Urban Reconstruction in the Developing World: Learning Through an International Best Practice*, pp. 309–332, Johannesburg: Heinemann.

Bonnin, D. (2000) "Claiming spaces, changing places: political violence and women's protests in KwaZulu-Natal," *Journal of Southern African Studies*, 26(2): 301–316.

Bonthuys, E. and Albertyn, C. (eds) (2007) *Gender, Law and Justice*, Cape Town: Juta.

Buur, L. and Jensen, S. (2004) "Introduction: Vigilantism and the policing of everyday life in South Africa," *African Studies*, 63(2): 139–152.

CMDA (2009) Cato Manor Development Association. Available at www.cmda.org.za (accessed 10 August 2009).

Cresswell, T. (2000) "Falling down: resistance as diagnostic," in Sharp, J., Routledge, P., Philo, C. and Paddison, R. (eds), *Entanglements of Power: Geographies of Domination/Resistance*, pp. 256–268, London: Routledge.

Goodenough, C. and Forster, C. (2004) "Community safety and public security," in Robinson, P., McCarthy, J. and Foster, C. (eds), *Urban Reconstruction in the Developing World: Learning Through an International Best Practice*, pp. 333–346, Johannesburg: Heinemann.

Haraway, D. (1988) "Situated knowledge: the science question in feminism and the privilege of partial perspective," in McDowell, L. and Sharp, J. (eds) (1997), *Space, Gender and Knowledge: Feminist Readings*, pp. 53–72, London: Arnold.

Harris, B. (2001) "As for violent crime that's our daily bread: vigilante violence during South Africa's period of transition," in *Violence and Transition Series*, Volume 1, May, pp. 1–90, South Africa: The Centre for the Study of Violence and Reconciliation.

Holloway, L. and Hubbard, P. (2001) *People and Place: The Extraordinary Geographies of Everyday Life*, Harlow, Essex: Pearson Education.

Holston, J. (1998) "Spaces of insurgent citizenship," in Sandercock, L. (ed.), *Making the Invisible Visible: A Multicultural Planning History*, pp. 37–56, Berkeley: University of California Press.

Holston, J. (1999) *Cities and Citizenship*, Durham and London: Duke University Press.

Holston, J. (2009) "Dangerous spaces of citizenship: gang talk, rights talk and rule of law in Brazil," *Planning Theory*, 8(12): 12–31.

Hornberger, J. (2004) "'My police – your police': The informal privatisation of the police in the inner city of Johannesburg," *African Studies*, 63(2): pp. 213–230.

Jewkes, R., Dunkle, K., Koss, M., Levin, J., Nduna, M., Jama, N. and Sikweyiya, Y. (2006) "Rape perpetration by young, rural South African men: prevalence, patterns and risk factors," *Social Science and Medicine*, 63: 2949–2961.

Khosa, M. assisted by Naidoo, K. (2003) "LED in the Durban Metropolitan Area: Informal Trading in Warwick Avenue." Available at www.local.gov.za/DCD/ledsummary/durban/dbn03.html (accessed March 2006).

Lund, F., Nicholson, J. and Skinner, C. (2000) *Street Trading*, Durban: School of Development Studies, University of Natal.

Maharaj, B. (1999) "The integrated community apartheid could not destroy: the Warwick Avenue Triangle in Durban," *Journal of Southern African Studies*, 25(2): pp. 249–266.

Maharaj, B. and Makhathini, M. (2004) "Historical and political context: Cato Manor, 1845–2002," in Robinson, P., McCarthy, J. and Foster, C. (eds), *Urban Reconstruction in the Developing World: Learning Through an International Best Practice*, pp. 27–40, Johannesburg: Heinemann.

Maylam: (1996) "The changing political economy of the region 1920–1950," in Morrell, R. (ed.), *Political Economy and Identities in KwaZulu-Natal: Historical and Social Perspectives*, pp. 97–118, University of Natal, Durban: Indicator Press.

Meagher, K. (2007) "Hijacking civil society: the inside story of the Bakassi Boys vigilante group of south-eastern Nigeria," *Journal of Modern African Studies*, 45(1): 89–115.

Meth, P. (2003) "Entries and omissions: using solicited diaries in geographical research," *Area*, 35(2): 195–205.

Meth, P. and Malaza, K. (2003) "Violent research: the ethics and emotions of doing research with women in South Africa," *Ethics, Place and Environment*, 6(2): 143–159.

Meth, P. and McClymont, K. (2009) "Researching men: the politics and possibilities of a mixed methods approach," *Social and Cultural Geography*, 10(8): 909–925.

Miraftab, F. (2009) "Insurgent planning: situating radical planning in the global south," *Planning Theory*, 8(32): 32–50.

Miraftab, F. and Wills, S. (2005) "Insurgency and spaces of active citizenship: the story of Western Cape anti-eviction campaign in South Africa," *Journal of Planning Education and Research*, 25: 200–217.

Naidoo, K. (1993) "Work and life of women in the informal sector: a case study of the Warwick Avenue Triangle." Master's thesis, Durban: Department of Geographical and Environmental Studies, University of Natal.

Nesvåg, S. I. (2000) "Street trading from apartheid to post-apartheid: more birds in the cornfield?," *International Journal of Sociology and Social Policy*, 20(3-4): 34–63.

Oomen, B. (2004) "Vigilantism or alternative citizenship? The rise of Mapogo a Mathamaga," *African Studies*, 63(2): 153–171.

Posel, D. (2004) "Afterword: Vigilantism and the burden of rights: reflections on the paradoxes of freedom in post-apartheid South Africa," *African Studies*, 63(2): 231–236.

Rangan, H. and Gilmartin, M. (2002) "Gender, traditional authority, and the politics of rural reform in South Africa," *Development and Change*, 33(4): 633–658.

Robins, S. (2008) "Sexual politics and the Zuma rape trial," *Journal of Southern African Studies*, 34(2): 411–427.

Rogerson, C. (1996) "Urban poverty and the informal economy in South Africa's economic heartland," *Environment and Urbanization*, 8(1): 167–197.

Routledge, P. (1997) "A spatiality of resistances: theory and practice in Nepal's revolution of 1990," in Pile, S. and Keith, M. (eds) *Geographies of Resistance*, pp. 68–86, London: Routledge.

Roy, A. (2009a) "Strangely familiar: planning and the worlds of insurgence and informality," *Planning Theory*, 8(1): 7–11.

Roy, A. (2009b) "Why India cannot plan its cities: informality, insurgence and the idiom of urbanization," *Planning Theory*, 8(1): 76–87.

Sandercock, L. (1995) "Making the invisible visible: new historiographies for planning," *Planning Theory*, 13: 10–33.

Sandercock, L. (1998) "Introduction: framing insurgent historiographies for planning," in Sandercock, L. (ed.), *Making the Invisible Visible: A Multicultural Planning History*, pp. 1–33, Berkeley: University of California Press.

Sandercock, L. (1999) "Introduction: translations: from insurgent planning practices to radical planning discourses," *Plurimondi*, 1(2): 37–46.

Sandercock, L. (2003) *Cosmopolis II: Mongrel Cities in the 21st Century*, London: Continuum.

Scott, J. (1990) *Domination and the Arts of Resistance: Hidden Transcripts*, New Haven, CT: Yale University Press.

Skinner, C. (2009) "Street traders threatened with eviction for new mall in Warwick Junction," presentation to public meeting, 26 May 2009. Available at www.abahlali.org (accessed 10 August 2009).

South African Policing Services (2009) "Statistics per province: stations." Available at saps.gov.za/statistics/reports/crimestats (accessed 6 August 2009).

Walker, C. (1991) *Women and Resistance in South Africa*, Cape Town: David Philip.

Watson, V. (2003) "Conflicting rationalities: implications for planning theory and ethics," *Planning Theory and Practice*, 4: 395–407.

Wirka, S. M. (1998) "City planning for girls: exploring the ambiguous nature of women's planning history," in Sandercock, L. (ed.) *Making the Invisible Visible: A Multicultural Planning History*, pp. 150–162, Berkeley: University of California Press.

Urban transformations

Chapter 5

Constructing "world-class" cities

Hubs of globalization and high finance

Louise David and Ludovic Halbert

This chapter provides an understanding of the way financial capital shapes the built environment of large cities of so-called "emerging" countries. This is done by looking at the complex interdependencies between the circulation of global finance capital invested in real estate and the production of the built environment of two cities that appeared on investors' radars in the late 1990s, Bangalore, India and Mexico City. The chapter first sheds light on the instruments, rationalities and behaviors of finance capital investors. Then, it focuses on how finance capital is permitted to reach a given tract of land thanks to the intermediation of local facilitators. This leads to a discussion of the notion of commutation as a central process to explain the interactions between financial investments and the production of cities.

Introduction

The terms of the articulation or interaction between economic and financial systems and the actual built environment of cities are still unknown as regards their magnitude, shape and implications for sustainability. In this chapter, the process of materially producing and reproducing urban spaces will be examined. The

Submitted by the Association pour la Promotion de l'Enseignement et de la Recherche en Aménagement et Urbanisme (APERAU).

First published in French in Pierre Jacquet, Rajendra K. Pachauri and Laurence Tubiana (dir.), *Regards sur la terre. Villes changer de trajectoire.* © 2010 by kind permission of Presses de la Fondation nationale des sciences politiques.

cities of Bangalore in India and Greater Mexico City[1] in Mexico will serve as two points from which to observe contemporary economic and political transformations (Box 1). Both cities are bound up with the economic globalisation of the 1990s, through their particular positions in the new, global division of labour – information technology services in Bangalore and the populous business and logistics centre of Greater Mexico City – and because of their status as beachheads for access to fast-growing "emerging markets." According to international real estate consultants, both cities are "opportunity spaces" for local and especially international investors who seek to profit from the strong demand for the material development of urban areas and their infrastructure, facilities and real estate. We will discuss how cities in emerging market countries may be increasingly shaped in terms of their built environment, their landscapes and their spatial organisation by a more globalised, and more "financialised" form of capitalism. By "financialisation," we mean the process of evaluating and managing real estate investments through the same types of financial tools that investors use to evaluate stocks and bonds (Orléan 1999). Using Bangalore and Greater Mexico City as examples, we set out to examine the mechanisms by which international capital flows are "fixed" (Harvey 1985) or "land" (Torrance 2008) in urban regions and by doing so, modify the regions' spatial organisation and landscapes. We will analyse the role of investors who, whether in tandem or in competition with other public and private investors, such as developers, business tenants, individuals or local and national governments, contribute to the material development of metropolises. The chapter thus first discusses how financialisation may have effects on all levels of the urban environment, from its physical shape to the evolution of a city's governance, and is thus central to an analysis of sustainable development.

That said, by taking a closer look at Bangalore and Greater Mexico City, we intend to discuss the conventional wisdom that international real estate investors act alone and independently, and with the same facility as they invest in stock markets. Rather, urban societies affect investors' strategies by making the integration of international capital in real estate projects possible. In this process, which we call "commutation," local urban societies have a direct influence on the practices investors use to access the metropolitan territory they covet. In other words, governments and local societies have an unrealized power over international investors because the investors need to negotiate their place at the table with other international, national and regional actors who have local relationships and expertise, and who are part of social networks without which nothing can be done.

Box 1. Bangalore and Greater Mexico City: becoming globalised cities

Greater Mexico City in Mexico, and Bangalore, India, are fast-growing economic beachheads to their countries. Bangalore's population has grown from 4 to 8.4 million in the last 20 years. The growth is based on high employment rates for skilled labour, moderate salaries and the internationalisation of the software industry since the 1990s (Parthasarathy 2004), as well as for outsourced information services, such as call centres, teletranscription, business processes, etc. (and certain high-tech activities such as research and development and biotechnology since the 2000s). This way of developing through exporting services has created wealth that, through a trickle-down process, can be seen optimistically as benefiting a large part of the local population. The growth model is probably not as inclusive as all that, and feeds strong tensions in the metropolitan area (Halbert and Halbert 2007). In particular, there is competition for land from salaried technology workers and the activities of a "globalised" economy versus the citizens of poor or modest means who work as bureaucrats, shopkeepers and artisans, or in services and light industry (Benjamin 2000). The whole is constantly energised by the attraction the urban region has for both qualified workers and poor migrants, despite the numerous criticisms that erupt from all sides against a rampant urbanisation that the local public authorities and the Karnataka State have a difficult time managing.

The result is a transformation of the metropolitan region's spatial organisation (Figure 5.1). The historical layout, where densely populated working-class neighbourhoods are adjacent to the government's administrative district and the bungalow city left over from the British cantonment that has since been renovated into the Central Business District, has been totally reconfigured by dynamics that combine multiple business clusters and sprawl. The city, which was not very densely populated originally because of the military's and state enterprises' hold over the land, has become a sprawling metropolis that stretches out 30 to 50 kilometres at its rural edges. There are also new "modern" business clusters emerging: because of competition for land, the business cluster around the M.G. Road is getting denser, and the secondary business cluster of Koramanala is expanding, while new business centres emerge on the periphery of the city around business parks and shopping malls. Since the beginning of the 2000s, several dozen kilometres of high-tech-oriented zones have been built up, ranging from new, additional

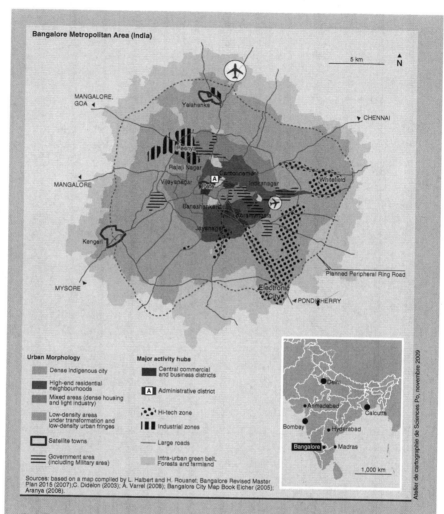

5.1 Transformation of Bangalore's metropolitan region's spatial organisation

Sources, Based on H. Rouanet (2010), Bangalore Revised Master Plan 2015 (2007), Didelon (2003), Varrel (2008), Eicher (2005), Aranya, (2008)

development for Electronic City in the southeast to Whitefield in the east, with its emblematic International Technology Park, Ltd. (which was developed by Karnataka State and the Singaporean investor, Ascendas (Halbert and Halbert 2008). Other recent dynamics, such as the opening of a new international airport 50 miles to the north of the metropolitan

area, or development projects for five new private cities around Bangalore, reinforce their "multipolar" spatial organisation, i.e. their organisation around several centres of activity. The "local productive economies"
which include light manufacturing, such as of textiles, for instance, in
the factories inherited from the socialist era, as well as the informal
economy, suffer from competition for land which forces them and their
workers to move to peripheral areas. In doing so, there is a break-up
of agglomeration economies that cannot be reproduced at a distance
(Benjamin 2000).

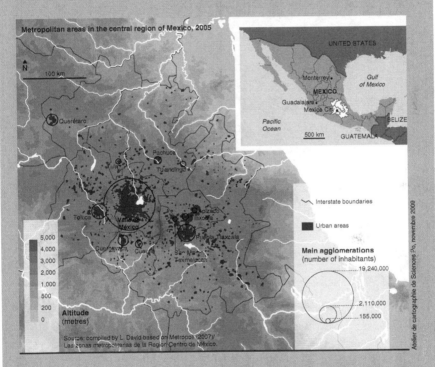

5.2 Greater Mexico City is being linked to a network of secondary cities

Source: Based on Metropoli (2007)

The metropolitan area of the Valley of Mexico which we call Greater
Mexico City has also experienced transformation recently in terms of its
spatial organisation because of population growth, which has massively
increased since the 1950s. By 2010, Greater Mexico City had nearly 22

million inhabitants. The urban area has extended beyond the Federal District which was originally designed to manage the metropolis, towards two neighbouring federal states. The region is undergoing a vigorous process of "megalopolisation"[2] (Garza 2000). The metropolitan area of Greater Mexico City is being linked to a network of secondary cities that surround it (Figure 5.2). This spatial dynamic is working even inside Greater Mexico City, where new business clusters are emerging along the main transportation routes. These transformations are linked to the economic evolution of the metropolitan region. Before the economic crisis of the 1980s and the liberalisation of Mexico's economy in the 1990s, Greater Mexico City was the largest industrial centre in the country. With the signing of the North American Free Trade Agreement (NAFTA) in 1994, and the construction of *maquiladoras* (export assembly plants) along the US–Mexico border, Mexico City deindustrialised to the extent that the retail and services sectors, which include a large share of informal, undeclared jobs, have become predominant. However, the retail and service sectors are not really economic engines because of their low level of wealth creation (Herniaux-Nicolas 1999).

However, despite a slowdown in economic growth, Greater Mexico City is the main area of consumption in the country. This factor, associated with the fact that it is the political capital of the country, encourages most multinational companies to set up headquarters or an office there. It explains why office-building construction was so strong from the end of the 1980s to the first half of the 1990s. In fact, beyond the effects of the 1985 earthquake that destroyed many government buildings, the arrival of foreign businesses stimulated the emergence of new business centres. The business centres contributed to a more vertically built environment and were organised along radial lines. In Greater Mexico City, real estate investors identify sites for office buildings all along the main roads, such as Paseo de Reforma, Santa Fe, which is on the road from Mexico City to Toluca, the ring road (El Periférico), and the Avenue Insurgentes. This trend is seen elsewhere in recent construction and the commercial and logistics activities that are undergoing extensive development. A recent census found no fewer than a hundred "service sector corridors" that feature concentrations of activities such as retail stores, banks and restaurants (Pradilla et al. 2008). Along with these services corridors there are several logistics corridors in the north of the Greater Mexico City area, such as near the municipality of Cuautitlan Izcalli, situated on the road to the metropolis of Querétaro.

The increasing power of international investors in the metropolitan areas of developed and emerging market countries

From the 1980s to the current, and probably temporary, slowdown due to the financial crisis, international investors have had ever-increasing amounts of capital to invest. A share of these funds was placed in so-called "alternative" investments, such as real estate, to diversify portfolios. Fund managers increased the number of locales acceptable for investment to spread geographic risk and profit from growth in emerging markets.

The increase in international investment in metropolitan areas

Since the 1990s, an increase and acceleration in investment flows into the construction of infrastructure and urban services in general, and in commercial real estate in particular, has been observed.

Financial services companies, such as banks, insurance companies, mutual companies, pension funds, hedge funds and private equity companies, collect huge amounts of capital, of their own and from third parties, which they invest in infrastructure and real estate projects. The funds are smaller than those invested in the largest stock markets, but are potentially very large in terms of the cities where they are invested. The financial institutions have become major actors, in the style of the Swiss pension funds that increased investments in the Swiss real estate market and are the primary holders of national real estate assets (Theurillat et al. 2006). International capital flows are made possible through the liberalisation of investments. They have picked up, in particular, with governments' frequent recourse to public–private partnerships, where private investment is used to finance urban development and the management of facilities, infrastructure and networked services.

Emerging market countries are not to be outdone, since their need for financing urban development is great. The participation of certain metropolitan regions in the redeployment of international value chains, for example the export assembly factories in Mexico; the IT industrial parks in India; the factory workshops of China; and the Russian and Eastern European engineering firms, come about because of and in support of international investors. The partial abandonment of protectionist policies after the 1990s in favour of policies of "opening" has transformed the processes by which urban spaces are produced in these countries.

Beyond global value chains to global investment risk management

The dynamics that led to "the city being seized by financial interests" (Renard 2008) are linked in part to the global fragmentation of production activities. Outsourcing and global supply chain management by multinationals and the participation of regional businesses in long-distance trade has fed a demand for infrastructure and especially adapted buildings and office parks. The existence and quality of infrastructure is a determining factor in a territory's ability to participate in the globalisation of production.

In the case of commercial real estate in newly opened countries such as South Africa, Brazil, India and Mexico, the entire value chain and local and national investment was turned upside down. This is partially because of multinational companies setting up offices in the largest cities in the world: cities drew international investors that built the office buildings they needed, which the investors very often rented to their peers. When the retail chain Walmart opened 180 stores in Mexico in 2008 and projected opening more than 250 stores for 2009, a whole set of specialist real estate developers found a rich vein to mine. Such is the case for Mexico Retail Properties (MRP), which was created in 2003 by two North American-based real estate investment companies, Equity International Properties (EIP) and Black Creek Group, and which develops shopping centres almost exclusively around Walmart group stores, including Aurrera, Suburbia, Sam's Club and Superama.

The increase in international real estate investments also results from the portfolio management strategies of fund owners. True to Markowitz's (1959) portfolio diversification theory for example, fund managers spread their investments among locations with complementary characteristics, which brings together infrastructure and facilities that are geographically very distant but which are evaluated and managed in a very similar way (Torrance 2008).

EIP does nothing further once finance capital is raised from financial institutions, namely banks, insurance companies and pension funds, and invested for third parties in funds where the principal features are defined by contract. Generally speaking, an investor relying on this type of real estate investment company commits his capital for a specific amount of time, whether 5, 10 or 15 years; a risk level, i.e. core, value added, or other targeted return; and very often a type of building or infrastructure, as well as a targeted geographic location or locations. For EIP, the first fund that opened in 1999 was mainly focused on Latin America and acquired minority shares in real estate companies according to country classifications, such as Mexico or Brazil; by type of property, such as industrial, residential, commercial or hospitality; and by type of activity, i.e. development, construction or financing. The two follow-on funds were extended to include Asia, the Middle East and Europe.

International investors are implicated to a greater degree in the reproduction of cities that they help finance and of which they hold a "piece of the action", or a share in a building or part of a city. There is talk of the financialisation of a city in that these investors apply financial tools which guide their investment decisions in terms of the production and holding of buildings, infrastructure and urban facilities. For example, N. Aveline-Dubach (2008) writes:

> With global finance, real estate has been financialised and has become an investment vehicle like any other financial asset. The real estate industry has adapted to this changing situation, inventing new forms of investment such as real estate securitisation (transformation of a real estate portfolio into financial stocks) and by borrowing analytical tools from finance.

Meanwhile, the territory does not function as a simple receptacle of external financial systems. Certainly, real estate fund managers have adopted so-called "dynamic" management methods, which, according to them, depend on three things: (1) the constant search for investment opportunities; (2) reinvestment in assets held to increase rental returns; and (3) a policy of "dynamic arbitrage" at the time of sale. To some extent, investment managers scrutinise the city on a quarterly basis (Clark 2000). However, if international investors are increasingly important in the financing of real estate projects, it is only because local urban societies enable them to be so, and in doing so participate in the shaping of international investors' strategies.

The financialisation of cities is a process of "commutation"

Even if it can be said that property development is in the process of being "financialised" with the arrival of international investors, the current dynamic can be more accurately analysed as a commutation process (Halbert and Rutherford 2008) in which various actors participate in long-distance networks that circulate international capital until it reaches the built environment. Looking at the actual form investments take helps explain the relationship between the deployment of a more financialised capitalism and the process of creating a metropolitan area. In Greater Mexico City, as in Bangalore, whether directly through one or more properties, or indirectly, through holding shares in a real estate investment trust or in land, fund managers are required to negotiate with the holders of the resources who have the ability to interconnect the financial systems to the urban territory: that interconnection is a function of commutation.

Local intermediaries prepare the terrain

Commutation implies friction between different spheres of activity, such as finance, property, buildings, regulations, etc. (and different scales, from a single building site to the distant office of a given investor). The deployment of new structured investment products and other practices by institutional investors is possible only when agents in the targeted territories *make themselves capable* of collaborating with international investors. This is because of the limited liquidity[3] of property markets and their relative opacity[4] (Clark and O'Connor 1998), particularly in cities in emerging markets where information is harder to access because of a lack of tracking tools and studies.

For example, Cuautitlan Izcalli's logistics corridor north of Greater Mexico City has become very desirable to investors since 2000. In the 1990s, national developers developed several logistics parks along its main highway. In turn, Walmart and other well-known multinational companies set up their logistics centres along the road. Following on the interest of the multinationals, international investors came to the Greater Mexico City region at the beginning of the 2000s and saw the municipality of Cuautitlan Izcalli as an investment opportunity. However, much of the land was part of the *ejidos*[5] system of cooperative agricultural land held in common tenancy by rural families and passed down through generations. While the privatisation and sale of such land has been authorised since 1994, the procedure for getting unanimous agreement from a community of owners is long and complex. After working on it for more than a year, a small developer was able to set up an agreement with one such community to put its land up for sale and offer it to international investors. This developer facilitated the articulation between financial and other constraints related to the real estate market in specific locales. This is an example of how diverse resources from more or less formal organisations can be mobilised by individuals who use their social networks, their personal knowledge and survey the metropolitan region to identify saleable land and clients.

Putting sufficiently large land parcels together requires knowledge that only local specialists have, particularly in countries where land titles are often contested. Local specialists are able to obtain cooperation from the owners – sometimes by force – and the clearing of titles and usages rights from local authorities. In political and administrative systems where economic agents openly acknowledge corruption, a system of local intermediation is crucial to facilitate the arrival of institutional investors who are less inclined to get involved in such arrangements. In all, no matter the extent of the investors' globalisation, the capital that has been collected on international markets can only effectively "land" on a given real estate project after testing the territory and negotiating with the organisations and individuals that control the targeted territory. This requires facilitators who provide valuable site-specific information and reduce entry costs.

Local facilitator is key

How the facilitator works can be illustrated by looking at a manager working in a large developer's land management office in Bangalore. His life story shows the importance of having a Rolodex of contacts and being part of a local metropolitan company. The manager comes from a Bangalorian family whose business was pharmaceuticals distribution. In the early 1990s, he started two businesses using his family's network with varied success. Benefiting from his Rolodex, he went into real estate just as it was going through its first boom years in the second half of the decade. Known for his entrepreneurial qualities, the manager made the most of his connections to property owners, politicians and local businessmen, which allowed him to identify real estate opportunities:

> I know a lot of landowners from Bangalore. I have a very large network of friends, who go back to grade school, my career as a businessman, and now my new job here (at the property company). I've met people in very different social circles and kept up good relationships with businessmen and landowners . . .
>
> [. . .] Everyone has political connections today, there's no doubt about that, everyone is very well connected. It's a small world, really. It's necessary if you want the powers that be to move our business ahead. It's about that simple, in fact: I know you and I can use your resources so you can get something for me. If you are well connected politically, so much the better – I can take advantage of your resources. All of that is important because some of the landowners are politicians. It's not the case for the majority of our transactions, but it happens sometimes. And then, often we buy land from the public authorities . . . in such a way that our contacts are very valuable. For example, when we go to a minister and we have to ask for an authorisation for, let's say, so much land, and we ask someone to please have a look at the file nicely, please, and to try to do what he can to move it ahead . . .
>
> Money is everywhere in this world . . . there is money made by everyone in real estate. But that's not all, it's also a question of contacts. It's your ability to use the tried and true services of people you know that matters for doing business.
>
> Authors' interview number 8 in Bangalore on 5 March 2008

Whose "world-cities" are they?

Key indicators and the "herd mentality"

A majority of real estate investment made by international institutional investors is concentrated in the major cities of industrialised countries and emerging markets. International real estate consultancies, such as CB Richard Ellis (CBRE) reveal an especially selective geography where, according to the depth of the analysis, the "useful" world for real estate investors is reduced to a few hundred, or even just a "Global 50" cities. This "geography of opportunities" is very selective, and when investors get interested in second- or third-tier cities like Mysore and Mangalore in India, or Lyon and Marseille in France, it is for the purpose of diversification. The "useful" world is thus represented as a hierarchy between places that are evaluated according to risk diversification strategies.

The rationale used by investors is to select spaces that have sufficiently large market volume to be relatively liquid. They also need to have a real estate consultancy industry that is up to international standards, and national legislation that is favourable to property investments. Taken together, the criteria create a sort of "herd mentality" or tendency to copy one another that some observers attribute to investors using the same comparison tools and data (Henneberry and Roberts 2008). The spatial concentration of investments is therefore a result of the process of commutation in which international investors can decide to invest only if and where actors and tools (such as CBRE's reports, legal authorisations, etc.) allow them to, as much as it is a result of agglomeration economies that are seen in contemporary geographic economics.

Take the case of Bangalore's metropolitan area. There has been an exponential growth in the demand for modern commercial real estate, despite two downward cycles during the crises of the late 1990s and 2008–09. Bangalore is regularly cited as one of the top five cities in the world in terms of having the most square metres of commercial space, alongside London and New York. Traditionally, development was supported by partnerships between the local land-owning notables and regional developers, using bank credit. International investors who are taking advantage of the opening of India's borders since 2005 have also financed the most recent real estate projects. They prefer certain neighbourhoods such as the central business district near M.G. Road or Koramangala, and smaller business centres, particularly in the southeast quadrant that links the two big high-tech parks, Electronic City and Whitefield.

The herd mentality of international investors is also seen in the type of projects in which they invest. Buildings developed with international funds follow so-called "international" standards and tend to be very large projects because of the large amount of capital funding available. In Bangalore, the preferred projects are those that offer very high internal rates of return of around 25 per cent, and are all five-star hotels, class A office buildings or office parks for prestigious tenant companies. In the wording of the marketing brochures, we find ourselves in "world-class" cities.

Financialisation and the production of two-tiered cities

The developers who partner with property investment funds are often asked to change the way they work to be able to collaborate with both the local real estate industry and the fund manager. They must produce comprehensive information about their company, their past activities, their financial situation, etc. They must also be capable of following the steps of property investing, of calculating and evaluating the criteria that the fund managers take into account when determining the interest of a project. In the case of Greater Mexico City, real estate development companies rush to find professionals who speak fluent English and who understand finance. Through the typology of buildings in which international investors invest, new "international" property standards spread to local real estate markets.

This dynamic affects the whole industry. The criteria for quality and security that are promulgated by so-called "institutional" developers who work with international finance capital quickly become the reference points for all other market participants. The buildings produced meet the needs of only the most creditworthy regional purchasers or renters. For businesses with lesser purchasing power, or for a whole host of activities such as slow-growing local industries or the informal sector, there is no option other than looking to other investors to finance the space needed, or, as in many cases, to suffer the consequences of a competition for land that is lost before it starts. Thus the process works in such a way as to be highly selective of "useful" places, and leaves aside a great number of urban areas that do not benefit from "global" capital accumulation (Harvey 1982), such as the precarious workers' encampments that spread out beneath high-rises or working-class neighbourhoods that are far from the beaten tracks of the financialised city. In Bangalore, real estate developers and investment companies produce a city of swimming pools in condominiums within a city of low-income families who must share one water source with 800 other families (Halbert and Halbert 2008), or a city where air conditioners run off their own generators in a city where there are chaotic cuts in electrical supply.

In the last five years, developers and investors seeking a "green" brand for themselves have led a trend toward a certain kind of "sustainable development" which was based on investments in buildings that were supposed to be more respectful of the environment during their construction phase and active lifetime (Nappi-Choulet 2009). The "green" image is still largely conceived as being about (predominantly) technological choices aimed at environmental preservation, rather than being aimed at any of the economic or social objectives of sustainable urban development.

The regional political and economic dynamics have also changed. The arrival of international finance capital allows certain market participants to gain influence over the real estate industry and in the political arena. The real estate industry developers and advisors who are well established in Bangalore had a strong influence on redrawing the lines of governance for the metropolitan region, notably by publishing their positions in the press (Halbert and Halbert 2007). In Mexico, the battle for transparency reinforces power struggles. On one side, international institutions who a priori lack local social contacts fight for a better circulation of information in order to facilitate the rapid development of the international market. On the other side, what are usually local actors, especially certain entrepreneurs and politicians, try to maintain their position by limiting the circulation of information, which really is their fundamental comparative advantage based on traditional social and economic systems. In this sense, if it is understood that "growth coalitions" can have an effect on the built environment of these metropolitan regions, the points of convergence follow relatively variable lines along which international investors are not systematically invited to participate. The hackneyed local power struggles behind the production of a city are repeated yet again with the arrival of international investors, even if the nature of the resulting conflict is not predetermined, especially because of location-specific factors that affect the commutation process of international real estate investment.

Conclusion

It may be appropriate to leverage international investors to finance urban development in the name of economic development. Particularly because of the way urban regions must adapt to the needs of companies that are potentially very mobile, or even, in a capitalist system, because of the urgency to provide investment vehicles to investment fund managers who contribute to collective functions such as insurance and retirement pensions, etc. And finally, because urban regions are faced with a reduction in national and local governments' funding abilities. However, the question of spatial organisation, of regulation and the negotiation of the type of "city" international investors help build is still open to debate. In formerly industrialised countries, as in emerging markets, the capacity to encourage sustainable development assumes that certain factors can be used as investment criteria, although with different terms. The criteria should include three elements:

1 Address economic issues by strengthening the ability of tenant companies to compete through the construction of buildings that are adapted to their activities, while still allowing industries or economies that are judged necessary, whether in the name of economic diversity or job creation, but not necessarily as economically viable to survive.
2 Take social considerations into account to avoid creating two- or several-tiered cities.
3 Address environmental issues by accepting the challenge of a type of development that is less predatory in terms of all resources, including natural resources.

A better understanding of the international real estate investment process is therefore necessary to identify at which times and by what means a government and an urban society can negotiate with international investors who often, and incorrectly, appear all powerful.

Notes

1   Greater Mexico City refers to the Metropolitan Area of the Valley of Mexico (*Zona Metropolitana del Valle de México*), an agglomeration which incorporates 59 municipalities in three states: the Federal District, the State of Mexico and the State of Hidalgo.
2   The term "megalopolisation" was coined in 1957 by the French geographer Jean Gottmann to describe the emerging city-region that today stretches from Boston to Washington, D.C. The term now refers to a vast urban region made up of several metropolitan areas that are interconnected by dense road networks and functional relationships.
3   A market is said to be liquid when its products can be traded rapidly, which depends on the volume of transactions. Property markets are considered to be relatively illiquid because of transaction costs and constructions delays. Their financialisation tends to increase property markets' liquidity through the creation of publicly traded property companies, the development of transparent information, comparative indicators, and the construction of standardized buildings.
4   The opacity reflects the private and specific nature of information related to a given property transaction, which makes it difficult to compare with similar goods. Opacity is often a comparative advantage for the party that possesses this private property-specific knowledge.
5   The *ejidos* or cooperative land system was created after the 1910 Mexican Revolution as part of a movement to restore land to Mexican peasants.

References

Aranya, R. (2008) "Location theory in reverse? Location for global production in the IT industry of Bangalore," *Environment and Planning A*, 40(2): 446–463.
Aveline-Dubach, N. (2008) *L'Asie, la bulle et la mondialisation*, Paris: CNRS Editions.
Bangalore Development Authority (2007) *Bangalore Revised Master Plan 2015*, Bangalore: BDA.
Benjamin, S. (2000) "Governance, economic settings and poverty in Bangalore," *Environment & Urbanization*, 12(1): 35–56.
CBRE (2007) "Global market rents. Office rents and occupancy costs worldwide," CBRE, November.
Clark, G. L. (2000). *Pension Fund Capitalism*, Oxford: Oxford University Press.
Clark, G. L. and O'Connor, K. (1998) "The informational content of financial products and the spatial structure of the global finance industry," in Cox, K. (ed.) *Spaces of Globalisation: Reasserting the Power of the Local*, New York: Guildford, pp. 89–114.
Cushman & Wakefield (2009) "Global commercial property investment volumes (excluding residential)," *International Investment Atlas*, Cushman & Wakefield RCA & Property Data.
Didelon, C. (2003) "Bangalore, ville des nouvelles technologies," *Mappemonde*, 70: 35–40.

Eicher (2005) *Bangalore City Map Book*, New Delhi: Eicher, ISBN No. 81–87780–50–3.

Garza, G. (2000) "La mégalopolis de la ciudad de Mexico, según escenario tendencial 2020" and "Ámbitos de expansión territorial," in Garza, G. (coord.), *La ciudad de México en el fin del segundo milenio*, Mexico: Gobierno del Distrito Federal.

Halbert, A. and Halbert, L. (2007) "Du modèle de développement économique à une nouvelle forme de gouvernance métropolitaine?," *Métropoles*, 2, Varia. Available at http://metropoles.revues.org/document442.html (accessed 12 June 2009).

Halbert, A. and Halbert, L. (2008) *La Ruée vers l'Est*, film DVD, 38 mins.

Halbert, L. and Rutherford, J. (2008) "Conceptualising city-regions as flow-places." Regional Studies International Conference, Prague, 27–29 May 2008.

Harvey, D. (1982) *Limits to Capital*, Oxford: Blackwell.

Harvey, D. (1985) *The Urbanization of Capital*, Oxford: Blackwell.

Henneberry J. and Roberts C. (2008) "Calculated inequality? Portfolio benchmarking and regional office property investment in the UK," *Urban Studies*, 45:5–6, 1217–1241.

Herniaux-Nicolas, D. (1999) "Los frutos amargos de la globalización: expansión y reestructuración metropolitana de la ciudad de México," *EURE*, 25:76, 57–78.

Markowitz, H. M. (1959) *Portfolio Selection: Efficient Diversification of Investments*, New York: John Wiley & Sons.

Metropoli (2007) "Crecimiento de la metrópolis de México de 1824 hasta 2005," published by ciudadanosen red.com, 11 September 2008. Available at http://ciudadanosenred.com.mx/mapas/este-mapa-muestra-c%C3%B3mo-ha-crecido-metr%C3%B3poli-partiendo-desde (accessed 16 January 2013).

Nappi-Choulet, I. (2009) *Les mutations de l'immobilier: De la finance au développement durable*, Paris: Éditions O. Jacobs.

Orléan, A. (1999) "*Le pouvoir de la finance*," Paris, Éditions O. Jacob.

Parthasarathy, B. (2004) "India's Silicon Valley or Silicon Valley's India? Socially embedding the computer software industry in Bangalore," *International Journal of Urban and Regional Research*, 28(3): 664–685.

Pradilla Cobos, E., Márquez López, L., Carreón Huitzil, S. and Fonseca Chicho, E. (2008) "Centros comerciales, terciarización y privatización de lo público," *Ciudades* 79, Puebla, Mexico: RNIU.

Renard, V. (2008) "La ville saisie par la finance," *Le Débat*, 148, 106–117.

Rouanet, H. (2010) "Bangalore (Karnataka, India): Promotion immobilière et construction de la world-city," Paris: Département de Géographie, Université Paris-1.

Theurillat, T., Corpataux, J. and Crevoisier, O. (2006) "Property sector financialisation: the case of Swiss pension funds (1994–2005)," Working paper in Employment, Work and Finance.

Torrance, M. I. (2008) "Forging glocal governance? Urban infrastructures as networked financial products," *International Journal of Urban and Regional Research*, 32:1, 1–21.

Varrel, A. (2008) "Back to Bangalore" – Etude géographique de la migration de retour des indiens très qualifiés à Bangalore (Inde), PhD thesis, Poitiers: Université de Poitiers.

Chapter 6

Are first-generation suburbs of Mexico City shrinking?

The case of Naucalpan

Sergio A. Flores Peña

The sole fact of population loss is not enough to apply the notion of "shrinking" to urban areas of metropolitan Mexico. As the case of Naucalpan illustrates, economic adjustments among sectors and (an apparent) population decline may be accommodated in a process of restructuring that is characterized by changing interrelations between several dimensions of social life at different levels. This analysis displays the ways in which changes in economic activity (global as well as local) are related to changes in urban form and spatial dynamics, always mediated by a changing institutional setting, the "spirit" of the time, demographic transition and the local (competitive) advantages produced by a particular historical trajectory. Non-economic factors such as symbolic values of regional leadership and prestige are historical and play a definitive role in the "capacity to adjust" in any given urban region.

Introduction

During the 1950s, import substitution policy was consolidating what would be known later as the "Mexican Miracle" of the 1960s. However, this economic bonanza was only one of a number of transformation processes affecting the basic structures of Mexican society. Along with industrial growth, there was the "Green Revolution" of the agricultural sector with technical modernization and export orientation. Simultaneously, an urban middle class highly tied to the expansion of white collar positions emerged as domestic markets consolidated.

Submitted by the Asociación Latinoamericana de Escuelas de Urbanismo y Planeación (ALEUP).

Originally published in Ivonne Audirac and Jesús Arroyo Alejandre (eds), *Shrinking Cities South/ North.* © 2010 by kind permission of Universidad de Guadalajara, Centro Universitario de Ciencias Económico Administrativas, Periférico Norte N° 799, 45100, Zapopan, Jalisco, Mexico.

Such changes were associated with an equally dramatic process of population growth and changes in its spatial distribution. While Mexico's population as a whole grew at a rate of about 3 per cent, the main cities, particularly Mexico City, Guadalajara and Monterrey, grew at twice that rate. It was in this context that suburbs appeared in Mexico City.

Naucalpan, State of Mexico, is a municipality located in the northwestern edge of the Federal District (DF), where Mexico City was founded and historically developed. It is traversed by the highway to the city of Queretaro, which continues on to Laredo, Texas, and the American east coast. This connection was the main link to the US economy and the main economic corridor between the two countries. Naucalpan's topography is a continuation of the hilly west side of Mexico City. As the home of the wealthiest Mexicans, it provided a great opportunity for middle to high-end real-estate development.

Historically, Naucalpan was the initiator of the industrial expansion of Mexico City's Metropolitan Area (MCMA) towards the north. It has enjoyed a position of leadership that has proven fundamental in achieving the desired economic development. Currently, five entities, one of them Naucalpan, account for 71 per cent of the industrial output of MCMA (PUEC 2006a).

As usually portrayed in the planning literature, shrinkage is understood as population loss resulting from the process of global economic restructuring, which implies the reorganization of political and social life in specific places. In some places the resulting population loss is tied to the demise of economic "relevance" – a spatial mismatch that devalues labor capacities and turns fixed capital (the "built environment") obsolete. In this chapter, I attempt to establish the conceptual base for understanding the transformations in urban form and activity patterns in the perimeter of Naucalpan as a component of Mexico City's Metropolitan Area.

The issue addressed here is how the combinations of changes in economic activity patterns combined with changes in urban form are shaping the future conditions of those who live in Naucalpan. However, the central question on this subject relates to what kind of restructuring processes result and their implications for the wider themes of social welfare and sustainability. In this respect, the case study of Naucalpan can help us understand a concrete manifestation of the process of economic restructuring.

The analysis of situations, actors and processes are presented in eight parts. I begin with a discussion of the conceptual approach used to situate the notion of "shrinking cities" within the Latin American context. In the second section, I present an overview of the national policy of industrialization through import substitution as the foundation for the industrial development of Naucalpan. Section three explains Naucalpan's urban form and the nature of its socio-spatial and economic dynamics, while sections four and five describe and discuss the social and economic trajectory of Naucalpan in relation to the national economic policies and the way alliances and specific interests worked to produce recent conditions and change processes. Section six presents the current conditions of development based on empirical data gathered from recent studies of manufacturing and services in order to address the initial question about shrinking conditions. Finally, sections seven and eight offer some conclusions about the notion of shrinking and the processes of structural change in Naucalpan.[1]

Conceptual approach: the city as a space of relationships

In this work, I frame the analysis of "urban shrinking" in the wider concept of social division of labor as the varying geography of opportunity offered by an economic arrangement in a given place (a city) at a given time. This site-specific economic arrangement – part of a general (global) economic space – is then characterized as a set of productive and distributive relationships, spatially organized for the appropriation of rent (Camagni 2004). The conceptual approach centers on the interrelation between processes of economic change and those of territorial transformation, whose particular combination in a given institutional setting and time period produces differentiated results of wealth generation and quality of life.

On the consumption side, Naucalpan shows a highly polarized situation. Around half of the developed area is occupied by middle to high income residents with patterns of consumption influenced by globally symbolic content. In this context, globalization can be observed in the consolidation of Plaza Satélite, which is the most important (shopping) center for luxury goods consumption in the whole northwest region of MCMA. It is also a source of "distinction" or social positioning (Bourdieu 1988) as a recognized place of centrality and prestige, in the metropolitan region. Together, the upscale residential areas, the Plaza Satélite retail and services center, and the commercial corridor along the highway to Querétaro form the older core of the most prosperous suburban section (the northwest) in the MCMA.

The other half of the developed territory is occupied by lower income groups, some of them living in poverty conditions. Their consumption patterns are hardly related to those previously described. They rely mostly on the traditional city center for retail and service provision and live in a place that looks like "anywhere" in Mexico, with a great diversity of small businesses, architectural styles and types of material; intense use of commercial signs, heavy traffic and congestion, with no noticeable architectural features other than the municipal government building (Palacio Municipal).

This very polarized pattern of socio-spatial differentiation defines in principle the urban form of Naucalpan and the possibilities for its evolution. The "modern, well-to-do" area that surrounds the main road towards Querétaro is basically dominated by "station wagon urbanism" (Weiss 1988), all formal (legal) subdivisions that are functionally organized south–north, following the highway, and enjoying access to Mexico City. It forms a well-known conservative agglomeration that perceives land-use change (density and activity) as a threat to the quality of the area. Zoning controls and single-family detached houses are key symbols of prestige, fiercely defended by the community.

The other half of Naucalpan, composed of middle-income to poor residential settlements – with many of them in the process of becoming formal neighborhoods – has traditionally relied on state support to improve their material conditions. As the mix of neighborhood economic activity expands with the growth of home-based businesses, residents' concerns are geared towards any public measure, service or work that will expand the likelihood of their access to opportunities for better or additional sources of income.

Therefore, public demands from poor settlements are for public services. They seek access to mainstream economic opportunities that are clearly restricted by the way they are functionally connected to the "modern, well-to-do" area (along the Querétaro highway), where industrial parks and residential subdivisions are located. The physical link between both groups is the highway to Toluca, which defines an east–west axis of territorial organization and intersects the south–north axis precisely in the traditional city center anchored by the Palacio Municipal. Along this road, one can see the extension toward the west of the traditional center as an urban corridor of lower profile, mostly made up of small retail shops and fewer services.

As a result of current urban socio-spatial conditions, Naucalpan demonstrates two different faces of consumption-based local economic activity. On the one hand, we can observe a prosperous, regionally based center of luxury goods and services, originally associated with upscale residential areas that explain its symbolic central positioning in Mexico City as a whole. This area's expansion is tied to the transformation of the entire northwest sector of the MCMA. Juxtaposed to it is a mix of low-income and basic consumption and production settlements, some of them still in the process of achieving adequate living conditions and relying on state support and on the expansion of the well-to-do neighborhoods in order to gain access to urban infrastructure, services and jobs.

Given this dual pattern of development of Naucalpan's urban fabric, its future also depends on the degree of obsolescence brought about by real estate life cycles and differences in the aging process of the housing stock. As they reach the fifth decade of existence, the oldest and most prestigious subdivisions – built during the late 1950s and 1960s – are likely to face demand for products in the real estate market that are more appealing to the lifestyle preferences of the upwardly mobile professional singles, new couples or smaller families (one or two children). Failing to do so may initiate a filtering down process to a new population of lower economic means and of diminished symbolic status (PUEC 2006a).

The consequences of aging are different for the lower income areas. Living in such conditions corresponds to an open process in time, where community organization and government programs combine with households' initiatives in a discontinuous, although sustained, process of house improvement and neighborhood upgrading. The most relevant disruption of this process comes from the reduction of public assistance, although community and households' contributions are maintained. Formal real estate markets are not much of a factor in a social context unattractive to conventional means of credit and finance.

During times of economic hardship, many small businesses turn to informality for survival, as this allows them to internalize income that otherwise would be dedicated to comply with fiscal obligations (Bustamante 2008). As part of this process of informalization, many businesses relocate to their owners' homes. New employees are hired locally, forming an economic network partially based on extra-economic relations, oriented to accomplish business survival and meet labor's basic needs. Problems and confrontations arise when actions from the market or the state (or both) block or restrict pathways to social mobility. This often leads to the appearance of resistance movements.

However, the poor and low-income areas are far from being homogeneous. They actually comprise a wide variety of material and mobility conditions very different in nature from the "landscapes of despair" of developed cities (Wallis 1991). Location – as access to services and opportunities of the formal city – plays an important role in the process of neighborhood improvement. Living conditions worsen in neighborhoods farther away from the road to Toluca, which is the organizing axis of the whole area. Although mobility cannot be considered as a rigid pattern, it is in general time–distance governed. Those who settled last are the outermost settlements, have the least accessible locations and the worst of conditions.

The production of Naucalpan's competitive advantages

Industrialization in Naucalpan started in the early 1940s as a result of the expansion of the local economy derived from external sources, and the consolidation of Mexico City as the most important market and the economic center of the country. In the drive for modernization of the country as a whole, central government officials considered the primacy of Mexico City inconvenient.[2] Since then, decentralization of (Mexico City) industrial activities has become an-ever present feature of the discourse and policies of economic development in Mexico. The first step in this direction was the introduction of the Law of Fiscal Exemption for Industrial Activities (1940), which offered tax breaks to all new industries located outside the Federal District (i.e. outside Mexico City's boundaries at that time).

The rapid expansion of the local market (Mexico City grew at an annual rate of 7 per cent during the 1920s and 3 per cent in the 1930s) led to a simultaneous expansion in most industries. For those in need of new space, Naucalpan was the best location. It was tax free and right on the outskirts of the city, and was easily connected with the main internal markets: old cities of the central region and the emerging regional centers of Guadalajara and Monterrey (Garza 1985 and 2003; Bustamante 2008). Furthermore, a more relevant and persistent advantage emerged for Naucalpan: its location in relation to Mexico City. The following year (1941), a new law was issued for the promotion of industry. The Law of New and Necessary Industries (LNNI) offered tax breaks of 5, 7 or 10 years to new industries in compliance with the criteria for being "new or necessary". Since there were no geographical restrictions for the law's application, more than 70 per cent of the industries that were created located in Mexico City and its surroundings, favoring Naucalpan and other similar locations.

However, the definitive consolidation of Naucalpan as a site of privilege for industrial development and living conditions came during the 1960s and 1970s, the height of the import substitution period. The final years of the 1940s witnessed the rise of the developmental state, responsible for steering the national economy toward modernization. Apart from the programs of fiscal incentives already mentioned, the federal administration of Miguel Alemán (1946–1952) carried out a number of major public works aimed at creating the conditions for development and modernization. Part of this endeavor benefited Mexico City and its surroundings, particularly Naucalpan.

By the end of this period, Naucalpan was linked to the wealthy west and southwest parts of the city – including the new university campus – by the Periférico (ring road) freeway and to downtown, the international airport and the city's east exit by the Viaducto freeway. As will be seen later, this process of construction of the competitive advantages of Naucalpan was not free from the interests of "political-entrepreneurs", represented by President Miguel Alemán himself and his predecessor Manuel Ávila Camacho. The relocation in 1947 of the bullring of the former Hacienda La Herradura, property of President Alemán, from the would-be parcels of Ciudad Satélite to the Cuatro Caminos site located on the jurisdiction's old highway, marked the beginning of Naucalpan's dominant position in the MCMA as we know it today. It launched the site planning for the development of the estates of both presidents (Camacho 2008).

In 1953, the new federal administration (1952–1958) widened the policies for industrial expansion with two additional instruments. The first one was a guarantee fund for the financing of new industrial initiatives of small and medium-sized businesses (FOGAIN), and the second was a trust fund for the creation of infrastructure in the form of industrial parks and cities (FIDEIN). In addition to these policies, there were other policies for industrial promotion that played a role in promoting concentration, such as lower prices in electricity and natural gas, better services of communication and transportation, and lower tariffs for railroad shipments.

These policies highlight how instruments – in principle oriented towards the promotion of the industrialization of the entire country – ended up allocating most of their resources in Mexico City and its immediate surroundings, as was the case of Naucalpan. Although by 1954 the law of fiscal exemption was canceled for the DF, the effect of such a measure was almost unnoticeable because of inefficiencies in small and medium sized targeted cities and because agglomeration economies more than offset the law's tax breaks. The only areas that benefited from the measure were DF's neighboring municipalities that could take advantage of both tax breaks and the access to markets, infrastructure and lobbying for key government decisions (Lavell 1972). From 1940 to 1970, two-thirds of the investment supported by the LNNI, 50 per cent of the credits granted by FOGAIN, and 95 per cent of all the industrial area built under FIDEIN was located in the MCMA (Garza 1985).

From the early 1960s to the end of the 1970s, the northern periphery of Mexico City was consolidated as the most important concentration of industrial activities in Mexico. In 1975, industry in the MCMA represented 54 per cent of the national industrial output, 25 per cent within the boundaries of DF and the remaining 29 per cent in the surrounding municipalities of the State of Mexico (Garza 1985). The rate of newly created firms illustrates the nature of the transformations: the national stock of industrial firms grew by 8,650 units between 1960 and 1970 and almost quadrupled the following decade, with 33,185 new firms that nearly doubled the output value. For Naucalpan, the number of industrial firms went from 7 units in 1956 to 1,158 in 1975, 2,034 in 1980 and peaked at 2,429 in 1989.

Urban population growth, vis-à-vis the expansion of urban markets, explains much of the evolution of Mexico City and its urban hinterland. From 1940 to 1970 the number of cities with a population equal or greater than 15,000 inhabitants more than tripled, from 55 to 174 and absorbed 66 per cent of the nation's population growth, equivalent to 18.8 million people. While it was one of the foundations of "success" for national industrialization, it was also the beginning of a transition from a monocentric system with Mexico City as the nucleus, to a polycentric one based on a few other cities with an older and dominant principal center (Garza 2003).

Industrial development, though important, is not sufficient to explain the dominant position of Naucalpan in MCMA and the comparative advantages upon which it was built. We have seen already how its relative position changed after a number of projects of infrastructure and public works were carried out during Alemán's administration. Those public works were the basis for Naucalpan's genesis – since a significant portion of its territory was family property of former presidents Alemán and Ávila Camacho – and were the living proof that modernization (through industrialization) was attainable and could be seen, felt and lived. Hence, Naucalpan then was endorsed with symbolic value and the associated public investments, which rendered it more dominant than the neighboring municipalities of Tlalnepantla and Ecatepec – also subject to similar pressures of industrial expansion. The symbolic endowment came in the form of urban developments which provided living space for consumption and social reproduction to all those families associated with the new (modern) way of living. For Cardoso and Faletto (1969), it was an "island of modernity" that legitimized a modernization project based on an alliance of foreign capital (multinational corporations (MNCs)), national industrialists, the elite labor movement and the (developmentalist) government. The construction of the Ciudad Satélite (Satellite City) development was the keystone of such vision.

The conversion of the road to Queretaro into a toll freeway by the federal government in 1957 created the continuity of the Periférico ring road through Naucalpan (and the rest of the northwest area of the MCMA), and consolidated this major artery into the backbone for development of the MCMA. Not surprisingly, the same road crossed through the middle of Ciudad Satélite. Ciudad Satélite, planned and designed by the leading planners and architects of the time, was modeled after the English Garden Cities, in its new towns version, but it was strongly influenced by the American suburbs of the late 1920s (the Radburn Idea of Wright and Stein) with its characteristic "superblocks," culs-de-sac and the differentiation between car and pedestrian traffic, with an extensive use of linear parks inside the superblock (Birch 1980). This new pattern of urban design became one of the symbolic characteristics of the well-off areas of Naucalpan and the whole northwestern region of the MCMA. By 1958 Ciudad Satélite was already a high-end real-estate market. It marked the beginning of a process that is still in existence and that clearly portrays the manner in which competitive advantages of a place within a region are produced.

Following the opening of Ciudad Satélite, there was a cascade of new land development projects in the area. In 1958, the Torres de Satélite (Satellite Towers) were built, including a new element of public art as the symbolic representation of a modern lifestyle in Naucalpan. With the start of development projects such as Lomas Verdes, Fuentes de Satélite, Bosques de Echegaray, La Florida, etc. a process was initiated, which is still ongoing, although somewhat reduced. This process has produced an unusual pattern of land occupation where more than half of the residential areas are inhabited by families with an economic profile of middle to high income. The symbolic capital attached to a place or locality, which provides the condition of dominance, was achieved with the construction of the Plaza Satélite shopping center oriented to luxury consumption. With this event, the image of a modern, progressive area emerges through the efforts of developers and settlers, who now proudly claim that the Satélite area (northern portion of Naucalpan) is distinguished by its culture (of consumption), its social status (as modern, well-off) and its appearance (of an American suburb). Distinction in this sense becomes a self-reinforcing process, with concrete material implication for the settlers. As argued Bhagwati (2004) and others, the wage of an individual depends not only on education, experience and talent, but also on where he or she lives and works.

Public works and infrastructure projects were always present and contributing to the competitive edge of Naucalpan. In 1973/74, the municipality was endowed with former *ejido*[3] land, where two important recreational areas (Naucalli and Los Remedios Parks) and the UNAM campus of Acatlán were built. These major urban development projects reinforced the competitive aspects of Naucalpan's quality of life and knowledge economy. By 1980, Naucalpan had gained a solid position as the leading and most prestigious area of the northern periphery of the MCMA, with clear and identifiable elements of symbolic representation in housing and land development, economic activity (industry), luxury consumption, education and recreation. By blending all these attributes, Naucalpan developed as the ideal living place for many of the upwardly mobile "techno-bureaucrats" of northern Mexico City. This also raised its economic profile due to an unusually large percentage of residents with middle to high income. Similarly, the Asociación de Industriales de Naucalpan – a very strong guild of industrial entrepreneurs of national relevance – a large number of universities for human capital formation, a scientific research center in Acatlán (UNAM), and a reasonable supply of recreational options – an outstanding example among these being the Naucalli Park, equally contributed to Naucalpan's leading position.

In 1982, the Cutzamala Aqueduct started to operate, supplying water at the rate of 16 cubic meters per second to Mexico City; such flow was intended to assure the MCMA of water availability in an effort to stop the overexploitation of underground waters (Legorreta et al. 1997). Owing to the topographic conditions, the main supply line to the metropolitan municipalities of the State of Mexico in the MCMA was accessed through Naucalpan's mountains, thus giving a privileged position to the municipality in terms of water availability. Until today, in a context characterized by chronic problems of water supply, Naucalpan boasts having ample water availability for new developments (PUEC 2006a).

In 1984, Naucalpan's link with the central areas of Mexico City was improved with the extension of the Metro System (subway) and the building of the Cuatro Caminos terminal station (Camacho 2008) precisely on the border with DF, along the Periférico freeway. Because of the intensity of the exchange,[4] the site became one of the "gates" to downtown Mexico City and an intermodal site for traffic flowing from many different places of the northern periphery.

The effects of the structural adjustment programs

Reorganization of industry and economic activities at the end of the twentieth century in Mexico cannot be explained without considering the effects of the Structural Adjustment Programs (SAPs). SAPs were designed by the World Bank in 1980, based on the experience gained from the stabilization programs of the IMF implemented during the 1970s. Their ultimate aim was to overcome the international crisis of profitability by allowing new waves of economic growth through the opening of new markets. The purpose of the SAPs was to achieve two main conditions. First and foremost was the opening of internal markets to the world economy. The second involved the cancellation of import barriers, price controls and all kind of public subsidies. These two conditions were the basis for the next tier of developmental policies aimed at:

- reducing domestic expenditure (subsidies to imports, manufacturing and consumption) in favor of production and savings;
- changing the pattern of exports and imports substitution, to open up the economy to international competition;
- restructuring international debt and attracting international investment for new development.

These programs were a precondition for SAPs' debt restructuring and payment guarantee, and resulted in a global set of new, market-based economic development policies emphasizing incentives to economic agents, selective subsidies, and limited regulations. Incentives were mainly applied to capital investment, better governance for economic growth and the relaxation of institutional structures hampering new investments, technology development and acquisition, merger of firms and economic restructuring. Subsidies were mostly for labor training and productivity, but also in the form of the state's supply of key productive assets. The strengthening of regulations was required for environmental protection, health and security (SAPRIN 2002).

SAPs' strategies applied to improving cities' competitiveness resulted in a decentralization policy based on the privatization of public functions and the cancellation of social programs (markets, clinics, schools, nutrition, families, etc). However, privatization had to redefine the functions of local governments (devolution) to allow for market intervention in the provision of public facilities, and for change in the legal and regulatory framework (deregulation) that would allow private firms to construct and operate public utility systems. The expected outcome was a rapid expansion of the local private sector through the reduction of expenditure on social programs and public services. This, in turn, would generate economic growth and new public revenue; improve local government's general financial health and the capacity to attract new development. Lastly, recommended use of labor-intensive technologies would improve formal employment levels.

Far from a "win–win" solution, SAPs produced cycles of economic contraction and currency devaluation for more than a decade (1982–1995). During those years, the Mexican economy was characterized by a drastic reduction in local investment for development, because the available capital was mostly dedicated to (international) debt obligations. A series of currency devaluations were implemented to promote exports and foreign direct investment for the improvement of the balance of payments. This resulted in extremely high rates of inflation and the impoverishment of a significant part of the population. Under this set of conditions, although severely restricted by the same SAPs' conditional ties, foreign investment was the only available source of financing development (Bustamante 2008).

By the end of the twentieth century, the cumulative effects of all these conditions highly favored places with an inherited capacity to support new economic development as result of the investment policies exercised during the import substitution period, particularly those bolstering important agglomeration economies.[5]

The processes of territorial transformation of economic activities can be considered slow and intergenerational, and are mostly determined by inherited circumstances and support initiatives (policies) from the state (Marshall, in Bustamante 2008)[6]. The impediments for the geographical transfer of agglomeration economies (Scott 2001) turn the central places of major urban areas into "innovation transmitters" that favor the creation of new firms and businesses. Those are places of greater prosperity where accumulated infrastructure, knowledge and opportunities for social exchange create the highest potential benefit. The localized nature of agglomeration economies – at specific points or areas in space – makes certain places more productive than others and, in time, these places become capable of supplying services and products to other regions and cities.

Therefore, Naucalpan can be said to be located at the core of the positive effects produced by industrialization policies because of the added powering effects of two fundamental factors: the spillovers (locational externalities) derived from its proximity to Mexico City, of which it became a part since the metropolization process started in the 1950s, and the interest of key actors – former presidents – that were in a position to favor the consolidation as the major center of the new suburbs of MCMA. The question to be seen now is whether all these advantages, historically given, were insufficient to provide Naucalpan with the basis to compete in the new global economy, and hence, corroborate if Naucalpan, a symbol of modernity through industrialization, is currently undergoing a process of general decline that we have characterized as "shrinking."

Is Naucalpan actually shrinking?

There are several ways in which the question of urban shrinking has been addressed. Most of them consider deindustrialization and disinvestment in general as the basic forces that drive a sustained and progressive situation of decline, which is ultimately expressed in a cumulative process of urban decay and outmigration. We have already established the historical trajectory of Naucalpan in the framework of wider processes of development in Mexico and the role of the MCMA within it. In the following pages we will review some empirical facts of the last two decades in an attempt to understand recent performance conditions in the municipality and to identify mismatches and dysfunctionalities that allow us to confirm or reject the applicability of the "shrinking city" notion to our case study.

Demographic changes[7]

As mentioned in the introductory section of this chapter, one of the reasons to analyze Naucalpan from the shrinking city perspective was the fact that in the last decade, the municipality showed a negative population balance, with a 7 per cent population loss. Naucalpan's growth initiated during the 1950s when it almost tripled its population, passing from 29,876 to 85,828 inhabitants. It was the beginning of a sustained period of growth that peaked during the 1970s, reaching up to 730,170 inhabitants in1980. This growth was 8.5 times larger than that registered in 1960, and equivalent to a sustained annual rate of 11.12 per cent. However, it came to an end in the final years of the 1990s, and during the early years of the new century (2000–2005) the population fell from 858,711 to 821, 442, with a loss of 37,269 inhabitants and a negative growth rate of -0.88 per cent. Although at first glance, this seems to be the start of a process of decline and population loss, when seen through the lens of more detailed demographic change in Mexico, another dimension emerges. The fact that Mexico's population is growing at smaller increments and growing into an older age structure is an essential part of our interpretation.

Fertility rates in Mexico are rapidly declining, from more than 6 children per adult female in 1960 to 2.7 in 2000 and an estimated 2.1 in 2005. Such a change involves the adoption of new social attitudes and values. It is the result of the wide use of contraceptives and the consequence of a growing improvement in the position of women in society, who now aim for smaller families. While in

1960 43.9 per cent of women of 40–49 years old had 6 or more children, in 2000 that percentage decreased to 25 per cent. The probability of having more children clearly diminishes between the second and third child. This behavior gets stronger in cities bigger than 100,000 inhabitants where the probability of having more than three children is sharply lower than in the rest of the country, mainly because more than 62 per cent of fertile women use contraceptives (Paz 2001).

Economic and welfare conditions are also contributing factors and are directly related to the case of Naucalpan. Human development conditions for the MCMA are similar to those of Cyprus and Spain in Europe (Tuirán 2001). If the current trend in fertility rates is maintained, it is soon expected to reach a condition of stability with only two children on average (Paz 2001). Improvements in the educational profile of population shows important changes equally biased towards the larger population centers. The rate of inhabitants attending a classroom of any kind is of 88.1 per cent for places of 2,500 inhabitants or less, and of 95.2 per cent for places bigger than 100,000 inhabitants (Gutiérrez 2001). Permanent single status is a growing reality, which is also associated with the improvement of the social condition of women and to urbanization. Its effect was observed to be twice as large in places with more than 100,000 inhabitants than in those with 2,500 or fewer inhabitants. New lifestyles are also emerging: in the same decade (1990–2000), single person households grew from 4.9 per cent to 6.4, showing a 30 per cent increase (López 2001).

Changes in labor markets are a part of this new reality. The rate of participation of women rose by 5 per cent between 1990 and 2000 (from 31 to 36 per cent), although still at a level substantially lower than men (71 to 78 per cent). The most important changes in this respect included the rise in the proportion of households headed by women, from 17.3 per cent in 1990 to 20.6 per cent in 2000. For metropolitan areas, this number is estimated to be around 25 per cent (one in four). Although census data is not sufficiently conclusive in the causes of such trends, it is generally presumed that they are driven by a growing tendency toward marital breakdown (stimulated by outmigration), the single mother phenomenon, and most important, the growing acceptance of new forms of households with women in the role of heads or unique provider (ibid.).

A 1999 survey on family roles showed that 32 per cent of women and 24.5 per cent of men accepted males in the role of housekeeper and females in the role of provider. It is clear that this is a societal trend because the level of acceptance grows with the decrease in the age of the subjects interviewed. Among young people, the percentage of approval are 60 per cent for women and 58.9 per cent for men (ibid.)

Economic crises have demographic impacts too, and families adjust to economic stress by increasing the number of income earners in the family as a defensive measure for the maintenance of welfare conditions. From 1977 to 1999, the average number of income-receiving members grew from 1.53 to 1.79 (Cortés 2001). The cumulative effects of all these changes and trends has been a shift towards smaller size households, decreasing from 5.8 members in 1970 to 4.4 in 2000 (Schteingart 2001). Furthermore, we observed a generalized decrease in family size with an increased number of working members, where women partici-pate in an unknown manner. New forms of household and household composi-tion, e.g. single parent and permanent single (one-member household), are now part of daily life scenarios.

In the case of Naucalpan, these changes are more accentuated. While at the national level, the average size of the household decreased from 4.86 to 4.4 mem-bers, in Naucalpan, it decreased from 4.83 to 4.18 members for the 1990–2000 period, reaching 3.94 members in 2005. When we apply these household-size figures to population registered in 2005, we observe a situation of growth in the number of households from 206,006 to 208,488. Based on this evidence, we can conclude that demographic changes in Naucalpan are the expression of general trends of demographic transition of the country as a whole, and that a more detailed analysis, considering new conditions of family size and structure, would give an alternative picture. In this respect we can state that Naucalpan still attracts new population, but primarily to the new medium-to-low income settlements, worsening the problem of social differentiation and segregation.

Manufacturing sector

During the early years of the current century (1999–2003), Naucalpan seemed to enter a process of economic decline, with a loss of 185 units from a high of 2,104 to a low of 1,919 firms. At the same time, a reduction in the average number of employees in a firm, from 38 to 34, could also be considered a symptom of dis-investment, which is part of an ongoing shrinking process (PUEC 2006c). How-ever, a closer look at the same figures suggests changes in the local organization of manufacturing that may appear to cause an apparent loss of economic drive while activities related to manufacturing could be actually growing. In this respect, Bus-tamante (2008) reported that changes brought about by globalization and the opening of the local economy to international markets have produced a rather complex pattern of interrelationships between firms that can hardly be registered by conventional statistics. This, in fact, challenges the widely accepted view of deindustrialization in metropolitan areas. According to Bustamante, industrial reorganization in the textile industry of MCMA, which is one of the leading industries of Naucalpan, involves the simultaneous occurrence of all kind of com-pany relationships through the global space economy. These relationships may be of various types: vertical, horizontal, internal to the firm and internal to the region or local area, between all kinds of firms (classified as big, medium, small and micro).

One of the main features of these changes is an ongoing process of technology transfer and assimilation, directly tied to the opening to international competition, which enables local firms to compete and contract in the international market. Most of the usual company innovations are related to internal reorganization for the adoption of flexible labor practices and software for "just-in-time" performance, complemented by a selective acquisition of machinery. In a competitive world governed by project-based contracts – uncertain and discontinuous – the purchase of new machinery or technology requires a very large commission; otherwise, the international contractor may resort to outsourcing or subcontracting. The result is a new economic ecology of manufacturing firms, where all kinds of firms collaborate in all sorts of ways. Big firms may subcontract smaller ones but the opposite holds equally true. In this process of adaptation to the new economic environment, some firms specialize in certain tasks and may find it unnecessary and expensive to operate in a formal manner, withdrawing to informality. Informality, in this context, becomes an adaptive strategy to the conditions of increased competition – national as well as international – in a local environment. It keeps firms from being overregulated by rules inherited from the import substitution period. It also shelters them from the disadvantages associated with the drastic reduction of the role of the national government as provider of incentives and new infrastructure (ibid.).

Becoming informal is one of the most frequent survival strategies in manufacturing. Firms and workers themselves make administrative changes to be written off from official records and fiscal registries to avoid payments of taxes, social security and the like. Usually, these kinds of adaptive measures involve relocating the business, mostly to the house of the owner (of the workshop) or to that of one of the workers. Another frequent strategy is the already mentioned practice of subcontracting. These changes produced an apparent deindustrialization because productive units are no longer visible and fragmented tasks are often very difficult to differentiate from services. Changes in manufacturing are also derived from the reorganization of MNCs, which redistribute production and administrative tasks in order to strengthen their competitive position in the world markets. This kind of vertical redistribution (the multi-plant arrangement) looks for places with privileged conditions for production, such as Naucalpan.

Being in Naucalpan allows MNCs to have direct access to the central region of Mexico, which represents a market of around 25 million people. In addition, it is a location (MCMA) with an easy access to federal government decision-making levels, well connected to all the main ports and regions of the country by highways, and to the USA and the rest of the world by plane. In addition to good connectivity, the MCMA offers a good stock of well trained people, high-quality areas for living and leisure, and a well developed culture of entrepreneurship dominated by the manufacturing sector. A very illustrative behavior of this type of MNCs' adaptive behavior is given by KORES, an Austrian company producing stationery products, which is located in Naucalpan. For KORES, Mexico is the most important market in Latin America, and their biggest factory is located in Naucalpan (around 1,000 employees) from where they supply other western hemisphere markets: Argentina, Peru, USA and Central America. In 2008, despite the decline in the use of stationery products and the beginning of the ongoing crisis, the company grew at an attractive rate of 8 per cent. Several factors explain such a positive situation:

- relocation of the production of all mechanical pencils and correction fluids to Mexico;
- substitution of materials: liquid correction products changed to solid and dry;
- supply of old products: banks and public institutions in Mexico and other Latin American countries still use carbon paper and typewriter ribbon.

In this new pattern of corporate organization, headquarters in Austria make design innovations to be manufactured in Mexico. For KORES, "[Naucalpan] is the strategic place to be in" (*Reforma* 2009). This example illustrates the benefits offered by Naucalpan to MNCs and why, rather than withdraw from the region, current conditions may make them expand existing operations to maintain, and even improve, their competitive position.

The same logic behind the reorganization of the KORES Corporation is changing Naucalpan into a logistics and distribution center: it is a privileged position to produce and distribute throughout Mexico's central region. A consequence of this emerging role is the growing pressure on former industrial facilities to be used for storage and distribution. A convenient location for these activities means significant savings and increased profits, because it implies shorter and more fluid routes of access and delivery leading to reduced expenditure on fuel, auto parts, maintenance, etc. Therefore, logistics firms bid higher for the use of premises located inside the industrial parks, all of them adjacent to the Periférico ring road, which gives them excellent accessibility to central Mexico City and the rest of MCMA. The move from manufacturing to storage and distribution thus produces an upward push in the value of property inside industrial parks that are no longer used for manufacturing. Because of this, fragmentation of tasks and the new conditions of competition have camouflaged the reorganization of industry by firms moving to informality in less visible areas. Between 1999 and 2003, wholesale firms grew from 163 units to 264, with an increment of 61 per cent. Along with additional pressures for space from growing professional and corporate service sectors, Naucalpan has become the most expensive industrial corridor of the whole MCMA.

Physically this transformation appears as the growing presence of tertiary activities inside the industrial areas. The trend is clearly illustrated with changes observed in the two biggest industrial parks, Alce Blanco (AB) and Fraccionamiento Industrial Naucalpan (FIN), from 1993 to 2005. In AB, the reduction in land occupied by industrial activities (basically manufacturing) was from 88 per cent to 42 per cent, while commerce and services grew from 12 per cent to 58 per cent. For FIN the reduction in industrial land use was from 73 per cent to 56 per cent, while tertiary increased from 27 per cent to 44 per cent (PUEC 2006c).

One of the points to be noted is the essential role played by agglomeration economies in supporting the new economic ecology brought about by exposure to international competition, with special reference to the impact on the survival of small firms.

Agglomeration economies offer advantages to firms (positive externalities) that originate in the concentration of activities and population. They have three basic sources of origin (Camagni 2005):

1 Scale: usually internal to the firm; they appear when a minimum level of demand (threshold) is reached and develop regions specialized in particular products.
2 Location: external to the firm but internal to the sector; they are based on proximity and complementarities between firms, causing reduced transaction costs, a more productive labor pool, specialized services and informal knowledge diffusion that facilitates innovation; in addition, they promote the consolidation of complexes (e.g. petrochemical, auto industry, food processing, tourist resorts).
3 Urbanization: external to the firms and to the sector; advantages stem from the presence of generic infrastructure and public services that generate benefits to all firms, from a good relationship between government and firms and a consolidated structure for consumption; advantages that promote larger concentrations of people, businesses and complementary activities (metropolization).

While scale and location economies can be associated with the consolidation of industry, it is clear that in the global economy, an adequate combination of location and urbanization economies is the key to competitiveness. Naucalpan has benefited from location economies generated by public policies in support of industrialization (import substitution). Urbanization economies were originally attained from urban spillovers into the MCMA, from Mexico City's central areas, but increasingly they now appear as a regular part of the local economic life. Concentration of firms and services in space represents the opportunity for sharing the cost of adjusting – reorganizing and restructuring – to the global/open economy, while proximity creates the conditions for everyday contacts that enable firms to have access to technical knowledge and strategic alliances for the reduction of risk, uncertainties and cost. Proximity in this respect works as a kind of low-cost "just in time" technology.

The critical role played by agglomeration economies – since small and medium-sized firms are willing to pay relatively high rents in exchange for that "bundle" of positive externalities associated with preferred locations – in combination with the impossibility of relocating to other places and the scarcity of well-equipped areas, aggravated by the reduction of public investment in social and productive infrastructure, has resulted in a heightened desirability of those few good locations. A survey in the textile industry of MCMA reveals the importance given to good location. Regardless of the size of firms, owners and managers interviewed consistently stressed the value of keeping an advantageous place, even if the actual manufacturing was moved to other premises. Entrepreneurs may look for alternative uses or complementary activities for the specific location, but giving it up is out of the question (Bustamante 2008).

The point to be made in this discussion is to highlight the crucial role of the historical trajectory of a city or place, which, given favorable economic and institutional circumstances, unchains a reinforcing cumulative process that makes certain places increasingly more appealing for the attraction and creation of new activities. Based on all these arguments, we can conjecture that Naucalpan's loss of industrial units and employment, as recorded between 2000 and 2005, may be inaccurate. On the other hand, it is very clear why big multinational businesses have maintained and, at times, expanded their operations in Naucalpan. From these facts, it is doubtful that manufacturing in Naucalpan is undergoing a process of contraction.

Tertiary sector and the spatial concentration effect

We have already observed the evolution of the industrial sector and the way it has been reshaped by the effects of its integration into the international markets. Next, we will analyze the performance of the tertiary sector of Naucalpan by emphasizing the services component. We note that the reorganization of economic activity has made it increasingly difficult to differentiate manufacturing from services performed inside the firm or outsourced, such as administrative and clerical work, marketing, research and development, among others (Méndez 1997). While outsourcing explains a good proportion of the expansion of service activities it has also meant that traditional statistical forms have become increasingly ineffective to properly track changes in the behavior of such activities. The boundary between manufacturing and services has grown ever more blurred.

Although producer services, in general, refer to those products of labor of intangible nature, they have become detached (spatially and functionally) from the actual line of production, becoming businesses in their own right. Parallel to what we saw in manufacturing, the integration into the global networked economy redefines the importance of services to a higher level and promotes the emergence and consolidation of urban centers. Working with international markets raises the complexity of business transaction, which also increases the type of high-level functions to be performed by the branches of the MNCs and the need for highly specialized services. This sequence of effects demonstrates a constantly growing demand for services, such as insurance, banking and financial, accounting and professional associations (Sassen 2005). In Naucalpan, the four most important services in terms of output value are: finance and insurance, information processing, transportation and real estate (PUEC 2006b).

The location of services tends to favor the principal urban concentrations inherited from the industrial phase (as in the MCMA and Naucalpan), promoting the consolidation of megalopolis or multicentered urban regions (Sassen 2005). Certain locations – cities and urban agglomerations – emerge as urban economic nodes of specific services (banking, professional) because they take advantage of the inherited complementarities with the industrial economy, demonstrating a path-dependent evolution, as shown in the case of Naucalpan.

Business location is determined by conditions produced by earlier economic activities, to which the incoming ones have to adjust. This concentration effect becomes a self- reinforcing process, because services tend to concentrate in cities of greater size. As urban concentrations grow, services grow in diversity and degree of specialization. Competition varies in scale according to the specific product and the kind of firm producing it. Certain aspects go beyond the city boundaries and reach the scale of what is called the "urban region." However, it is important to address the relevance of local governments, because they are essential in coordinating key local actors. Competitiveness in this respect is a highly localized process that promotes specialization and efficiency (Scott 2001).

The way that local authorities manage inherited advantages plays an important role in shaping the business climate that allows firms to operate – and compete – in a collaborative environment – as opposed to a condition of isolation – in a city or in an urban region. Such conditions of collaboration work mostly to the advantage of firms with a very low level of competitive resources, i.e. firms of small size. In such cases, actions by the local authority producing a system of positive externalities for the firms (i.e. promotion of new investment, new development, collaborative institutional environment, etc.) leverage a quasi-market for small firms that liberates them from restrictions associated with their size. They expand the organizational and technical capacities of the firms, reducing the bias effect against economic units of smaller size, and thus favor the creation of new ones. In

Naucalpan, business support services grew at spectacular rates from 1999 to 2003. The number of firms in this category grew from 116 to 332, while the number of employees multiplied by almost tenfold, from 1,314 to 12,923, making the average number of employees per firm increase from 11 to 39 (PUEC 2006b).

In the global networked economy, capital flows to a limited set of fixed destinies in cities, establishing a hierarchical order according to the local capacities to leverage this kind of global financial operations. These fixed destinies are intraurban nodes that change the economic dynamics of the city or urban region, intensifying the centrality of such places but also affecting the functional qualities of the local urban space (Camagni 2005). While analyzing these effects, one has to bear in mind that not every economic activity is global, and that such effects are limited to certain strategic activities linked globally by the MNCs, networks of firms or exchange terms.

This recentering of functions requires highly developed systems of information and communication technologies (ICTs). As ICTs are not readily found everywhere, the articulation of the network, for the continuous flow of information, gives rise to a differentiated concentration of functions in the nodes that makes the operation of a system possible. This system is simultaneously physically dispersed and functionally concentrated. Global corporations spatially distribute manufacturing and services operations, while they concentrate command and control functions as well as all kind of "dematerialized" (intangible) production, such as financial operations, and specialized corporate services. Therefore, demand derived from these new (networked) forms of operations has a direct impact on the urban pattern and the consolidation and distribution of urban centers. Though urban centers are basically a bundle of agglomeration economies, and a massive concentration of information, they make CBDs and nodes within an urban region the principal components of the system. Moreover, they are more diversified by the effects of ICT and the control functions over large territorial areas because physical proximity is no longer the determinant factor. (Sassen 2005).

Node location is not random. Their location takes advantage of local infrastructure, particularly that which increases network connectivity to other cities (fast trains, freeways, airports) reinforced by digital connectivity that allows integration into the global network. The "trans-territorial center" based on ICT has emerged from these conditions, where the hierarchy involves principal financial centers and international business that integrate other cities to form the emerging regional hierarchies. Such is the case of Mexico City (MCMA) where Naucalpan seems to perform the function of a complementary node. At the intraurban level, emerging locations favor the possibility of key face-to-face personal contact, which facilitates collaboration in scales and places of economic interest. For this reason, the consolidation of intraurban nodes needs to provide a number of privileged areas of high environmental quality that combines working facilities with entertainment and leisure in proximity with areas of high-end residential areas. This, as we have seen in the case of Naucalpan, gives a symbolic value much appreciated by people and firms involved in the global economy. The symbolic capital that once determined the dominant position of Naucalpan in the northwestern expansion of Mexico City, has transformed it into a second-tier center of services linked to the persistence of manufacturing and the emergence of logistic operations.

Permanence and consolidation of a reduced number of urban centers in the economic space comes from the fact that nodes are the connection between wealth produced in a place and the world market. The limited number results from the networked organization of production and economic activities, which no longer depends on proximity, but on consolidated markets (areas) controlled by way of ICT. These nodes need to take advantage of the existing infrastructure concentrated in old industrial centers. Nodes with better conditions of digital connectivity and positive externalities for the new operational forms (agglomeration and locational economies) may turn into urban centers with a renewed drive as a result of the following three processes (Sassen 2005).

Performance of financial and control/command functions

All coordination operations require highly talented and innovative services in areas such as advertising, marketing, accounting, legal services, economic forecasting and all other kind of corporate services. Naucalpan, regardless of its proximity to Mexico City's CBD, is unquestionably a center for subcontinental corporate services; data about services performance reveals an increase in the local importance of these types of trades.

Capacity for information processing is also another important aspect of central functions. It allows corporations to take advantage of specialized information but demands highly skilled people capable of judging, interpreting and assessing the value of different sources. When well performed, this capacity can give a quality of leadership to an urban center. It strongly depends on the local capacity to produce the type of skills demanded by the particular mix of economic activities in that central location, the social infrastructure that facilitates formal and informal ways for capturing valuable information, and the degree of global connectivity that puts these capacities to work at the global economic level.

For Naucalpan, the capacity for information processing may be one of its best inherited advantages from the import substitution period. As observed earlier, it has consolidated as a higher education center with 21 universities and has a long tradition of businesses association, local leadership and collaboration with local authorities, built by the very influential Naucalpan Association of Industrialists

(Asociación de Industriales de Naucalpan). Central coordination also requires technical capacities, state-of-the-art premises, and infrastructure that facilitates knowledge transmission to make the best use of ICT advances in competing with other centers. Naucalpan offers a good pool of qualified labor with 24 per cent of its population having college education or more, a percentage almost 50 per cent higher than the average for the whole MCMA (PUEC 2006a).

Consolidation of markets

The merger, acquisition and strategic alliances among firms require the support of very specialized services in legal, fiscal and governmental affairs. In relation to other corporate services previously mentioned, these specialized services are only found in places where the control functions of firms cluster within a sufficiently large market region. In this sense, MNCs not only compete but also collaborate among themselves. Otherwise, they would have to bear on their own the extraordinarily high cost of this kind of (in-house) services. Naucalpan in this respect is basically a manufacturing location and depends completely on the capacities of Mexico City's corporate services, where most of the corporations have their local headquarters. However, strategic alliances, and merger and acquisition decisions are mostly made at the highest level of the hierarchy, usually hiring services of the same level located in the few global centers of the world (London, New York, Tokyo, Frankfurt, and others).

Subculture of global businesses

MNCs promote loyalty to the corporation over and above notions of local (national or regional) identity, which becomes relevant only when it serves the interest of the company. They need to facilitate – and at times provide – opportunities to disseminate corporate practices and values among the local managerial employees, who will later promote them in the lower echelons of the workforce. A good part of this global business environment is based on the possibility of accessing entertainment and cultural events at equally global level, such professional sports, rock concerts and musical events, avant-garde performances and art, first-class museums, and so on. In this respect, Naucalpan benefits again from its location inside the MCMA, taking advantage of the wide range of events offered in Mexico City. With regard to the provision of courses and conferences of "global leaders," the Industrial Association, in combination with the pool of universities, locally complement their wide availability found in Mexico City. From this perspective, is not surprising that one of the main proposals of Naucalpan's Plan for Economic Development of the city is the promotion of the city as a world-class cultural and entertainment center (PUEC 2006a).

Conclusion

The central question that this study attempted to address was whether the city of Naucalpan, Mexico, is undergoing a process of urban shrinkage. We considered the term "shrinking" as a general situation of decay that results from local deficiencies in adjusting to the global functioning of the economy. Such failings make the city ineligible as a place for production, capital investment, or as a pool of skilled labor that may be attractive to the leading economic forces. The loss of economic significance is then generally manifested in a progressive loss of population, and ultimately, in the abandonment of the place. Extreme examples of this condition are dying mining towns that are no longer viable to operate.

However, our contention is that such a notion is conceptually too broad and strongly rooted in the experience of western capitalist countries. In this respect, we argue that negative economic and demographic balances, in a given period and place, are not sufficient indicators of a general process of urban decline. Doing so in developing country societies risks missing the real dynamics of change, and most certainly would lead to the application of inadequate development policies.

The Naucalpan case study helped us understand concrete expressions of local change under conditions of global economic restructuring and provided us with new elements to refine, or even rethink some theoretical formulations. To overcome a biased application of the notion of shrinkage we framed the analysis in the wider concept of the social division of labor and its spatial expression, which enabled us to characterize productive and distributive relationships in time and space. From this perspective we depicted the spatial organization of the city and its processes of transformation – social, economic and cultural – in terms of specific relations of power and economic interest, for the purpose of social control, income generation and the appropriation of ground rent. A key to this interpretation was the understanding of the city as a space of relationships and the product of changing specific social relationships, which accumulate in time as the built environment.

Thus, the city or urban territory that we analyze or inhabit is hence described as a composite of social relationships with economic (functional–productive) and organizational (hierarchical–distributive) attributes. We also examined the trajectories and key actors essential in understanding the nature of local adaptive strategies to the restructuring of the world economy. In the case of Naucalpan, we illustrated how historical circumstances promoted its emergence as one of the leading centers of industrial production of the country, supported by its advantageous location in relation to Mexico City, and by the symbolic capital derived from an institutional environment that allowed prominent politicians to personally benefit from public investment. These conditions in the long run produced a collection of competitive advantages that played a major role in Naucalpan's adaptation to the new global economy. The process of adjustment to global competition was built on the economic strengths inherited from the previous phase of industrialization.

In Mexico, as in most developing societies, well-equipped competitive locations are few and became even more so due to the effect of SAPs which drastically reduced public expenditure on social and economic infrastructure. As private investment is highly dependent on public works and services, successful places that thrived during Mexico's industrialization have inevitably become platforms of the new economy. Naucalpan is an example of this evolution towards a second-tier central node for the northern part of MCMA. Without losing its original role as an industrial center, it has fostered the growth of new business support and professional services activities. Most recent studies on the evolution of manufacturing in Naucalpan have revealed a transformation of economic units and the reorganization of production as a result of international competition, assimilation of new technologies, and growing practices of flexible employment and outsourcing. Combined with a rigid regulatory framework, this transformation makes informality one of the frequent strategies of adaptation for the survival of small firms.

Becoming informal expands the division of labor as small firms tend to specialize in tasks rather than in full processes of production. As these small firms are not visible, official statistics show signs of deindustrialization while in reality manufacturing is expanding through the effect of the permanence – and expansion – of the operations of MNCs. The selective relocation of production operations in places like Naucalpan supports this restructuring of production. Thus, informality then is not a deviation or inadequate response to globalization, but a direct consequence of the transformation of the spatial division of labor according to the capacities of different territories. The appearance of new specialized services in Naucalpan can be explained by these facts. Advanced specialized services are tied to production areas even though their location is based on positive externalities derived from good access to relevant transportation and communication networks, agglomeration economies, and high quality –prestigious – living conditions. The very factors that promote renewed centrality attract logistics operations of storage and distribution which along with the new wave of services, occupy old industrial areas and increase the prices of land.

The restructuring of Naucalpan is inevitably affected by other structural conditions such as population trends, which, when observed in relation to the general process of Mexico's demographic transition, reveals a situation where Naucalpan is attracting new households. The case study of Naucalpan's evolution, described here, suggests that the term "urban shrinking" as a category for defining the condition of decline of a city or region based on some negative statistics may miss some of the critical aspects of the nature of restructuring processes. Therefore, an approach based on the notion of the social division of labor and its spatial expression is suggested for attaining a more accurate assessment of the processes of transformation and that may provide a more solid foundation for policymaking.

Notes

1   All economic and demographic data were taken from *Mexico's Population and Economic Census*, edited by Instituto Nacional de Estadística y Geografía (INEGI).
2   Many of the pacific *caudillos* (former regional military leaders from the Mexican Revolution of 1910) were now living in their states of origin and claiming a share of the benefits derived from this new (modern) form of prosperity.
3   The *ejidos* or cooperative land system was created after the 1910 Mexican Revolution as part of a movement to restore land to Mexican peasants.
4   Mexico City is still highly centralized in terms of specialized services and commerce (Reyna 2008).
5   In this respect, the process of territorial evolution of economic activities can be observed from the recognition and adaptation to possibilities (advantages) offered by different places as a result of a particular historical trajectory. Each level of territorial organization (scale) – country, region, city-region, metropolitan area, municipality, city, district, zone or place – strives to make the best possible use of its inherited (accumulated in time) stock of common wealth and goods (Harris 1997). In this sense, the location of economic activity and manufacturing, in particular, is a far more complex process – not economically determined – than has been considered in conventional "Location Theories" (Richardson 1978; Goodall 1977).
6   Common interest and values (culture) are major components of the accumulated knowledge and capacities of organizations, which are essential for the pursuit of economic endeavors. A given social organization, responsive to a set of values, becomes a crucial screening device for the processes of innovation and progressive accumulation of knowledge. It is the basis of collaborative strategies – between individuals, groups and firms – for the active defense of symbolic aspects of power (politics and policies) and culture, which may determine local innovation capacity.
7   All population figures are from official INEGI documents.

References

Bhagwati, J. (2004) *In Defense of Globalization*, Oxford: Oxford University Press.
Birch, E. L. (1980) "Radburn and the American Planning Movement," *Journal of the American Planning Association*, 46 (4). Available at http://repository.upenn.edu/cplan_papers/31 (accessed 9 December 2012).
Bourdieu, P. (2002) *La Distinción*, México: Taurus.
Bustamante, C. (2008) *Actores urbanos y políticas públicas*, UNAM-Porrúa.

Camacho, F. (2008) "El proceso de urbanización e industrialización de la zona Metropolitana de la ciudad de México en el siglo XXI. El caso de los municipios de Naucalpan y Nezahualcóyotl." Unpub. Thesis, Lic. en Urbanismo, UNAM.

Camagni, R. (2005) *Economía Urbana*, Spain: Antoni Bosch.

Cardoso, F. H. and Faletto, E. (1969) *Dependencia y Desarrollo en América Latina*, Mexico: Siglo XXI.

Cortés, F. (2001) "Hogares y desigualdad. Crisis, miembros del hogar e ingresos," *Demos* no. 13.

Garza, G. (1985) *El Proceso de Industrialización en la Ciudad de México (1821–1970)*, Mexico: El Colegio de México.

Garza, G. (2003) *La Urbanización de México en el siglo XX*, Mexico: El Colegio de México.

Goodall, B. (1977) *Economía de las zonas urbanas*, Spain: Instituto de Administración Local.

Gutiérrez, F. (2001) "Educación: los niveles educativos de la población y su distribución en el año 2000," *Demos* no. 13.

Harris, N. (1997) "Cities in a global economy: structural change and policy reactions," *Urban Studies*, vol. 34, no. 10.

Lavell, A. M. (1972) "Regional industrialization in Mexico: some policy considerations," *Regional Studies* 6.

Legorreta, J. et al. (1997) "Agua y más agua para la ciudad." Available at www.planeta.com/ecotravel/mexico/ecología/97/0897agua2.html.

López, M. (2001) "Los hogares. Cambios sobresalientes en la composición de los hogares," in *Demos* no. 13.

Massey, D. (1984) *Spatial Division of Labor*, UK: Macmillan.

Méndez, R. (1997) *Geografía Económica*. Spain: Ariel.

Paz, L. (2001) "La fecundidad y el crecimiento de la descendencia," *Demos* no. 13.

PUEC-UNAM. (2006a) "Programa de Desarrollo económico de Naucalpan." Unpub. Report.

PUEC-UNAM. (2006b) "Programa Parcial de Corredores Comerciales y de Servicios en Naucalpan." Unpub. Report.

PUEC-UNAM. (2006c) "Programa de Reconversión de Zonas Industriales de Naucalpan." Unpub. Report.

Reforma (11 March 2009) "Descubre Kores nuevos mercados," Sec. Negocios, p. 2.

Revista Demos (2000) "Carta Demográfica sobre México. no. 13," UNAM-INEGI-ONU, 2001.

Reyna, F. (2008) "Efectos de las tecnologías de información y comunicación en la estructuración de las ciudades. El caso de la ciudad de México." Unpub. Thesis; lic. en Urbanismo UNAM.

Richardson, H. (1978) *Urban Economics*, Hinsdale, Ill: The Dryden Press.

SAPRIN (2002) "Las políticas de ajuste estructural en las raíces de la crisis económica y la pobreza. Una evaluación participativa multi-nacional del ajuste estructural." Available at www.rrojasdatabank.info/saprin/SAPRI_Findings_Esp.pdf (accessed 9 December 2012).

Sassen, S. (2005) "Situando ciudades en circuitos globales," in Macías, C., Cabrero, E.

and Ziccardi, A. (eds) (2005) *Ciudades del siglo XXI ¿competitividad o cooperación?* Mexico: CIDE-PORRÚA.

Schteingart, M. (2001) "La vivienda: evolución reciente de la situación habitacional," *Demos* no. 13.

Scott, A. (ed.) (2001) *Global-City Regions: Trends, Theory, Policy,* Oxford: Oxford University Press.

Tuirán, R. (2001) "Crecimiento y composición de la población. Estructura por edad y desarrollo humano," in *Demos* no.13.

Wallis, B. (ed.) (1991) *If You Lived Here,* Seattle: Bay Press.

Weiss, M. (1988) "What your address says about you," *Washingtonian.*

Chapter 7

Towns in the forest

The "large objects" as expressions of the informational
technical-scientific environment in the Amazonian space

Saint-Clair Cordeiro da Trindade Júnior

This chapter analyzes the "towns in the forest", created to support the great
projects or "great economical objects" implanted in the Amazon in the
second half of the twentieth century; such towns were also called company
towns or *cidades-empresa*. The discussion is inspired in theoretical reflections elaborated by the Brazilian geographer Milton Santos when conceptualizing a theorization of space and, similarly, when proposing a reflection
able to comprehend the Brazilian territory of the period, which he called
informational technical-scientific, and its respective spatial correspondent,
the informational technical-scientific environment.

Introduction

It has been unusual to see theories proposed in the Brazilian academic geographical literature that adequately address the Brazilian spatial dynamic in all its territorial diversity. Milton Santos was one of the few to concern himself with this
theoretical dimension. He sought to do so both within the scope of the so-called
peripheral countries, particularly in Latin America, as well as to illustrate the
famous theory of the "two economic circuits" (Santos 1979).

Submitted by the Associação Nacional de Pós-graduação em Planejamento Urbano e Regional
(ANPUR).

First published in Portuguese under the title "Cidades na floresta: os 'grandes objetos' como
expressões do meio técnico-científico informacional no espaço amazônico," in *Revista do Instituto
de Estudos Brasileiros*, n. 51: 113–137, Mar./Sep. 2010, São Paulo: Institute of Brazilian Studies
(IEB). Used here with kind permission.

Santos's contributions are often mentioned in reference to the urban econ-omy of Brazilian and Latin American cities, principally to discuss issues involving the totality of Brazil's territory, as seen when discussing Brazilian urbanization (Santos 1993) or when the configuration of the totality of Brazil's territory at the outset of the twenty-first century is analyzed (Santos and Silveira 2001).

The question, however, is to what extent does an analytical framework made for Brazil as a whole apply to the regional peculiarities of a space such as Ama-zonia? This question was brought up while conducting our studies focused on eastern Amazonia.[1] Although Amazonia as a space is not so frequently mentioned as a specific reality for study in Santos' work, several of his contributions aid in understanding, to a large degree, the more recent dynamics that have character-ized the eastern portion of this region.

One article in his work deserves special mention, as it also served as an initial systematization of a reflection on the space in this present historical period. It was also included as a chapter in one of his books published in the 1990s (San-tos 1994). The article addresses large-scale projects in Amazonia (Santos 1995), wherein he placed these enterprises as part of global action systems that force new territorial dynamics on the region.

During the seminar Grandes projetos: Desorganização e Reorganização do Espaço (Large Projects: Disorganization and Reorganization of Space) held in Belém city (Brazil) in 1991,[2] Santos presented these reflections on Amazonia, wherein large-scale enterprises were given the adjective "large objects." These reflections were part of a collection of articles (Castro, Moura and Maia 1995) pre-sented during the said seminar. In that article, Santos takes a fresh look at the space theory and thinks of it as a system of actions and of objects and as a level of society itself, even questioning the pertinence of thinking of the possibility of "disorgani-zation of space," as suggested by the seminar, inasmuch as space is a material aspect of society and therefore an expression and condition of its relations. Advancing along these lines, he speaks of a new spatial materiality that is being configured in contemporary times, seeing it, in the case of Amazonia, through "large objects" commanded by a "system of actions" increasingly foreign to the place itself.

We will look again at this problem here, considering certain studies con-ducted on cities and the urbanization process of eastern Amazonia, and more specifically, the State of Pará. Here we emphasize the "towns of the forest" which, in our understanding, are quite appropriate characterizations of the urban centers created to attend to the economic "large objects," implemented in the region, beginning in the 1960s, and which were also denominated as company towns (Becker 1990) or *cidades-empresa* (Piquet 1998).

This chapter, therefore, seeks to address these towns. We shall do so, inspired by Milton Santos' most recent theorizations regarding space and, likewise, in pro-posing a reflection that seeks to understand Brazilian territory based both on what he calls the technical-scientific informational period and its corresponding spatial component, the technical-scientific informational environment.

"Towns in the forest"

Some explanation is needed regarding the title of this chapter. The expression "city in the forest" is not new. The title of Oliveira's PhD thesis was *Cidades na Selva (Towns in the Jungle)* (Oliveira 2000); Browder and Godfrey (1997) also called their book *Rainforest Cities*, and most recently Castro (2008) organized a collection of articles on cities and urbanization of Amazonia called *Cidades na floresta (Towns in the Forest)*.

Here, however, we use this expression as inspired in the discussion Santos (1993) had regarding Brazil when he distinguished "towns of the rural areas" *(cidades do campo)* from "towns in the rural areas" *(cidades no campo)*. We say "inspired by," because the meaning given by this author, when discussing Brazilian urbanization, is not literally transposed with the same equivalence when addressing Amazonia. When we use the expression "towns in the forest," we seek to distinguish between the other type of city, the "towns of the forest," which, in our opinion, were predominant until the mid twentieth century, and since then Amazonia has experienced a process of effective territorial integration with the Northeast and Central-South regions of Brazil. Therefore, defining the city in Amazonia, in addition to the landscape, must initially take account of other elements that consider their relation with their respective surroundings and with processes the region has undergone since its effective national integration.

Until the 1960s, "towns of the forest" were the most common in the region. Their characteristics of small towns, often associated with river travel, caused them to create strong links with the dynamics of nature, with basic rural conditions and the forest space still largely undisturbed. Moreover, these towns always established strong links with their respective surroundings and nearby locations (villages, settlements, riverbank communities, etc.). Although many towns have been losing these characteristics, considered as rural, effectively speaking, they have not completely disappeared, and are still marked in certain subregions of Amazonia.

"Towns in the forest," on the other hand, are those that tend to create links principally to demands from outside the region, making the forest an element of little integration with new values of urban life, their denial mainly seen as a space for economic exploitation (timber, ore, fragrances, animal and plant species, tourism, etc.).

Forms of linkages and interactions of "towns in the forest" largely occur much more with realities outside the region than with the internal reality. The majority of these towns that became logistics bases for economic relations following a logic from outside the region, for example, company towns (Carajás-PA, Porto Trombetas-PA, etc.), which provide support to large economic projects, established in the region to meet resource demands of the foreign market.

Unlike company towns, some towns are not urban enclaves, but rather local towns with strong connections to the surrounding spaces. This is the case of the riverbank towns, which conduct intense interactions with their immediate surroundings, and are thus considered prime examples of "towns of the forest," due to the role they play in interacting with their surroundings and with the rhythm of the forest.

Thus, riverbank towns have strong roots, strong socioeconomic and cultural connections at the local and regional geographical scale. These roots translate into a close relationship to the river, not only through its physical location, as they are on the riverbank, but primarily because of a functional interaction with the natural elements. Examples of this are river navigation, material subsistence (source of foodstuffs, domestic use, etc.), leisure (use of the river for leisure) and symbolic use (importance of the river in the sociocultural imaginary). Therefore, they are normally: (a) small towns in terms of population size and extent of territory; (b) located on banks of rivers, and, generally speaking, large rivers, considering both their size and water volume and also size of their navigational course – this is even an important physiographical characteristic to be considered; (c) traditional, in the sense of spatial organization of the spatial group of which it is part, the pattern of their intraurban organization, economic production, and local and regional sociocultural relations (Trindade Jr., Silva and Amaral 2008).

As a result of these latter characteristics, these towns have undergone little economic and territorial modernization, and the so-called technical-scientific informational environment is only barely present, compared to other towns in Brazil and even specifically in the Amazon Region.

In any case, we cannot think of these realities as being mutually excluding and dual. On the contrary, the cohabitation of relationships, temporalities, spatialities and territorial patterns makes us recognize the complexity of the urbanization process in the region.

Towns and urbanization in Amazonia

Understanding the presence of "towns in the forest" in the current context presupposes a consideration of the regionalization process that marked regional differentiation in Brazil and the dynamic that defined the role of Amazonia in the Territorial Division of Labor.

In this sense, Santos and Silveira (2001), supported in the understanding of the expansion of the technical-scientific informational environment in Brazil, spoke of the existence of "four Brazils," in an attempt to describe the new regions of the country, as follows: the Concentrated region, the Midwest, Northeast and Amazonia.

There are certain elements cited for Amazonia to characterize it in terms of the dynamic of territorial modernization, including: low demographic and technological densities; the importance of new technological networks, for example, highways and waterways; weak centrality of transportation and communication; inventory of resources not tallied; the possibility of gaining knowledge of its resources and potentials through modern satellite and radar; the coexistence of modern, high-speed systems with slow and traditional systems; connections between the most important towns, established notably with extra-local spaces; the interwoven relations between these towns and their respective hinterlands; the existence of globalization links in export-oriented productive areas; the presence of modern cities as support sites to these same activities.

This new regional profile, which also adds a new characteristic to the dynamics of the urban network, tends to establish a set of relations that hinders the rigid structure of a traditional urban network, in large part, based on "towns of the forest," and also establishes connections that form a network of "short-circuits" and which take precedence over the logic of traditional urban hierarchy. Furthermore, they are towns that offer prime examples of elements of the technical-scientific informational environment and territorial modernization, which began to take place particularly after 1960.

Therefore, these are new objects inserted into a new territorial dynamic, such as modern industrial plants, new circulation systems, established by highways and railways, modern telecommunications networks, new port systems, expansion of the electrical grid, supported by the building of modern hydro-power plants; the presence of large economic and infrastructure projects, and, similarly, their modern cities.

It would be fair to point out, however, that this advance of modernization of the territory that has occurred over the last few decades has not taken place in a homogeneous manner throughout the entire regional space. There is intra-regional differentiation to be considered, in terms of the unequal and differentiated manner in which modernity has advanced throughout Brazil, addressed by Santos and Silveira (2001).

In western Amazonia, for example, which has Manaus as its principal expression of the urbanization process, the metropolis grows faster than the region, reaffirming the trend towards urban, population and economic concentration, which is different from the urbanization seen as a trend throughout the rest of Brazil. This is specifically the result of the establishment of the industrial complex of Manaus, which favored economic and demographic concentration. In this case in particular, territorial urbanization, understood as the spreading of these connections of modernization of space (Santos and Silveira 2001), did not occur at the same pace as the urbanization of society, marked by the spreading of the urban lifestyle, which is increasingly present throughout the region.

The principal urban centrality in eastern Amazonia, on the other hand, is the city of Belém. Its metropolitan space is officially formed by six municipalities, but at least two others not officially included are also part, and the region tends to grow more than the metropolis. This dynamic in eastern Amazonia is largely due to the strong presence of economic expansion fronts and large projects that, unlike in western Amazonia, do not result in excessive economic, urban and demographic concentration in the metropolis and its immediate surroundings. Instead, it spreads labor, investments, capital, technological networks, etc., throughout the interior of the region, following strategies of settlement, resource exploitation and economic and territorial integration, established by the Brazilian state since the latter half of the twentieth century.

For this reason, territorial urbanization is much more present in eastern Amazonia than in western Amazonia, which follows more closely the urbanization process of society that occurs equally throughout the region.

Statistics from the Brazilian Institute of Geography and Statistics (IBGE) partially express this trend, when we compare the degree of population concentration in Belém (27.77 per cent) with that of Manaus (51.78 per cent) in terms of the population of each respective state, Pará and Amazonas, in 2010. Belém actually saw a drop in this concentration from the 1970s onward, when the population of the capital represented 30 per cent of the state population, quite unlike Manaus, where concentration has increased steadily over the last decades.

We can thus observe that the spread of the urban phenomenon does not occur in the same manner, when comparing the two major subregions of Amazonia. Our interest here is to analyze one of these expressions of urbanization of territory and society, taking account of the importance that "large objects" represented to eastern Amazonia and to the state of Pará in particular.

"Systems of objects" in the globalized world: thinking of the Amazonian space

In addressing this current historical moment, called the technical-scientific informational period, Santos (1996) seeks to characterize it by analyzing the processes of modernization of society and their effects on the territorial plane. It shows, for example, that these new geographical objects increasingly become not only more technical and endowed with scientific knowledge, but that they also become bearers of information that coordinate differentiated and widely separated parts of the territory itself. They therefore bring with them a nature of ubiquity and universality; since they are present everywhere, even in regions of low technological density, such as Amazonia, compared to other regions where modernization of the territory occurred not only earlier but in a more intense manner.

For this very same reason, they are more artificial objects, which respond to the need to modernize both society and territory, especially production needs directed towards new market demands, as seen in the "large objects" established in Amazonia, in terms of modern industrial plants, power plants, port systems, in general, accompanied by modern cities (Table 7.1), conceived and built to meet the new technological demands of production, circulation and consumption. They are therefore distant from the natural systems that exist in the region, or the already existing technological systems, although they show themselves to be intrinsically associated with the dynamics of the *pre-existing* systems.

These technological and more artificial systems are conceived and built to serve the hegemonic members of society. In Amazonia, this intention is made clear when observing the agents (Table 7.1) who control or who took control, after the privatization process, most notably in the 1990s, of these technological systems and the wealth they produce. Generally speaking, these are large companies, predominantly in the field of mining, and tend to reaffirm the role of a region such as Amazonia in the current Territorial Division of Labor on the global scale:

> The center of power is in those regions where the system of objects and system of actions is densest. The center of dependency and incapacity for self-direction is in those areas where the system of objects and system of actions is less complex and less intelligent. The word region means *regis* – rule. Today, however, more and more regions only engage in doing, and fewer and fewer regions engage in ordering, in *regis* – ruling. Those regions involved in doing are increasingly regions doing things for others.
>
> (Santos 1995: 17)

Table 7.1 Companies and their towns in Amazonia

State	Town	Company
Amapá	Serra do Navio	ICOMI
	Vila Amazonas	ICOMI
Amazonas	Vila de Pitinga	Mineração Taboca
	Vila de Balbina	Eletronorte
Pará	Monte Dourado	Cia. do Jari
	Carajás	Vale
	Vila de Tucuruí	Eletronorte
	Porto Trombetas	Mineração Rio do Norte
	Vila dos Cabanos	Albras/Alunorte
Rondônia	Vila Cachoeirinha	Mineração Oriente S.A.

Organizer: Saint-Clair Cordeiro da Trindade Júnior.

Directly linked to a command unit, the "large objects" suggest a response, on the territorial plane, to the needs of the new spatial arrangement on the regional scale, and to the new configuration of relations to which space is subject in this technical-scientific informational period. They therefore represent a set of objects, and, in the case of Amazonia, of "large objects," commanded by a set of actions increasingly foreign to the location. The command unit is provided by large corporations that set the rules, pace, form and logic of production, based on their command spaces. They, *ipso facto*, transfer technologies, forms of organization of labor, production systems and values in the internal order of the company, as well as in the plurality of social relations necessary to the dynamics of this new production. One prime example of this are urban values, reproduced in new forms of urban organization, to provide support to the new technological objects:

> Spatial arrangements, under these conditions, do not occur only as they did in the past, figures formed of continuous points. Nowadays, there are also, side by side with or superimposed on them, constellations of discontinuous yet interlinked points, which define a space of regulating flows.
>
> (Santos, 1994: 104)

This territorial logic, which combines horizontalities and verticalities, continuity and discontinuity, regulation and hierarchy, complementation and domination, command and obedience, also defines the new roles of regions in the current technical-scientific informational period. These regions are increasingly linked to a global order, whether in their logic of command, or in their logic of obedience to these command units:

> The objects are created with precise intentions, with a carefully and previously established purpose. Likewise, each object is also located in a suitable manner to produce the results expected of it. In the past, objects obeyed us in the place we were, and where we created them. Now, objects no longer obey us where we are, because they are established according to a logic that is foreign to us, a new source of alienation. Their functionality is extreme, yet their ultimate purposes are beyond us. This intent is mercantile, yet it is also often symbolic. Actually, in being mercantile, it often needs to have been symbolic beforehand.
>
> (Santos, 1995: 15)

Verticalized relations establish a type of unity in the new spatial arrangement, which does not reject the horizontalities of the past, yet attributes new content to them and adds new geographical elements to regional spaces, which previously were held as more autonomous and with groups of more distinct and

Table 7.2 The dynamic of regions according to Milton Santos

Elements	In the past	In the present
Relation society–space	Direct relation between local society and its space.	Result of relations between a given place and distant factors.
Territorialities	Places as absolute territorialities of a social class or group.	Places seen as supports to hegemonic relations and agents on the global scale.
Regional "framework"	Associated to characteristics of identity, exclusivity and limits.	Relating to a large number of mediations.
Objects	They were held as collections of localized objects, of a relatively autonomous nature.	They tend to function as systems of objects or technological systems.
Actions	Actions at a lower level were not necessarily subordinate to others.	Hierarchization and subordination of actions and their command objects.
Relations between actions and objects	Relative equilibrium in point of view of functioning of system of actions and of objects.	Existence of tension between system of hegemonic and non-hegemonic system of actions and of objects.
Role of objects	Objects used as means of action.	Means of action and information.
Spatial arrangements	Spatial systems technologically and socially denser, with little overlapping of interests.	Spatial systems more complex from technological point of view and that of the plurality of relations and interests.

Source: Santos (1996).

Organizer: Saint-Clair Cordeiro da Trindade Júnior.

differentiated objects (Table 7.2). This logic becomes clearly visible when one focuses on the spatial arrangement of Amazonia in terms of the insertion of these "large objects."

These objects justify, by their location, a certain precise intent that is both commercial and symbolic in nature, and therefore a certain discourse regarding said objects is necessary. The need to translate this discourse is what determines those professionals qualified to address the functionality of the global dynamic at the local level.

It is also the discourse that transmits the symbolic weight of the new spatial arrangement in Amazonia, invariably associated to the idea of modernity and which counterpoints tradition and existing lifestyles to a new rationality full of innovation, and which convinces through this idea of modernity, progress and economic growth:

> When they tell us that hydro power plants will provide hope of salvation of the economy of the country and to the region, integration with the world, ensuing progress, all of these are symbols that allow us to accept the rationality of the object, whose advent, in reality, will ruin our relationship with nature and impose unequal relations.

> (Santos, 1995: 15–16)

Consequently, it suggests an illusory positive change for the new inhabitants who arrive, attracted by the new objects, and for those who were there prior to its advent. This discourse is very present in regards to the establishment of company towns, also called herein "towns in the forest," which are conceived, designed and equipped to provide logistical support to the "large objects."

Company towns: objects and actions

This dynamic imposed on the space of Amazonia has certainly caused a reorganization of the region, where organizational solidarities tend to replace organic solidarities of the past, or even incorporate them in a new rationality:

> In the past, the solidarity that held regional life together was an organic solidarity, while today solidarity is organizational. Before, elements of an area related with each other where they were and their unity occurred through exchanges of energy. Nowadays, their relation is with an organization that is increasingly foreign to them. Before, the organization of life was local, near the individual; now this organization is increasingly distant and foreign. Before, life itself was the reason, today rationality is without reason, with no objective, no teleology, to rule the existence of spaces.
>
> (Santos 1995: 18)

This solidarity is strategic from the standpoint of the globalization of economic processes, responsible for connecting different places of space unified by technology. In other words, the connections of energy are replaced by connections of information, which protect the interests of certain actions (Santos 1995).

This is why, in Amazonia, the most relevant technological network is linked mainly to mining and smelting projects, in which isolated companies, or companies organized in joint ventures, establish efficient spaces of production and management connected to the global market. These verticalizations are only made possible through horizontalizations, the most evident example of which was the Greater Carajás Program (*Programa Grande Carajás* – PGC), a post-Fordist option of production that conceived the creation of economies of scale in order to integrate projects (Castro and Marin 1993).

The response of the region to this new period is clearly seen in the fragmentation of its space. The new regionalization under way in Amazonia represents the overlapping of actions and flows that result in differentiating the territory. Eastern Amazonia, particularly the space defined for PGC implementation, has become a good example of this. In this case, the technical bases of globalization express themselves through the system of objects – as per the example of the "large objects" – and respond to demands for a new division of labor imposed through regulations employed by structures of global power:

> The current technical systems are endowed with a huge capacity for invasion, but this invasion is limited precisely because these objects are in the service of actors and forces that only apply them if they are ensured a return on their investments, whether these investments be economic, political or cultural. These technical objects are the transmission lines of the objectives of the hegemonic actors, of culture, politics, economy, and can only be used passively by non-hegemonic actors. Active use is increasingly reserved for the few and passive use is left to all other actors, who thus take on a subordinate role within society.
>
> (Santos 1995: 17)

This is how the existence of processes that are placed on a global scale can have consequences of integration, tension, fragmentation and imposition at the local scale. Consequently, they are interactive, integrating and yet also antagonistic and subordinating processes. As a receiver of these technologies, space within this arrangement appears as rendering the totality functional, and therefore, the process of globalization and materialization of different forms of power on the global scale.

The other side of this issue is represented by the sequence of events experienced in the region and the overlapping of technological systems, which select, suppress and/or replace territorial configurations, and social relations and processes established therein. A clear example of this dynamic is, without a doubt, the alteration of the regional urban network, in which a decisive role has been assumed by the company towns.

In large part, the urban centers of eastern Amazonia are configured as operational bases for the economic projects from the 1980s, participating in the creation of a transnational space not only as ports and industrial centers, but also as places where electronic communication takes place and also as management offices of different projects. Therefore, all of the "large objects" have planned towns as bases of their "self-sufficient" territories within the regional space, efficiently linked to the transnational space of command of their actions (Becker 1990).

Participation of the state has been decisive in this context, contributing to their establishment. A prime example of this took place in Vila dos Cabanos, jointly built with the Port of Vila do Conde, both in the Municipality of Barcarena, with the purpose of fulfilling the needs of the Albras/Alunorte, an industrial complex for alumina and aluminum production. Moreover, there are other examples, including the residential villa of Tucuruí, built by Eletronorte (Electricity Company of the North of Brazil S/A), for electricity generation; and the urban center of Carajás, built by the then-state run company Vale do Rio Doce Company, for iron mining operations.

The establishment of these company towns in the region is not recent. Even prior to the process of regional integration, which received support at the beginning of the 1960s, there had already been a number of experiences, for example, the establishment of towns such as Fordlândia and Belterra on the Tapajós River.

For the most part, these experiences resulted in failure from the standpoint of local development, whether due to the failure of the projects (Fordlândia, Belterra); or due to the rationale for their existence, since these functioned as veritable enclaves within the region (Serra do Navio, Vila Amazonas, Monte Dourado).

The sum of these experiences has contributed little to a reassessment and advances of urbanization policies for the planned cities of Amazonia. A second era in the history of these towns seems to be becoming quite apparent, when one sees, for example, changes in regional development policies for Amazonia, not well accustomed to the establishment of new urban spaces in the manner as seen in the initial decades of the process of regional integration into the global economy.

Urban spaces, created by the companies to render their economic projects in Amazonia viable, retain several dimensions of urban production that associate objects and actions, landscapes and intents, visible materiality and symbolic representation. On the other hand, it is important to consider the manner in which spatial practices related to the companies' purposes are configured in the internal organizational plan of these urban settlements and how they establish relations with their surroundings, whether with nearby towns or also the regional urban network. Here we seek to highlight the significance of these towns to local and regional development.

Generally speaking, creation of these urban nuclei establishes: (a) a new urban standard for Amazonia, until recently almost unseen in the regional landscape; (b) a new rationality towards a new economic production, even associated with a logic of flexible accumulation, a landmark of contemporary reproduction of the capitalist system; (c) logistical support to the "large objects," whether related to mining, primary transformation industries or hydroelectric plants; (d) insertion of specialized labor, brought in to enable the functioning of this new rationality of regional space, previously with low technological density and poorly integrated into modern forms of production; (e) denial of the existing regional urban network; (f) regional connection with new global circuits of production.

The initial experiences, Fordlândia and Belterra, on the Tapajós River, were established in the 1940s, and already heralded the need for this new rationality, at the time focused on rational planting of rubber trees, in order to supply the US market. Thus, the characteristics of these towns were quite different from the regional context in which they were established: US standard; strong integration company–town; proximity to production site; denial of multiple potentials of local space, based on the river and rainforest. Other towns established in the following decades, especially in the state of Pará, further enabled this differentiation.

This differentiation denotes the condition of "towns in the forest," which defines quite well the role of urban centers established to meet the demand and economic rationality of the "large objects." These towns become bearers not only of dissemination of new technical elements, highly visible on the landscape they have configured, for example, in the architectonic standard of housing, and urban standard of streets, but also of new lifestyles that these same objects bring with them.

The facilities provided also demonstrate the standard of living these small towns in Amazonia seek to ensure within them. They are different from their surroundings due to the qualities of these facilities, conceived to ensure that the inhabitants remain who meet the labor demands, generally for qualified labor, of the major companies and those that provide services to them, or in some manner are linked to the functioning of the large project:

> These new objects, which bring with them the system of current technology, require a discourse. Until recently, they could speak to us directly; now, we see them, but they tell us nothing, as if any translation were impossible. Therefore, the towns, even those in the hinterland, receive a large number of translators, people trained in reading technical systems and using technical objects.
>
> (Santos 1995: 15)

At the same time, these facilities disseminate new habits of consumption and new forms of sociability, as can be seen in towns such as Vila Permanente de Tucuruí, with the existence of a large supermarket and ecumenical center. This latter, due to its very nature, was conceived within a rationality that does not necessarily consider the peculiarities of diverse religious creeds, in a clear sign of lack of consideration for social diversity, which is quite typical of towns on the economic frontier, where migrants converge from different points of origin and cultural traditions, including religious ones, quite distinct from one another.

The existence of these towns, with their urban standards and exclusive facilities cannot occur without first a discourse, present, for example, in regional development plans and even in specific urban plans for establishing the "large objects":

> The objects have a discourse, a discourse that comes from their internal structure and which reveals their functionality. The discourse is pragmatic, but also persuasive. There is also the discourse of actions, on which its legitimacy depends. Actions require prior legitimacy to be more passively and actively accepted in social life and thus more quickly repeated and multiplied.
>
> (Santos 1994: 1030).

The results of these discourses occur at different levels. The first is regarding the regional insertion of the "large objects." It must be reaffirmed, according to the logic of the companies, that this new technological arrangement imposed on the regional space has its pertinence, even though concretely speaking the relation of the "large objects" with the region is moving in a different direction. This is because there is a rejection of the already existing spatial arrangement, denying it as part of the new imposed logic. The rainforest and rivers, for example, which, in the previous logic were seen in their multiple dimensions (resources, circulation, leisure, play, domestic use and symbolic-cultural representations), in a relationship where man–nature interaction tended to express a strong organic nature, in the new urban logic of the "large objects" become mainly sources of resources, and, secondarily, spaces for leisure and contemplation.

This is why the "towns in the forest," when located on riverbanks, are nothing more than towns near a river, losing the sense of riverbank towns quite specific to the "towns of the forest." This takes place because the new rationality imposed on the region its organizational solidarity, directly linked to the logic of the companies and their respective hierarchies, which is considered as foremost, defining the importance of the verticalizations of spaces in detriment to their horizontalities, even though still availing themselves of the latter.

Another dimension of this discourse that precedes the new objects is regarding the idea of polarization to be disseminated by these economic spaces, held as truly dynamic and bringing modernization and development. Within this perspective, towns are thought of as centers for disseminating innovation and as true central localities, as proposed in the Urban Plan of Barcarena for the case of Vila dos Cabanos:

> The idea of dimensioning facilities of an industrial complex solely in function of its own inhabitants does not seem correct. The belief is that, with the characteristics of the population to be attracted, there will be a new center, a demand for more complex services than in the rest of the region and that these services will become economically viable within a given scale that will certainly be reached, when said services are extended to the micro-region.
>
> (SUDAM 1980: 122)

Unlike what is proposed in the planning discourse, then, the role of the new center as central locality shows itself to be quite distant from the actual reality of the micro-region referred to and even of the municipality in which the large project is located. This same situation is reproduced for the other large-scale economic projects in Pará. These facilities are primarily for the benefit of the "*insiders*," those who live inside the planned town, reinforcing a type of isolation from the towns, villas and traditional or more recent settlements surrounding them. The interconnections that do occur do so much more intensely with dynamic regional and extra-regional urban centers, reinforcing the configuration of new verticalities in detriment to horizontalities.

The proposal of regional development is linked to this idea of polarization, which is primarily expressed with the presence of companies in local communities through assistance programs and incentives to one activity or another, reinforcing their participation in local life, or even discourses that support the image of the "open town," without guard posts, such as in Vila dos Cabanos, or "semi-open" town, such as the Vila Permanente de Tucuruí. The latter, even with its guard posts, establishes a more flexible control on the coming and going of "*outsiders*," that is, non-inhabitants and whoever has no association to the companies or large project installed in the location.

All of this is despite changes in the companies' strategies, to increasingly transfer maintenance responsibility of company towns to the municipal authorities. This responsibility is added to that previously assumed regarding surrounding settlements, which sprung up or expanded due to the waves of migrants attracted by the discourse of development, economic modernization, income generation and improved quality of life in the local space.

Increasing even further the density of these surrounding settlements are the workers laid off from projects as labor vacancies diminish, whether due to the end of the construction phase of the project, or due to new forms of labor organization adopted by the companies that make up the "large objects."

Reducing this burden also leads the companies to stop investing in or creating new planned spaces to settle their more qualified labor force. Instead, they seek to provide incentives for this same labor force to settle within the surrounding towns, as can be seen in the strategies of Vale do Rio Doce Company for projects under its control, in an attempt to resolve two problems at the same time: the former, reduction of costs from the necessary investment in urban infrastructure; the latter, the affirmation of an image that denies the segregationist nature of the companies regarding their most qualified workers, who, settled around the project, theoretically will be more integrated into local life.

The relation between socio-spatial development, the sum of the Gross Domestic Product (GDP) and level of tax collection enabled by large projects reveals, however, their perverse side, inasmuch as these municipalities where the "large objects" are located are the ones that stand out, in terms of production and tax collection, in the state of Pará.

On the outskirts of many of these planned towns, however, one sees the presence of urban settlements that emerged or sprang up spontaneously. These settlements represent the other side of the large economic enterprises and are characterized by extremely poor quality of life, not unlike most urban centers in the region.

It is therefore important to understand not only the spatial elements directly related to the establishment of company towns, but also the repercussions of these elements on regional urban dynamics and local development, considering a geographical perspective of analysis, which in turn considers the socially produced space as well.

The importance of company towns to the process of territorial organization of Amazonia appears to have been duly considered in many analyses that address recent dynamics of spatial structuring of Amazonia. After the initial phase of operation of the "large objects" in the region, we believe it is extremely important to also assess not only the repercussions of these projects on regional spatial organization, but also to situate how new elements of spatial arrangement are inserted, in terms of assimilation or coexistence of the same in local dynamics.

The logic of these projects, which is reproduced throughout the region, defines the segregationist character of these "large objects". Of all of the realities visited during this analysis, the precarious surrounding settlements (Table 7.3) draw our attention as they reveal the contradiction of the planned towns.

This is what makes us consider these spaces, however differentiated, as a whole, since their dynamic is explained only by the set of actions, the command of which is located outside the regional space. Therefore, these are two sides of the same spatial reality, which takes on markedly segregationist spatial differentiation. In this case, the provision of urban infrastructure and facilities is primarily the logic of the companies and their units of global command, and are, *ipso facto*, prime examples of corporate towns (Santos 1993); serving the local population only residually, which normally inhabit the surrounding areas.

Table 7.3 Company towns, "large objects" and their surroundings: state of Pará

Towns	Large projects	Company	Population	Surrounding territory
Monte Dourado	Jari	Cia. Jari	12,000	Laranjal, Água Branca, Vitória
Vila de Tucuruí	UHT	Eletronorte	3,200	Tucuruí Breu Branco
Porto Trombetas	Trombetas	Mineração Rio do Norte	6,000	Boa Vista, Caranã
Carajás	Carajás	Vale	4,240	Parauapebas
Vila dos Cabanos	Albras Alunorte	Albras/Alunorte, PPSA, RCC, Soinco, CDP, Eletronorte, etc.	7,600	Barcarena, V. do Conde, Itupanema, Vila Nova, Laranjal, Pioneiro.

Organizer: Saint-Clair Cordeiro da Trindade Júnior.

Final considerations

The presence of "towns in the forest," directly linked to large economic enterprises, reveals certain elements important to an understanding of the new spatial arrangement designed for Amazonia beginning in the second half of the twentieth century, as well as to understand the process of regional urbanization in light of the so-called technical-scientific informational environment, which characterizes spatiality of the current period of human history. Thus, company towns linked to "large objects" indicate:

- denial of a regional past, its spatial arrangement considered as poorly suited to dissemination of new connections of globalization within Amazonian space, and affirmation of the discourse of construction of a technical-scientific informational environment better suited to corporate interests, which consider the forests and rivers primarily as spaces where economic resources are located to be exploited;
- an urbanization process of the territory, enabling the region, where they are located, particularly in eastern Amazonia, to grow more than the metropolis (capital); a process following the same trend as the rest of Brazil, that is not present in western Amazonia, for example;
- a profile of "economic towns" and "corporate towns" (Santos 1993), either from the break with traditional patterns, related to urban lifestyles and the presence and importance of the company as an organization of labor and production aligned to new demands of the global market; or due to the importance of their staffs to the very existence of the town and to the dynamic of local social and political life, including dissemination of new patterns of consumption;
- forms of connection of places where communication and dependency schemes predominate in relation to extra-regional space, and where the weight of organic solidarities – which follow the hierarchy and interdependency of corporate organization – define verticalities much more than horizontalities. This relationship characterizes them as small towns, but not local in character, inasmuch as they share few links and centrality with their respective surroundings, in counterpoint to their high degree of connection with command spaces, located outside the region;
- insertion of Amazonia in global production and information connections with the technical-scientific informational period, reaffirming the role of the region in the new Territorial Division of Labor as a "region of activity," or a region that obeys commands defined in the "regions of ordering" (Santos 1995).

Given the logic enabled by the existence of the "large objects" in the region and their implications for regional space and life, they impose a new regional planning. Therefore, the dimension of place appears to be considered pre-eminent, as it is considered as a space of immediate relations and strong organic links from the social and spatial standpoint. Previous knowledge of these links and potential of their relations and logic of their spatial arrangement appear to be a condition for any socio-spatial development project. This suggests a new practice of regional planning:

Everything begins with knowledge of the world and is expanded with knowledge of place, a joint task that is currently even more possible because every place is the world. From this arises a possibility of action. Knowing the mechanisms of the world, we perceive why foreign intents come to establish themselves in a given place, and we arm ourselves to suggest what should be done in the social interest.

(Santos 1995: 19)

Local social life, therefore, appears as an assumption for a new practice of urban and regional planning, as long as it is duly situated in the new conditions required for its existence. On the other hand, a critical analysis is needed of the towns of the "large objects," regarding the relations that they have historically had with these local realities. This would lead to an understanding of the meaning of this type of town for regional policies, as well as how the same are inserted in the urban network of the region and, especially, the significance they assume to local development, in order to capture, from the experiences themselves, proposals that can lead to a rethinking of urban management and policies for the regional space of Amazonia.

Notes

1 Region of direct and indirect influence of Belém city, which, in addition to the state of Pará, includes Amapá, part of Maranhão and Tocantins States.
2 Organized by the Center for Higher Amazonian Studies (NAEA) of the Federal University of Pará (UFPA) and by the National Association of Post-Graduate and Research in Urban and Regional Planning (ANPUR), Brazil.

References

Becker, B. (1990) *Amazônia*, São Paulo: Ática.
Browder, J. and Godfrey, B. (1997) *Rainforest Cities: Urbanization, Development,*

and Globalization of the Brazilian Amazon, New York: Columbia University Press.

Castro, E. (ed.) (2008) *Cidades na floresta*, São Paulo: Annablume.

Castro, E. and Marin, R. E. (1993) *Amazônia oriental: territorialidade e meio ambiente*, in Lavinas, L. et al. (eds), *Reestruturação do espaço urbano e regional no Brasil*, pp. 121–148. São Paulo: Hucitec.

Castro, E., Moura, E. and Maia, M. L. (eds) (1995) *Industrialização e grandes projetos: desorganização e reorganização do espaço*, Belém: EDUFPA.

Oliveira, J. A. (2000) *Cidades na selva*, Manaus: Valer.

Piquet, R. (1998) *Cidade-empresa: presença na paisagem urbana brasileira*, Rio de Janeiro: Jorge Zahar Editor.

Santos, M. (1979) *O espaço dividido: os dois circuitos da economia urbana dos países subdesenvolvidos*, Rio de Janeiro: F. Alves.

Santos, M. (1993) *A urbanização brasileira*, São Paulo: Hucitec.

Santos, M. (1994) *Técnica, espaço, tempo: globalização e meio técnico-científico informacional*, São Paulo: Hucitec.

Santos, M. (1995) "Os grandes projetos: sistema de ação e dinâmica espacial," in Castro, E. Moura, E. and Maia, M. L. (eds), *Industrialização e grandes projetos: desorganização e reorganização do espaço*, pp. 13–20, Belém: EDUFPA.

Santos, M. (1996) *A natureza do espaço: técnica e tempo, razão e emoção*, São Paulo: Hucitec.

Santos, M. and Silveira, M. L. (2001) *O Brasil: território e sociedade no início do século XXI*, Rio de Janeiro: Record.

SUDAM (1980) *Plano Urbanístico de Barcarena*, vol. 2, São Paulo: Guedes e Associados.

Trindade Jr., S., Silva, M. A. and Amaral, M. D. (2008) "Das 'janelas' às 'portas' para os rios: compreendendo as cidades ribeirinhas na Amazônia," in Saint-Clair Trindade Jr. and Tavares, Maria Goretti (eds), *Cidades ribeirinhas na Amazônia: mudanças e permanências*, pp. 27–47, Belém: EDUFPA.

Rethinking metropolitan frameworks

Chapter 8

Urban forms, mobility, and segregation

A Lille–Lyon–Marseille comparison

*Dominique Mignot, Anne Aguiléra, Danièle Bloy,
David Caubel, and Jean-Loup Madre*

This chapter presents the main results of a research programme that looked at the impact of specific polycentric urban forms on commuting patterns and intraurban segregation. Our results highlight three different models; Lyon presents an extended monocentrism, with visible and diffuse inequalities; Marseille is characterized by a consuming duocentrism that is doubly unequal in social terms; in Lille, we highlight a polycentrism that is parsimonious with regard to individual motorized travel but that is socially excluding. Further research will test the relevance of these models on other European cities with over a million inhabitants.

Introduction

The growth in mobility and, above all, in car traffic, poses problems of three types: social (inequalities in terms of accessibility), economic (infrastructure costs, costs caused by congestion) and environmental (incompatibility with sustainable development objectives). The issues associated with growth in traffic are essentially concentrated in two types of space, namely major European (and even global) corridors, for land-based freight and passenger transport (in particular during peak summer periods with regard to the latter), and major metropolitan areas which concentrate local and through traffic (passengers and freight). It is within

Submitted by the Association pour la Promotion de l'Enseignement et de la Recherche en Aménagement et Urbanisme (APERAU).

This chapter was originally published in French in a slightly different version in the journal *Recherche Transports Sécurité (RTS)*, edited by INRETS (IFSTTAR since 1 January 2011) and published until December 2010 by Éditions Lavoisier: 2009, "Formes urbaines, mobilités et ségrégation. Une comparaison Lille, Lyon et Marseille," *Recherche Transports Sécurité (RTS)* no. 102 (Special issue: Mobilité, accessibilité en milieu urbain et périurbain): 47–59.

these metropolitan areas that the stakes in terms of managing individual mobility and reducing mobility and accessibility-related inequalities are greatest, although the potential inequalities between metropolitan and non-metropolitan populations would also make for an interesting analysis.

This chapter explores, in the cases of the three French cities of Lille, Lyon and Marseille, the extent to which urban form – in terms of the geographic and functional distribution of intraurban location choices (for housing and jobs) – can meet these environmental and social objectives. The review of the literature made here is only partial and principally targets work concerning the relation between polycentrism and urban form. For a more comprehensive review of the literature on the issues of urban forms and mobility, please see: Pouyanne 2004; Aguiléra 2005; Charron 2007; Mignot et al. 2007.

The analysis of interactions between urban form and daily mobility characteristics constitutes an important an area for study (Cervero 1996; Peng 1997; Priemus et al. 2001) as any "regular" subject of controversy in the literature (Giuliano and Small 1993). In particular, the differences in characteristics (in terms of distance, duration or choice of transport mode) of daily mobility, and specifically of commuter migrations (between the home and the workplace), generated by a monocentric – or, alternatively, polycentric – organization of jobs are at the heart of the debate (Aguiléra 2005; Aguiléra and Mignot 2010; Cervero and Wu 1997). The empirical results obtained are, it is true, often contradictory in terms of impact on motorized journeys (Aguiléra and Mignot 2004; Schwanen et al. 2004). For certain authors, these contradictions are in large part due to divergences in methods: for example, some studies take into consideration only internal migrations within the urban spaces analysed, whereas others also consider migrant flows out of these spaces – which, on average, involve much longer journeys. Other authors view divergences in results as a consequence of the fact that the form of location choices is not a determining explanatory factor for commuting patterns with regard to socio-economic factors, as well as with regard to the size of the urban spaces considered. Finally, yet other authors rightly underline the impact of the diverse forms of urban polycentrism, which appear to be associated with a certain diversity in the effects on daily mobility (Schwanen et al. 2004). This difficulty in reaching a conclusion is confirmed by a recent theoretical model on the relation between urban form and commuting distance (Charron 2007). The author accordingly confirms that a given urban form may generate contradictory effects.

Furthermore, questions on the subject of the sustainable city cannot be reduced to the environmental dimension alone (we will examine this aspect in terms of mobility), but inevitably leads to the social dimension also being taken into account. Indeed, socio-spatial segregation is increasing at all spatial levels in France, particularly within metropolitan areas themselves (Bouzouina 2008;

Bouzouina and Mignot 2009). The issue of daily travel and accessibility within spaces that are becoming more and more segregated and functionalized is therefore of central importance. A number of recent studies (for a summary of these works, please see Mignot and Rosales-Montano 2006) lead us to the conclusion that there are few inequalities in terms of daily mobility from the moment that individuals have access to a private vehicle. However, considerable inequalities can be highlighted for those without cars, who make up a not insignificant proportion of populations living in underprivileged neighbourhoods, as well as in terms of accessibility to jobs by social category (Orfeuil and Wenglenski 2002). In addition, certain studies have shown that recent policies promoting the development of "heavy" public transport lines have favoured access to the city centre for the least underprivileged neighbourhoods and populations (Caubel 2006).

The studies cited above, together with our own work, do not therefore enable us to come to a definitive conclusion at this stage regarding the advantages of one urban form (monocentric or polycentric) over another. After a brief reminder of the environmental and social factors associated with the issue of links between urban forms, mobility, accessibility and socio-spatial segregation, we shall evaluate the three major characteristics of contemporary urban forms, namely urban sprawl, polycentrism and segregation. Then, on the basis of a recent study (Mignot et al. 2007), we shall examine the influence of urban form on social and environmental sustainability in Lille, Lyon and Marseille, using the two following indicators: commuting patterns, and the level of socio-spatial segregation.

Environmental and social issues surrounding urban mobility

The issues at stake

While it is true that the issues that affect metropolized spaces are the local manifestation of national phenomena and trends, the most recent changes with regard to local traffic suggest that margins for manoeuvre exist locally.

Environmental issues

An analysis of changes in traffic over the last 30 years and the modal distribution observed for both passenger and freight transport does not make for terribly optimistic reading. Car travel represented 83 per cent of the inland passenger transport market in 2007, which puts France in line with the European average. Similarly, road freight represented 87 per cent of inland transport (measured in tonne-kilometres, excluding oil pipelines) (MEEDDAT–SESP 2008).

However, recent changes in road-based passenger traffic in France reflect a stagnation, since 2003, in the total number of passenger-kilometres. Moreover, fuel consumption has been falling since 2002. At the same time, a doubling of regional rail traffic in the space of 15 years has been observed, as well as significant growth in urban public transport traffic levels, both in Paris and other French cities. It would therefore appear that the vicious circle of inevitable growth in road traffic has been broken. However, are these related developments due to local and regional policies that promote public transport, or rather to the initial effects of rising fuel prices, despite the extensive dieselization of the car fleet in France in the 1980s which, at least for a certain time, helped to mitigate the price effect?

Although this price effect is altogether real, research has nonetheless shown that the elasticity of car traffic in line with income is less marked in the most densely populated areas (Madre et al. 2002; Berri and Madre 2003) and that, with regard to daily mobility, the income effect was not a distinguishing factor from the moment that access to a car was guaranteed (Diaz Olvera et al. 2004). It is clearly too early to draw any conclusions on the sustainability of this change in trend; however, we can confirm that policies led at a local level, in terms of the public transport on offer, at least, are not ineffective and are likely to contribute to sustainable territorial planning.

Beyond these broad tendencies, the question that interests us, however, is how to determine whether there are certain urban forms that are more sustainable than others (monocentric cities, sprawling cities or polycentric cities). However, it soon becomes obvious that the numerous research projects on this issue do not enable any conclusions to be drawn.

Concerning the relation between urban form and mobility, certain empirical studies confirm that a multicentric configuration means greater use of private cars in subcentres (secondary centres and suburban employment centres), which is the case in the polycentric metropolitan areas of the Netherlands (Schwanen et al. 2004), while other studies show the opposite, for instance in the case of subcentres close to metro lines in Toronto (Pivo 1993), which clearly highlights the role of heavy public transport services in this regard. As we can see, it all depends on the type of polycentrism, the size of the city, and the range of public transport on offer between the different centres of attraction studied.

Concerning the relation between urban form and commuting distance, Mathieu Charron (2007) demonstrates and confirms that a given urban form can generate contradictory effects. After performing a theoretical simulation, Charron (2007) concludes that *"we are not in a position to conclude that the monocentric form offers shorter or longer commuting possibilities than the polycentric form."*

There is therefore no obvious relationship that can be assumed, and simply denouncing sprawling cities does not help to provide a model for more sustainable cities. This is a question we shall return to later.

Social issues

At a time when numerous studies are revealing an increase in socio-spatial inequalities at all spatial levels, and particularly within cities, the issue of accessibility to the city and its services is quite clearly at the heart of urban sustainability.

A number of studies highlight these issues. For example, concerning accessibility to the job market in Île-de-France (the Paris region), Orfeuil and Wenglenski (2002) observed that the average commuting times (36–38 minutes) and distances (14–15 km) do not reveal inequalities according to social scale, but that "*accessibility to the job market for manual workers [. . .] is considerably less than for executives*". Still in Île-de-France, Beaucire et al. (2002) showed that, with regard to accessibility to city resources in the periurban fringes of the region, the results were not terribly clear; the only difference observed was that travel-to-work times were noticeably shorter for manual workers (34 minutes on average) than for executives (46 minutes on average). In fact, there are few differences or inequalities in terms of daily mobility from the moment that households have access to a private car (Diaz Olvera et al. 2004).

Furthermore, bearing in mind the location of areas where jobs and housing are concentrated, recent policies promoting the development of heavy public transport modes have tended to benefit the most affluent neighbourhoods (Caubel 2006). We can therefore conclude, following our various studies of the issue, that if public authorities wish to reduce inequalities in daily travel, then there are a number of possible policies, which are not mutually incompatible:

- help the poorest households to gain access to a car; this is not incompatible with more controlled car use in general;
- develop "social" travelcards, at highly reduced rates, for use on public transport for all types of journey;
- make a particular effort to invest in public transport on the outskirts of cities – or to provide other sorts of transport services for residents in commuter-belt areas (e.g. demand-responsive transport, new services).

Issues that also raise the question of scales of public action

Raising the issue of sustainable mobility within a given territory inevitably also raises the question of what is the best spatial scale for observing and/or analysing this sustainability. In France, the metropolitan area (*aire urbaine*) is, by its very definition, the scale generally used to analyse commuter migrations, since it is essentially a travel-to-work area (more specifically, the *aire urbaine* of a given city centre is composed of all municipalities where more than 40 per cent of workers travel into the city centre – or the continuously built-up urban hub surrounding the city centre – for employment). However, these metropolitan areas are purely statistical constructions and in no way represent spaces for action or even policymaking.

The example of Lyon is a good illustration of this contradiction between the necessarily large size required for effective urban planning and transport policies and the nervousness of local authorities, and even economic bodies, at the prospect of the emergence of an administrative entity of such a size. The results of a survey carried out in the Lyon area (Mignot, Rosales-Montano, and Toilier 2007) showed that the right scale for taking account of metropolitan issues was the Lyon Urban Region (LUR, with 810 municipalities). During the interviews for this survey, most respondents agreed that the current local administrative boundaries – *Grand Lyon* (the urban community with 55 municipalities) or the *aire urbaine* (metropolitan area with 296 municipalities) – were too restrictive, and the Rhône-Alpes region too large. These existing boundaries would seem to be unsuitable for taking account of the way a metropolis functions. The LUR, on the other hand, would appear to be the most suitable area for this purpose, and this was backed up by the responses of those interviewed for the survey: the LUR was the "ideal" scale for 46 per cent of those who answered the question. The relatively high score obtained by the Rhône-Alpes region (36 per cent) can be explained either by a desire to include the important city of Grenoble within the Lyon metropolis, or by the fact that it represents a sort of default response when one is unsure where to draw the boundary.

However, the LUR was called into question as soon as the issue was approached from the angle of public intervention, whether in terms of land-use planning or transport policy. Those interviewed who were involved in the economic or government sectors generally rejected the idea of any action at this level. In summary, these respondents judge that the relevant territorial scales for understanding economic development are the largest (LUR), but apparently "refuse" or "have reservations about" them when it comes to considering actual tools for action (e.g. single business tax, unified land-use planning) at this scale. This is a strong contradiction, and a potentially very significant obstacle to producing relevant policies on the subjects of housing and transport.

Intrametropolitan spatial dynamics: urban sprawl, polycentrism and segregation

Urban sprawl and polycentrism

The concentration of economic activity and population in ever larger metropolises always seems to be an essential characteristic of the metropolization process, as shown or suggested by a number of studies concerning different spatial scales (Fujita 1994; Krugman 1995; Arthur, 1995; Lacour and Puissant 1999). Furthermore, services – and more specifically business services – play an active role in the phenomenon of metropolization (Léo and Philippe 1998; Aguiléra 2003; Huriot and Bourdeau-Lepage 2009), as a result of both their concentration and their diversification. Our research (Mignot et al. 2004, 2007; Mignot and Villarreal Gonzáles 2009) shows first and foremost that the concentration of such services always favours the highest level of the urban hierarchy, which thus continues to develop. These studies also show that within these metropolitan areas, the historic city centre continues to play an important role, and that the concentration of jobs also occurs in privileged areas (in the city centre and elsewhere), including well-located suburban centres situated along transport routes (Mignot 1999; Gaschet 2001; Million 2009).

In French cities – unlike in large American cities, which have given rise to research into the phenomenon of edge cities (Garreau 1991) – the historic core has not tended to lose jobs or population (or the losses have been minimal). While it is true that many French city centre areas did suffer significant losses in the 1970s and 1980s, these areas have since recovered, as confirmed by data from the first round (2004–2008) of the rolling census in France (INSEE 2009). Furthermore, in most cases, city centres have seen their area of attraction (defined as places at the origin of commuter journeys into the city centre) increase (Mignot et al. 2004). Although the "market share" of core cities, in terms of the number of inhabitants and jobs, is highly variable from one city to another, in particular because of the considerable differences in municipality size from one region to another, these core cities nonetheless still have more than a 50 per cent share within their respective urban areas. In all cases, jobs are more concentrated than population. Indeed, this "spatial mismatch" (Gaschet and Gaussier 2005) is, for certain authors, the key explanation for persistent unemployment or difficulties in accessing employment for a whole section of the population, and thus for spatial segregation. Moreover, the important role still played by core cities today is the result of a certain control over jobs related to decision-making being maintained. Indeed, these historic cores are still the preferred location for business services (Aguiléra 2003; Villarreal Gonzáles 2007; Huriot and Boudeau-Lepage 2009).

Analyses also highlight the development of suburban centres of attraction in all French cities (Gaschet 2001; Mignot et al. 2004). The influence of transport infrastructures (especially motorways) and business parks is vital with regard to the location of these centres.

Spatial segregation
An analysis of changes in the distribution of average tax revenue by municipality shows a regular increase in dispersal within French metropolitan areas in the period

between 1985 and 2004 (Bouzouina and Mignot 2009). A more detailed analysis (concerning the Lyon area in particular) shows that it is the poorest municipalities that are experiencing the weakest growth in average tax revenue, whereas those that are better off to start with are also those enjoying the strongest growth. Poor municipalities that "lost out" were also widely "abandoned" by their populations between 1990 and 1999; these towns are located in the inner eastern suburbs of Lyon. The rich municipalities that were "winners" were also the preferred residential locations for executives, as well as for service-sector and high-technology companies; these towns are located in the western suburbs of Lyon.

The growth in spatial disparities can therefore be analysed as an increase in segregation – driven above all by the most advantaged areas. Bresson, Madre and Pirotte (2004) also show, through an econometric study concerning the link between urban sprawl and changes in average household tax revenue by municipality, that this segregation increased between 1986 and 1999. They even go so far as to talk about a ghettoization process affecting the richest towns on the one hand, and the poorest towns on the other.

These different studies show that spatial segregation is a reality and that it is developing not just within metropolises, but also at other spatial levels. Attention should also be drawn to the fact that growth in spatial segregation is driven above all by those areas that are most affluent. This result has been demonstrated at different spatial levels, and in different national contexts: within metropolises (Buisson and Mignot 2005), within infra-regional spaces, such as the Swiss cantons (Maillat and Quiquerez 2005) and between French regions (Carrincazeaux and Lung 2005) and Mexican regions (Mignot and Villarreal Gonzáles 2009).

In conclusion

The permanence of forms of metropolization, having been verified and characterized, leads to a rather "pessimistic" observation, namely that strong trends towards concentration, urban sprawl and intraurban segregation continue to thrive. These major trends would therefore appear to leave little room for manoeuvre for public action that seeks to reduce the attendant environmental and social consequences.

What influence can urban form have? A comparative analysis of three French cities: Lille, Lyon and Marseille

Despite these common trends mentioned above, different cities present contrasting urban forms, particularly in terms of the number of centres they have and the relative size of the urban core. Are these differences also reflected in terms of mobility and segregation? This is the question that we have tried to answer through an analysis of the cities of Lyon, Marseille and Lille (defined according to the boundaries of their metropolitan areas), as these three cities present three contrasting forms of polycentrism (Mignot et al. 2007). Lyon has an extended monocentric form with a large urban core (formed by the municipalities of Lyon and Villeurbanne) that is surrounded by the main suburban centres. The Marseille area has two main centres, although the size of the primary centre (Marseille) is very much greater than that of the secondary centre (Aix-en-Provence). It therefore exhibits an unbalanced duocentrism. In Lille, on the other hand, we can talk of balanced quadricentrism, as the primary centre (Lille) and the three main secondary centres (Villeneuve-d'Ascq, Roubaix and Tourcoing) are all of a relatively comparable size.

We shall first consider the relationships between urban form and home–work migrations, and then between urban form and socio-spatial segregation. The data used is taken from the 1990 and 1999 French censuses.

Urban form and commuting

We shall first carry out an overall analysis of all three metropolitan areas; we shall then try to highlight the impacts of different forms of polycentrism on commuting trips (distance and modal share).

The impact of the size of the metropolitan area and territorial organization on travel-to-work distances

The average distances[1] between home and work are significantly higher for the Marseille–Aix metropolitan area (13.6 km), despite the fact that it is not the largest territory in terms of area. In addition, the average distances travelled by migrants (i.e. commuters who work outside the municipality in which they live) are considerably greater than in other metropolitan areas, not because the distances travelled by commuters living in outlying suburbs are higher, but because the distances covered by workers travelling between the city centre and the suburbs, and vice versa, are particularly high – more than 24 km (see Table 8.1): this is principally due to journeys between Aix-en-Provence (or neighbouring municipalities) and Marseille. It would therefore appear to be the considerable distance between the primary and secondary centres of the metropolitan area that generates such long travel-to-work distances, and therefore the high total number of kilometres travelled.

These results show that the surface area of the metropolitan area alone does not explain the relative values of travel-to-work distances (in particular, those of

Table 8.1 Travel-to-work distances in 1999

	Lille	*Lyon*	*Marseille*
Total km for all workers	2,898,811	6,415,821.9	6,748,277.8
Total km for migrants*	2,543,743	5,226,223.0	3,522,605.2
% of km for migrants*	87.8%	81.5%	52.2%
Metropolitan area radius (km)	17.5	32.4	30.0
Average journey distance (km)	7.6	9.7	13.6
Average journey distance for migrants* (km)	9.9	13.0	21.0
Average distance/MA radius	0.4	0.3	0.5
Average distance/MA radius for migrants*	0.6	0.4	0.7

Source: INSEE data

* Migrants: people who work outside the municipality in which they live

migrants), and that a lower proportion of migrants does not necessarily lead to lower travel-to-work distances: indeed, the case of Marseille–Aix demonstrates the exact opposite, as there are (proportionately) fewer migrants, but these migrants travel much further, owing to the distance between the primary centre, Marseille, and the secondary centre, Aix-en-Provence.

Greater proximity to employment for workers living in the city centre

Residents of the city centre are, on average, located closer to their workplace than workers living in the suburbs both for 1990 and 1999 (see Table 8.2). This is due to both the very high proportion of people in stable employment in the city centre, and the fact that migrants travelling out of the centre – except in the case of Marseille–Aix – live relatively close to their workplace. These shorter journeys can be explained by the fact that the suburbanization of jobs benefits first and foremost municipalities in the inner suburbs. Relocations tend to favour the shortest distances possible, as shown by the survey carried out in the Lyon area (Belanger et al. 1999). The migrants who have the longest average travel-to-work distances are, generally speaking, those who live in the suburbs and whose jobs are located either in the city centre or a different suburb.

Table 8.2 Average travel-to-work distances (in km) according to place of residence in 1990 and 1999

	Lille			*Lyon*			*Marseille*		
	1900	*1999*	*Change*	*1990*	*1999*	*Change*	*1990*	*1999*	*Change*
City centre	5.3	6.0	13.6%	7.0	7.3	3.6%	12.3	12.9	4.6%
Suburbs	6.8	7.4	8.8%	9.6	11.2	17.1%	12.9	14.3	10.3%

Source: INSEE data

Increased distances and modal share

The increase in average distances is another characteristic of all three metropolitan areas (see Table 8.2). This change was particularly pronounced in the Lille and Lyon areas, and a little less marked in Marseille. Workers who live in the suburbs saw their travel-to-work distances increase at a much faster rate than those living in the centre, with the notable exception of Lille, where travel-to-work distances from the centre had increased much more sharply than those for journeys originating in the suburbs. Among suburban residents, the increases in average distance were highest for those who live and work in the suburbs. This confirms the assertion that the suburbanization of jobs and workers over the decade was accompanied by an increase in the distance between their respective locations.

The modal share in the three metropolitan areas is, overall, around 10 per cent for public transport, compared with 67 per cent for private vehicles. Furthermore, in terms of distance travelled, private cars cover 8 to 10 times more kilometres than public transport (see Table 8.3).

From these initial observations, we can see that the Lille metropolitan area is notable for its lower total distances compared with Lyon and Marseille: around half as many kilometres travelled by car, and around a third as many kilometres travelled by public transport – despite having a similar number of workers to the other two metropolitan areas. Although the modal share for public transport in Lille is low overall, it can be seen that the key metro lines that link the primary centre (city centre) with the secondary centres generally work well. This is the case in particular for the line linking Lille, Villeneuve-d'Ascq and Tourcoing (modal share of over 20 per cent).

Polycentrism and commuter migrations

The objective here is to refine the analysis, in particular by identifying flows to or from the different centres. We shall analyse in turn the geography of home–work flows, the implications in terms of travel-to-work distances, and finally the impli-

Table 8.3 Modal share in the three metropolitan areas of Lille, Lyon and Marseille in 1999

	Lille	Lyon	Marseille
Total distance by car (km)	2,261,455	4,693,968	4,929,053
Total distance by public transport (km)	232,169	621,262	674,400
No. of journeys by car	264,004	421,256	336,214
No. of journeys by public transport	33,316	80,167	54,840
Total no. of journeys	383,804	638,296	496,808
Modal share: private car	68.8%	66.0%	67.7%
Modal share: public transport	8.7%	12.6%	11.0%

Source: INSEE data

cations in terms of transport modes. The changes observed between 1990 and 1999 are linked with the way concentrations of workers and jobs are distributed within the metropolitan areas.

Comparative analysis of the geography of commuter flows in 1999

In all three metropolitan areas, the most significant flows are: internal flows within the city centre; flows between employment centres; and flows between these centres and the city centre. The employment centres in the Lille area host more than 60 per cent of workers and jobs, which is considerably more than in the other two metropolitan areas; however, it is also in the Lille area that internal migrations within the city centre are lowest: just 9 per cent in 1999, compared with 25 per cent in Lyon and 45 per cent in Marseille – where the city centre hosts half of all workers and almost 60 per cent of jobs. This means that, in the Lille area, almost half (49 per cent) of commuter migrations take place between employment centres, which is by far the highest score of the three metropolitan areas (for comparison: 30 per cent in Lyon and 26 per cent in Marseille). Journeys from the employment centres towards the city centre represent the third most significant flow in Lille. Dependence on jobs in the city centre, including for those living in employment centres, is therefore significant everywhere, and is linked to the fact that there are many more jobs than resident workers.

In all three metropolitan areas, the volume of migrations from employment centres towards the city centre is greater than the volume of journeys in the opposite direction, as the vast majority of workers who live in the city centre also work in the city centre, and very few in suburban areas. In Lyon, the phenomenon of nearby employment centres exerting an attraction on residents in the urban core is amplified by the fact that the centres closest to the urban core are also the most important in the metropolitan area in terms of the number of jobs they represent.

As expected, the marked imbalance between the number of jobs and the number of residents in the other municipalities in the metropolitan areas (outside the city centre and key employment centres) implies that a high proportion of workers in these municipalities work in the city centre and key employment centres. However, in all three metropolitan areas, the employment centres combined account for more jobs than the city centre, highlighting the attractiveness of these centres.

The impact on transport modes

The private car dominates, except for journeys within core cities, where public transport enjoys a high modal share – with the notable exception of Marseille, where, owing to its large surface area, public transport performs well only in the city centre.

The modal share of public transport is not insignificant on routes from employment centres into core cities. In Lille and Lyon, the main centres concerned by these journeys are essentially located close to, and enjoy good public transport links with, the city centre. In Marseille, on the other hand, the employment centres are further away and public transport performs less well than the car. Finally, to get to and from the other municipalities in the metropolitan area, car use is very much predominant, as the range of public transport on offer is not highly developed and is poorly suited to the dispersed nature of the origin–destination pairings concerned.

Changes in the geography of commuter flows since 1990 and increased distances between home and work

For all three metropolitan areas, the main observation is a marked decrease in internal migrations within the city centre and within key employment centres, while migrations between different centres have, correlatively, increased greatly: +16 per cent in Lille, +23 per cent in Lyon, and +32 per cent in Marseille (see Table 8.4). In other words, workers who live in employment centres are increasingly working in other employment centres than the one they live in. The greatest increase in commuter flows originating in municipalities in the rest of the metropolitan area concerns migrations towards employment centres – and, to a lesser extent, the city centre – thus reinforcing their attractiveness.

Accordingly, Table 8.4 shows that the average travel-to-work distance increased in all three metropolitan areas, at similar rates in Lille and Lyon, but around half as quickly in Marseille.

In all three metropolitan areas, the sharpest increases concern journeys between municipalities in the rest of the metropolitan area; this increase is linked to the marked drop in intra-municipal journeys, i.e. due to workers in these places increasingly working outside the municipality in which they live. The number of such workers has increased significantly in all three metropolitan areas: +35 per cent in Lille, +51 per cent in Lyon, and +83 per cent in Marseille.

Aside from these journeys, the growth in travel-to-work distances in Marseille was more restrained for all types of journey than for the other metropolitan areas,

Table 8.4 Changes in travel-to-work distances according to place of residence in all three metropolitan areas between 1990 and 1999

Place of residence	Lille	Lyon	Marseille
City centre + city centre	+11.3%	+9.5%	+4.7%
Centres outside city centre	+13.0%	+12.5%	+7.9%
Rest of metropolitan area	+18.9%	+19.5%	+10.6%
Total for metropolitan area	+16.0%	+15.3%	+7.5%

Source: Calculations made by LVMT based on the 1999 and 1990 French censuses

as the distances concerned were already very long from the outset in Marseille. In addition, in the other two metropolitan areas, there has been an important development of commuter flows for long-distance journeys, e.g. between the city centre + city centre and relatively distant employment centres, or between municipalities in the rest of the metropolitan area and employment centres. Most of the flows that have developed concern journeys for which the car is the most popular form of transport, producing considerable gains in terms of speed compared with journeys made for other types of flow using public transport.

Urban form and spatial segregation

Research carried out in France in recent years into metropolization has above all given rise to studies concerning issues relating to urban sprawl and the concentration processes at play in these areas, as we have described above. However, the question of increasing social and/or territorial inequalities within urban spaces is very much at the heart of issues surrounding sustainable cities.

The question of the link between urban form and segregation – the subject of a recent thesis (Bouzouina 2008) – is discussed here on an exploratory basis in order to test the relevance of the three models of city, with respect to social cohesion within cities. In addition to the assessments that may be made in terms of socio-spatial segregation and/or travel-related inequalities, the development of journeys out towards the suburbs and within suburban spaces once again raises questions with regard to public policy on transport for urban spaces. Accordingly, it is possible to highlight, in the case of the three metropolitan areas studied, the issues at stake in connection with transport provision and segregation.

Suburban spaces with poorer public transport that are frequented by more and more people

The issue of mobility-related inequalities can be considered in terms of both individuals and territories. More specifically, the issues at stake concern the question of access for all to suburban spaces, as well as the question of special transport provision for certain underprivileged suburban areas.

Throughout the 1990s, a drop in the proportion of home–work journeys within central cities was a trend observed everywhere: by the end of the 90s, these journeys would represent less than 20 per cent of all home–work journeys in Paris, Grenoble and Bordeaux, for example (see Table 8.5).

By contrast, in all metropolitan areas, the proportion of journeys from the city centre out to the suburbs and from one suburb to another increased – journeys that typically involve the greatest distances. The development of private car use and the ease of ownership access – and, more specifically, ownership of a

Table 8.5 Distribution of home work journeys between the city centre and the suburbs in 1990 and 1999

Origin (home)	Destination (work)	Marseille	Saint-Étienne	Dijon	Lyon	Bordeaux	Grenoble	Paris
City centre	City centre	49.5% (54.7%)	48.4% (53.1%)	35.3% (37.0%)	24.5% (27.6%)	15.8% (18.4%)	17.2% (20.0%)	14.1% (16.0%)
City centre	Suburbs	5.1% (3.7%)	8.3% (7.0%)	11.0% (11.1%)	13.9% (8.4%)	7.2% (5.9%)	11.0% (10.3%)	6.0% (4.8%)
Suburbs	City centre	11.4% (10.4%)	18.7% (17.0%)	25.5% (25.5%)	17.9% (20.5%)	24.3% (26.5%)	21.6% (22.6%)	18.4% (19.3%)
Suburbs	Suburbs	33.9% (31.2%)	24.5% (23.0%)	28.2% (26.5%)	43.6% (43.5%)	52.8% (49.2%)	50.2% (47.0%)	61.6% (59.9%)
Total		100%	100%	100%	100%	100%	100%	100%

Source: Mignot et al., 2004, p. 58.

The figures in brackets correspond to 1990.

detached house on the outskirts of urban areas or in commuter-belt areas – have contributed to this increased distance.

These journeys are most often made by private car, as public transport either performs poorly or is nonexistent in the suburban areas concerned. Indeed, most public transport networks are organized according to a radial model that favours journeys within the city centre or between residential suburban areas and the city centre, where the concentration of jobs used to be highest. Journeys from one suburb to another via public transport, where the option exists, therefore often require users to pass via the city centre, weakening public transport performance. Between 1981 and 1993, public transport gained market share only within core cities; moreover, public transport lost market share on radial routes (city centre–suburbs). This is a point that will be analysed in greater detail with regard to changes between 1993 and 2007, once the processed data from the 2007/08 national travel survey is available.

Access to employment would therefore appear to be potentially unequal, depending on the geographical location of households, on whether or not they have access to a car, and finally on their income. In a context where the available public transport performs poorly, and where access to a private car is limited, the key issue for those populations with the lowest incomes is therefore transport.

Growth in intraurban spatial segregation

Analyses carried out regarding the Lyon area show that those in situations of poverty (15 per cent of households, a third of which are classed as extremely poor) are not all concentrated in central and pericentral areas, although the very poorest neighbourhoods are over-represented in the urban core (Lyon and Villeurbanne) compared with other types of household. Although concentric ring-based approaches fail to reveal "pockets" of poverty, more detailed analyses, in particular at local (i.e. municipality) level, point to the development of a phenomenon of intraurban spatial segregation, confirmed by a number of recent studies carried out at different spatial levels (Bouzouina and Mignot 2009).

In this context of urban growth and increased dispersal of residential areas, employment areas and commercial/leisure areas, the question of the role of transport policies within these urban spaces becomes a key issue: in particular, do these transport policies reinforce or compensate for this process of spatial segregation?

An analysis of the transport infrastructures in different areas (expressed in terms of kilometres of public roads and bus routes) clearly shows that not all municipalities benefit from the same levels of infrastructure. However, this simple measure, which does not consider the level of service provided (quality of roads, frequency of bus services, etc.), cannot be used in isolation to take account of transport needs within cities. Discussing transport and inequality means analysing the ways in which the transport system enables (or fails to enable) inhabitants to access their city and its various services and economic activities. This also means identifying and understanding the ways in which the transport system plays an integrating role, even if this role is merely symbolic.

Appropriate transport provision for suburban neighbourhoods

The symbolic – in addition to functional – dimension of the tramlines that have been (and continue to be) built in so many French cities is no longer called into question. The underlying reason for choosing to build tramways is that this mode of transport enables local authorities to redevelop the city at the same time and thus obtain the approval of residents. The transport function could, in certain cases, be provided in the same way, or even more effectively, by bus services running in dedicated lanes (COST action TU 603: "Buses with a high level of service," CERTU et al. 2005, 2009); however, the negative public image of bus travel does not capture the public imagination in the same way when it comes to urban redevelopment.

Indeed, one might question whether a neighbourhood that is served *only* by buses can truly be said to be served by public transport. The arrival of a metro or tram line in suburban areas is always the subject of heated political debate: for some people, it is a disproportionate response to the needs in question, or even a financial risk, while for others it is an essential tool for urban integration. Transport provision and the quality of service provided, particularly with regard to how well suited it is to social practices in the area concerned, become the outward signs of a demonstration of political acceptance of the right to inhabit and "live" the city in the way one wishes.

In addition to the symbolic dimension mentioned above, policies promoting the development of heavy public transport lines refer to another reality altogether. A recent thesis (D. Caubel 2006) has shown that, bearing in mind the location of city facilities (shops, services, leisure facilities, etc.), the recent development of heavy public transport modes (in Lyon) has tended to favour the *most affluent* neighbourhoods.

This shows that policies for developing heavy public transport lines may not have the effects initially expected with regard to the most underprivileged populations (but then were these effects truly expected?). On the other hand, one might question whether the inherent difficulty of suburban transport provision means that priority should be given to lighter, "on demand" transport solutions.

The social dynamics of urban sprawl

Although the years 1990 and 1999 do not correspond to notable points on any graphs (unlike, say, 1989 and, in some regards, 2002), it is clear that the comparative statics based on census data reveal a very poor image of changes, making interpolation dangerous, and extrapolation (with the slowdown in the early

2000s) or retropolation (with the strong increase in inequalities in the second half of the 1980s) even riskier. Should we deduce from this that censuses provide no dynamic information? Of course not, as the previous place of residence (at the time of the last census) constitutes a valuable source of data at individual level concerning the property moves that lead to urban sprawl. Regarding inequalities, no questions about income were included in the French census, but the recording of the "PCS" (*profession et catégorie socioprofessionnelle* – profession and socio-professional category) of the reference person in each household allows a different approach to be adopted with regard to inequalities.

Tables 8.6 and 8.7 therefore show the proportion of what French marketers call "CSP+" (households whose head is – or has been, in the case of retired people – an executive manager, freelance worker or member of an intermediate profession). In all the metropolitan areas and concentric rings considered, this proportion is higher among new arrivals than existing residents. The contrast is often more marked in provincial cities than in Île-de-France, and also in central cities compared to suburban areas; it is not, therefore, surprising that inequalities increase more rapidly in urban hubs (central cities + suburbs) than in commuter-belt areas.

Infra-municipal social fragmentations in Lille, Lyon and Marseille

The analyses carried out in the metropolitan areas of Lille, Lyon and Marseille highlight the structuring of the urban space at infra-municipal level according to the income and socio-professional category of the population. These analyses

Table 8.6 Percentage of households with a high PCS1 among new and old residents[1]

	Total		City centre		Suburbs		Commuter belt	
	New	Old	New	Old	New	Old	New	Old
Paris MA	59.0	50.0	66.2	58.1	54.9	48.0	58.1	46.8
Lyon MA	57.4	45.2	60.3	47.4	55.1	43.4	55.8	48.0
Marseille MA	52.9	39.4	48.9	36.4	55.1	43.4	60.2	47.9
Lille MA	58.1	38.4	60.6	36.2	55.2	38.0	62.3	43.8

Source: INSEE (processed by INRETS), 5% survey of households of the 1999 census.

Key: MA = metropolitan area (aire urbaine in France); Suburbs = urban hub (pôle urbaine in France) minus the city centre; Commuter belt = suburban rim (couronne périurbaine in France); Old = existing residents: households where the reference person was living in the same zone (i.e. same metropolitan area and same concentric ring) on 1 January 1990; New = new arrivals: households where the reference person was not living in the same zone on 1 January 1990.

1 Households where the reference person is a farmer, freelance worker, executive manager or member of an intermediate profession (former PCS for retired people).

  Households with a student at their head are excluded from the calculation (they represent more than 40% of households that come from outside the metropolitan area to live in the city centre, and 10–20% of households that come from the outer suburbs to live in the city centre or from outside the metropolitan area to live in the inner suburbs, but are present in very few other cases).

Table 8.7 Proportion of households with a high PCS1 according to the place of residence of their reference person in 1990 and 1999[1]

	Place of residence in 1999							
	City centre		Suburbs		Commuter belt		Total	
	Grenoble	Lille	Grenoble	Lille	Grenoble	Lille	Grenoble	Lille
Place of residence in 1990								
City centre	44	36	53	57	65	63	46	41
Suburbs	48	51	45	38	62	64	47	39
Commuter belt	58	60	55	42	46	44	47	44
Outside metro area	67	65	61	56	64	60	64	60
Total	49	46	48	40	50	48	49	42

Source: INSEE (processed by INRETS), 5% survey of households of the 1999 census

1 *Households where the reference person is a farmer, freelance worker, executive manager or member of an intermediate profession (former PCS for retired people).*
  Households with a student at their head are excluded from the calculation (they represent more than 40% of households that come from outside the metropolitan area to live in the city centre, and 10–20% of households that come from the outer suburbs to live in the city centre or from outside the metropolitan area to live in the inner suburbs, but are present in very few other cases).
  Interpretation guide: the proportion of non-student households of category CS+ is 46% in the outer suburbs of Grenoble and 44% in the outer suburbs of Lille among those households that already lived in this zone in 1990; this proportion rises to between 60% and 65% among households that moved into the outer suburbs of these metropolitan areas between 1990 and 1999.

confirm and clarify the results already obtained elsewhere with regard to the social inequalities of urban territories at municipal level. The municipality is no longer considered a homogeneous space, but a space differentiated by neighbourhoods, each of which is distinguished by concentrations of affluent and deprived populations, different standards of living, and distinct social structures.

We have been able to show that inequalities increased more rapidly in urban hubs than in the commuter belt, and that location strategies among the most affluent households played a role in reinforcing these inequalities. Conversely, households with the lowest incomes tend to migrate towards less segregated territories in outer suburbs or the commuter belt.

Conclusion: three city models

The results highlighted as part of this comparison of the three metropolitan areas in France with over one million inhabitants (excluding Paris) illustrate and confirm that, although links may exist between urban form and travel-to-work distances, these links are far from clear-cut, and the influence of polycentrism on commuter migrations depends on the type of polycentrism or the characteristics of the employment centres in question. Similarly, the analysis of socio-spatial segregation within the three metropolitan areas studied shows that no simplistic interpretation can be applied here. Ultimately, three models can be used to summarize the characteristics of the metropolitan areas of Lille, Lyon and Marseille:

- **Lyon: "an extended monocentrism, with visible yet diffuse inequalities."** In this case, there are clear centres of attraction that can be identified; however, they do not in any way compete with the urban core (Lyon and Villeurbanne) and, in reality, simply extend the influence of the core over a wider territory. The development of secondary centres helps not just to extend the role of the urban core in spatial terms, but also to create attractive "pockets" for affluent households, thus helping to keep social marking to a minimum across the metropolitan area as a whole.
- **Marseille: "a consuming duocentrism that is doubly unequal in social terms."** This model is described as consuming, as the considerable distance between the city centre and the main suburban centre is a key factor behind much greater average travel-to-work distances than the other two metropolitan areas. The city of Marseille proper appears to be split into two socially, while the secondary centre of the metropolitan area, Aix-en-Provence, is overwhelmingly dominated by concentrations of very affluent populations in almost all its neighbourhoods.
- **Lille: "a polycentrism that is thrifty, but socially excluding."** A number of secondary centres can be clearly identified; these centres compete to some extent with the city centre. The presence of these secondary centres is what gives Lille the lowest average travel-to-work distance of all three metropolitan areas. A dualism between rich and extremely underprivileged neighbourhoods is nonetheless clearly visible.

Conclusion

In all three cases, we have seen that the modal share of public transport is broadly the same. If we bear in mind that artificialization through housing is very low in periurban spaces (Tourneux 2006), it becomes clear that the implications of urban sprawl are therefore growth in motorized travel, and consequently increased pollutant levels.

The polycentric model organized around major centres therefore appears to be potentially "thrifty" with regard to individual motorized travel, provided that the centres of attraction at least partially structure the territories around them, and that the flows between these centres (including the city centre) – representing the greatest volume – are effected using less-polluting means of transport, i.e. public transport. Of course, this assumes that adequate transport and urban planning policies are put into place. If this is not the case, there would appear to be nothing that can slow down the growth in travel-to-work distances, and indeed journey distances in general.

And yet, for all this, does the polycentric model encourage a greater social mix? Here too, there is no simple answer at this stage. However, any urban forms that encourage mixed-income populations to settle in specific centres that do not correspond to "naturally" attractive places for the most affluent households would help to reduce, in part, spatial segregation within urban spaces.

Note

1   Calculations of travel-to-work distances are necessarily approximate, in view of the general nature of the census data available, i.e. municipality of residence and municipality of employment. For distances between municipalities, we have – as is often the case in this type of study – used the straight-line distance between centroids, weighted by a factor of 1.3. Concerning distances within a given municipality, we felt it was necessary to take account of the size of the municipality concerned, in particular because of the extreme variation that exists in this regard (for example, municipalities in the Marseille–Aix area are very large by French standards). We therefore represented each municipality as a circle, and took commuting distances within this municipality to be the radius of this circle, also weighted by a factor of 1.3 for consistency. As no data for *arrondissements* (administrative city districts in Paris, Lyon and Marseille) was available for 1990, travel-to-work distances within the municipalities of Lyon and Marseille were calculated in the way described above, i.e. using the radius of the circle representing the municipality. This means that we have probably slightly overestimated distances within the city of Marseille proper, which, although a very large municipality, registered high numbers of intra-*arrondissement* commuter journeys in 1999, and so probably also in 1990.

References

Aguiléra, A. (2003) "Service relationship, market area and the intrametropolitan location of business services," *The Service Industries Journal*, vol. 23, no. 1: 43–58.

Aguiléra, A. (2005) "Growth in commuting distances in French polycentric metropolitan areas: Paris, Lyon and Marseille," *Urban Studies*, vol. 42, no. 9: 1537–1547.

Aguiléra, A. and Mignot, D. (2004) "Urban sprawl, polycentrism and commuting: a comparison of seven French urban areas," *Urban Public Economics Review*, no. 1: 93–114.

Aguiléra, A. and Mignot, D. (2010) "Multipolarisation des emplois et mobilité domicile-travail: une comparaison de trois aires urbaines françaises," *Canadian Journal of Regional Science*, vol. 33, no. 1: 1925–2218.

Arthur, W. B. (1995) "La localisation en grappes de la «Silicon Valley»: à quel moment les rendements croissants conduisent-ils à une position de monopole?" in Rallet, A. and Torre, A. (eds), *Économie Industrielle et Économie Spatiale*, pp. 297–316, Paris: Economica.

Beaucire, F., Berget, M. and Saint-Gérand, T. (2002) "L'accessibilité aux ressources de la ville dans les franges périurbaines d'Île-de-France," Laboratoire Mobilité Réseaux Territoires Environnement, LADYSS, EOSYSCOM MRSH Caen. Rapport pour le PUCA et le PREDIT.

Belanger, A., Bloy, D., Buisson, M.-A., Cusset, J.-M. and Mignot, D. (1999) "Localisation des activités et mobilité," recherche effectuée pour la DRAST, rapport final, Décembre 1999, 300 pp.

Berri, A. and Madre, J.-L. (2003) "Forecasting the growth of car traffic according to different scenarios of urban sprawl and of economic growth," European Transport Conference, Strasbourg.

Bouzouina, L. (2008) "Ségrégation spatiale et dynamiques métropolitaines," Thèse de Doctorat en Sciences Économiques, Université Lumière Lyon 2.

Bouzouina, L. and Mignot, D. (2009) "La ségrégation spatiale à différentes échelles," in Gaschet, F. and Lacour, C. (2009), *Métropolisation et ségrégation*, pp. 63–77, Bordeaux: Presses Universitaires de Bordeaux.

Bresson, G., Madre, J.-L. and Pirotte, A. (2004) "Is urban sprawl stimulated by economic growth?" Tenth WCTR, Istanbul (CD-ROM of the conference).

Buisson, M.-A. and Mignot, D. (eds) (2005) *Concentration économique et ségrégation spatiale*, Brussels: Coll. Économie, Société, Région, De Boeck, 368 pp.

Carrincazeaux, C. and Lung, Y. (2005) "Configurations régionales des dynamiques d'innovation et performances des régions françaises," in Buisson, M-A. and Mignot, D. (eds), *Concentration économique et ségrégation spatiale*, pp. 127–143, Brussels: De Boeck.

Caubel, D. (2006) "Politique de transports et accès à la ville pour tous? Une méthode d'évaluation appliquée à l'agglomération lyonnaise," Thèse de Doctorat en Sciences Économiques, Université Lumière Lyon 2.

Cervero, R. (1996) "Mixed land uses and commuting: evidence from the American housing survey," *Transportation Research A*, vol. 30, no. 5: 361–377.

Cervero, R. and Wu, K. L. (1997) "Polycentrism, commuting and residential location in the San Francisco Bay area," *Environment and Planning A*, vol. 29, 865–886.

Charron, M. (2007) "La relation entre la forme urbaine et la distance de navettage: les apports du concept de « possibilité de navettage »," Thèse de Doctorat en études urbaines, INRS, Université du Québec à Montréal, March, 242 pp.

CERTU, INRETS, GART, UTP (2005) "Bus à haut niveau de service, concepts et recommandations," Édité par le CERTU, Dossier 166, Transport et mobilité, 111 pp.

CERTU, INRETS, GART, UTP, CETE (2009) "Bus à haut niveau de service: synthèse des projets et expérimentations en France," Édité par le CERTU, à paraître.

Diaz Olvera, L., Mignot, D. and Paulo, C. (2004) "Daily mobility and inequality, the situation of the poor," *Built Environment*, Guest Editor Dupuy, G., vol. 30, no. 2: 153–160.

Fujita, M. (1994) "L'équilibre spatial – L'interaction entreprises ménages," in Auray, J. P., Bailly, A., Derycke, P. H. and Huriot, J. M. (eds), *Encyclopédie d'économie spatiale*, pp. 213–223, Paris: Economica.

Garreau, J. (1991) *Edge Cities*, New York: Doubleday.

Gaschet, F. (2001) "La polycentralité urbaine." Thèse en Sciences Économiques, Université Montesquieu Bordeaux IV, December, 345 pp.

Gaschet, F. and Gaussier, N. (2005) "Les échelles du mauvais appariement spatial au sein de l'agglomération bordelaise," in Buisson, M-A. and Mignot, D. (eds), *Concentration économique et ségrégation spatiale*, pp. 221–241, Brussels: De Boeck.

Giuliano, G. and Small, K. A. (1993) "Is the journey to work explained by urban structure?" *Urban Studies*, vol. 30, no. 9: 1485–1500.

Huriot, J. M. and Bourdeau-Lepage, L. (2009) *Économie des villes contemporaines*. Paris: Economica, 366 pp.

INSEE (2009) INSEE Première, no. 1218, January 2009.

Krugman, P. (1995) "Rendements croissants et géographie économique," in Rallet, A. and Torre, A. (eds), *Économie Industrielle et Économie Spatiale*, pp. 317–334, Paris: Economica.

Lacour, C. and Puissant, S. (eds) (1999) *La métropolisation: croissance, diversité et fractures*, Paris: Anthropos–Economica.

Léo, P. Y. and Philippe, J. (1998) "Tertiarisation des métropoles et centralité. Une analyse de la dynamique des grandes agglomérations en France," *Revue d'Économie Régionale et Urbaine*, no. 1: 63–84.

Madre, J. L., Berri, A. and Pappon, F. (2002) "Can a decoupling of traffic and economic growth be envisaged?" in Black, W. R. and Nijkamp, P. (eds), *Social Change and Sustainable Transport*, Bloomington: Indiana University Press.

Maillat, D. and Quiquerez, F. (2005) "L'évolution des disparités régionales en Suisse," in Buisson, M. A. and Mignot, D. (eds), *Concentration économique et ségrégation spatiale*, pp. 105–125, Brussels: De Boeck.

MEEDDAT–SESP (2008) "Les comptes des transports en 2007."

Mignot, D. (1999) "Métropolisation et nouvelles pôlarités: le cas de l'agglomération lyonnaise," *Les Cahiers Scientifiques du Transport*, no. 36/1999: 87–112.

Mignot, D., Aguiléra, A. and Bloy, D. (2004) "Permanence des formes de la métropolisation et de l'étalement urbain," Rapport final, recherche financée par l'ADEME, LET–ENTPE et INRETS, Lyon, 114 pp.

Mignot, D. (dir.), Aguiléra, A., Bloy, D., Caubel, D., Madre, J. L., Proulhac, L. and Vanco, F. (2007) "Formes urbaines, mobilités et ségrégation: une comparaison Lille–Lyon–Marseille," LET–ENTPE, LVMT–INRETS, DEST–INRETS. Rapport final, recherche pour le GRRT, financée par la délégation régionale à la recherche et la technologie Nord–Pas-de-Calais, Lyon, 118 pp.

Mignot, D. and Rosales-Montano, S. (2006) "Vers un droit à la mobilité pour tous: Inégalités, territoires et vie quotidienne," Paris: La Documentation Française.

Mignot, D., Rosales-Montano, S. and Toilier, F. (2007) "De la «planification stratégique et volontaire» à la «planification interactive et incitatrice», quelles évolutions dans la territorialisation du développement économique lyonnais? Colloque «Les dynamiques territoriales: débats et enjeux entre les différentes approches disciplinaires»," 43ème colloque de l'ASRDLF, Grenoble–Chambéry, 11–13 July 2007, 25 pp.

Mignot, D. and Villarreal Gonzáles, D. R. (eds) (2009) *Les formes de la métropolisation: Costa-Rica, France et Mexique*, Paris: INRETS–Lavoisier.

Million, F. (2009) "L'impact des zones d'activité sur la localisation des entreprises," in Mignot, D. and Villarreal Gonzáles, D. R. (eds), *Les formes de la métropolisation: Costa-Rica, France et Mexique*, pp. 91–112, Paris: INRETS–Lavoisier.

Orfeuil, J. P. and Wenglenski, S. (2002) "L'accessibilité au marché du travail en Île-de-France: inégalités entre catégories sociales et liens avec les localisations résidentielles," Recherche pour le programme «déplacements inégalités» du PREDIT et du PUCA, IUP – Université Paris X, DEST–INRETS.

Peng, Z. R. (1997) "The jobs–housing balance and urban commuting," *Urban Studies*, vol. 34, no. 8: 1215–1235.

Pivo, G. (1993) "A taxonomy of suburban office clusters: the case of Toronto," *Urban Studies*, vol. 30, no. 1: 31–49.

Pouyanne, G. (2004) "The influence of urban form on travel patterns: an application to the Metropolitan Area of Bordeaux," Presented at the European Regional Science Association conference (ERSA conference papers ersa04p244). Available at www-sre.wu-wien.ac.at/ersa/ersaconfs/ersa04/PDF/244.pdf (accessed 18 December 2012)

Priemus, H., Nijkamp, P. and Banister D. (2001) "Mobility and spatial dynamics: an uneasy relationship," *Journal of Transport Geography*, vol. 9: 167–171.

Schwanen T., Dieleman, F. M. and Dijst, M. (2004) "The impact of metropolitan structure on commute behavior in the Netherlands. A multilevel approach," *Growth and Change*, vol. 35, no. 3: 304–333.

Tourneux F. P. (2006) "L'évolution de l'occupation du sol dans les franges franciliennes: des artificialisations concentrées plus qu'un étalement urbain?" in Larceneux, A. and Boiteux-Orain, C. (eds), *Paris et ses franges: étalement urbain et polycentrisme*, pp. 101–127, Éditions Universitaires de Dijon.

Villarreal Gonzáles, D. R. (2007) "Concentración del empleo y movilidad de la población trabajadora en la Zona Metropolitana de la Ciudad de México," in Villarreal Gonzáles, D. R. and Mignot D. (eds), *Metropolización, concentración ecónomica y desigualdades espaciales en México y Francia*, pp. 75–104, Colección Teoria y Análisis, Editor UAM–Xochimilco.

Chapter 9

Metropolitan strategic planning
An Australian paradigm?

Glen Searle and Raymond Bunker

This chapter describes the characteristics of a distinctively Australian paradigm of metropolitan planning which reflect circumstances of governance, infrastructure provision and concentration on suburban expansion into surrounding countryside. The resultant plans are detailed in their arrangement of land use and communications, comprehensive and long term. There are indications this paradigm may be changing as these dominating influences alter in character. Contemporary metropolitan strategic planning in Europe and America is overviewed to establish the distinctiveness of the Australian paradigm. Changes in plan-shaping forces are leading the emergence of a new European strategic spatial planning paradigm very different to Australia's. Strategic spatial planning in the United States, while heterogeneous, has examples that reinforce the idea of an Australian paradigm in terms of the influence of governance structure and infrastructure agency on the level of spatial plan detail.

Introduction

In an era of heightened inter-city competition, metropolitan strategic planning has assumed renewed importance. A central issue is how this might have changed the nature of city-wide spatial plans. Taylor has claimed that globalization is producing 'unprecedented [planning] similarities across leading cities of the world [that] are happening despite diverse public policies' (Taylor 2006: 1164). Conversely, Newman and Thornley (2005) argue that, while global forces and the response of a dominant global neoliberal ideology are important, highly variable institutional contexts produce fundamental differences in city planning responses.

Submitted by the Australia and New Zealand Association of Planning Schools (ANZAPS).

Originally published in *Planning Theory*, 9:3: August, 2010: 163–180. Used here with kind permission of Sage publications and the authors.

The aim of this chapter is to describe the main characteristics of a distinctively Australian style of metropolitan planning which has evolved over the years in response to the combined influence of the structure of governance, the challenges of metropolitan suburban expansion to accommodate the strong population and economic growth of the nation since the Second World War, and the way in which infrastructure has been provided to support this growth and change.

We argue that these conditions have given rise to a particular paradigm of metropolitan planning. In Australia, the state governments have constitutional authority for spatial planning. They are not legally required to gain the concurrence of national or local government to produce metropolitan strategies, which gives them freedom to insert much spatial detail to guide and direct development. In Australia also, where most infrastructure is provided by the state governments or their agencies, coordination of future infrastructure planning via state-led metropolitan plans is much more feasible. Indeed, recent metropolitan plans have been seen by the state governments as the blueprints for infrastructure planning and investment in each city. Hence the metropolitan strategic plans in Australia specify infrastructure size and location, and land use generally (as this is the major determinant of infrastructure need), in some detail. The proclivity to detailed spatial prescription that thus characterizes Australian metropolitan spatial strategies has been reinforced by the emphasis, until recently, on greenfields development, in which there are relatively few constraints (compared with intraurban development) on the ability of state planners to specify the structure and extent of new development in detailed fashion.

Five metropolitan strategies have been constructed for the largest cities of Australia since 2002. This chapter reviews how far they reflect that paradigm, framing this in terms of the basic circumstances that have generated the paradigm.

Contemporary metropolitan strategic planning in Europe and America is then overviewed to establish the distinctiveness of the Australian paradigm. The chapter considers how changes in plan-shaping forces in Europe are leading to a changing style of planning there and the emergence of a new European strategic spatial planning paradigm that is very different to the one that has endured in Australia. It also considers strategic spatial planning in the United States in terms of its potential similarities and differences with the Australian paradigm. Finally, the chapter notes some changes taking place in the Australian paradigm which could represent a move towards a more European style of planning.

Australian urbanization and strategic planning issues

Australia's urban system is dominated by the large metropolitan centres of Sydney, Melbourne, Brisbane, Adelaide and Perth, the capital cities of their respective states. The state governments are responsible for a wide range of economic, social and environmental matters including health, education, welfare services, security, recreation, environmental protection and enhancement, economic development and urban growth and planning. Local government is a junior participant in this process. It is created by state legislation and is not recognized in the constitution of Australia. Because of these circumstances, state governments have been able to bring a wide variety of policies and programmes to bear in planning and supporting the functioning of their respective capital cities. This has offered the potential for coordination and providing effective standards of service.

In particular, two components of this high level of urban authority vested with the state governments have been critical in shaping a distinctive Australian form of metropolitan planning. First, in contrast with both Europe and the US, constitutional responsibility for metropolitan planning in Australia lies entirely with the state governments (see Stilwell and Troy 2000). Local government and its planning are legally bound by state planning laws and controls. Metropolitan region authorities with legal powers that are independent of the state governments do not exist. Thus state governments make metropolitan plans with little need to take account of local government wishes, while national government has only intermittent and limited involvement at those times when it allocates significant funds affecting urban management.

Second, metropolitan infrastructure in Australia has traditionally been supplied by state government authorities. This is a legacy of the nineteenth-century Australian colonial governments, which provided the necessary infrastructure for development under a 'colonial socialism' system (Butlin 1982). The principal exception is Brisbane, where the city council has the bulk of the metropolitan population and provides water and sewerage infrastructure and bus services. While significant elements of metropolitan infrastructure have become privatized over the last 20 years, including electricity infrastructure in several cities, water and sewerage (Adelaide), rail and trams (Melbourne), and some motorways (Sydney and Melbourne), public control and regulation of infrastructure services is still strong via state regulatory tribunals and the like. Thus coordination of future infrastructure planning via state-led metropolitan plans is much more feasible. Indeed, recent metropolitan plans have been seen by the state governments as the blueprints for infrastructure planning and investment in each city. Thus the metropolitan strategic plans in Australia specify infrastructure size and location, and land use generally (as this is the major determinant of infrastructure need), in some detail.

A third aspect reinforces the distinctiveness of Australian metropolitan planning. Specifically, this planning has, until recently, been preoccupied with the organization of new suburban growth into greenfields at the rural–urban fringe. While Australian planning has not been unique in this respect, this emphasis has meant that the above two components – state government control of planning and of infrastructure provision – have resulted in metropolitan strategies with detailed spatial planning of urban expansion as a dominant element. These three features – state control of planning, state provision of infrastructure, and the importance of greenfields growth – can be seen as basic drivers of the distinctive character of Australian metropolitan plans, which are overviewed in the next section.

The nature of Australian metropolitan plans

In response to these drivers, Australian metropolitan plans have been characterized by:

- being based on state government planning legislation which marries strategic purposes and social objectives with supervision and direction of local land use zoning and control instruments and – as a result – considerable detail as to arrangement of activities, land uses and communications;
- long-range planning a generation ahead;
- being designed as a coordinating instrument for infrastructure provision;
- reliance mainly on private investment and the market to flesh out the strategy.

The first post-war plan for Sydney (Cumberland County Council 1949) expressed these principles and later metropolitan plans followed many of them, although there were important differences of emphasis. Early plans for Perth (Western Australian Government Printer 1955), Melbourne (Melbourne and Metropolitan Board of Works 1954), and Adelaide (Town Planning Committee of South Australia 1962) were very much in this idiom. The Sydney Region Outline Plan of 1968 (State Planning Authority of NSW 1968) was more conceptual and abstract in its form, though it specified the location and timing of future urban development in detail. It was so successful in acting as a coordinating instrument for massive suburban development – containing staging proposals and a short-lived policy of levying a betterment charge on development – that it was labelled a 'developers' guide'. All these plans were reasonably successful in planning and organizing suburban expansion, although help was needed from the federal government in one of its short periods of interest in urban affairs in 1972–5 in overcoming a sewerage backlog and other deficiencies in infrastructure provision. There were, however, difficulties in generating employment in the expanding suburbs and in providing adequate services, and many people, particularly women, found suburban life monotonous. There were also the beginnings of a social divide that has grown over the years.

After these plans, there was something of an interregnum including a flirtation with plans showing the influence of 'urban management' (Gleeson and Low 2000), such as the Sydney plans of 1994 and 1999 (Searle 2004).

They also included a more abstract plan for Adelaide (Premier of South Australia 1994) using broad arrangements of land use and communications which were heavily annotated to show policy initiatives in various spatial realms.

In the early years of the twenty-first century five major plans have been produced for Melbourne (Department of Infrastructure, Victoria 2002), Sydney (Department of Planning, NSW 2005), Perth (Western Australian Planning Commission and Department of Planning and Infrastructure 2004), Brisbane (Office of Urban Management 2005) and Adelaide (Government of South Australia 2006). This followed something of a public outcry among the public and developers about the lack of purpose and certainty in the direction of metropolitan growth and change, particularly shown in the run-down of infrastructure and above all of public transport. In Sydney, the Property Council of Australia (PCA) published two discussion papers recommending a number of measures that it considered should be in a plan for the city. These included spatially specific recommendations for the plan to direct where population and employment growth is planned to occur and expressing this through location-specific targets; to adopt PCA-nominated job targets for major centres; and to identify major urban renewal opportunity areas (PCA 2002, 2004).

These strategies are reviewed elsewhere in detail (Bunker and Searle 2009), but they illustrate interesting variations on the previous Australian paradigm, reacting in different ways to changing circumstances among the three leading drivers which are the subject of this review. The Melbourne and Sydney strategies have much in common and show a reversion to the classical model of Australian metropolitan planning with all the characteristics mentioned above. Sydney is the exemplar here and the plan is shown in Figure 9.1.

The Sydney and Melbourne strategies are marked by a high degree of blueprint-style detail that reflects the two drivers identified above, state control of planning and state provision of infrastructure, and which is particularly evident in Sydney's proposed new urban sectors. Sydney's strategy specifies the future size, location and role of all sizeable centres, the location of future rail and regional bus routes (for example, the planned rail lines to the new north-west and south-west urban sectors shown in Figure 9.1), intermodal centres for freight transport, and subregional housing and employment targets to be adopted in local planning. The boundaries of new urban sectors and new employment zones are precisely delineated, as indicated in Figure 9.1. The sites for redevelopment along the strategy's renewal corridor from the City to Parramatta are similarly mapped exactly in the main strategy document. In supporting documents the strategy sets out the spatial blueprint for new sector development. The plans for the new sectors have street layouts and zonings that are precisely specified.

Melbourne's plan has equivalent imperatives and spatially detailed responses. It has a hierarchy of centres, with the locations of over 100 different centres being

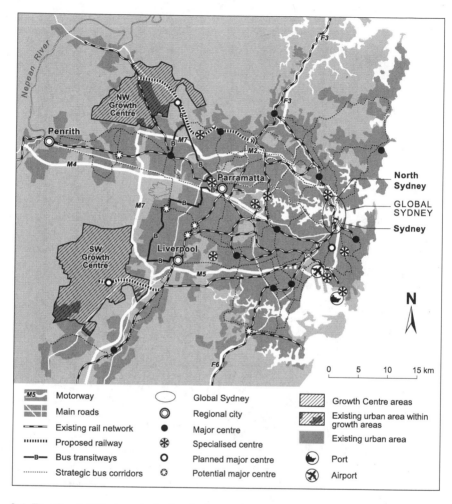

9.1 The Spatial Strategy in Sydney's City of Cities metropolitan plan, 2005.

Source: Bunker, R. and Searle, G. (2007) 'Seeking certainty: recent planning for Sydney and Melbourne', Town Planning Review, 78: 619–642

shown. Proposed new rail and tram routes are mapped. A central element of the strategy is a detailed urban growth boundary to shape and limit suburban expansion that can only be changed by Parliament. Within this boundary, new urban growth areas and proposed major industrial areas are delineated precisely. Housing targets are specified for each subregion.

Brisbane's strategy is in similar vein but has somewhat less prescription as a result of the reduced influence of the two main paradigmatic drivers and its ensuing long gestation. This weaker influence stems from the power of Brisbane City Council, the most powerful local government in Australia, with a population approaching a million. The city council has responsibility for a wide range of services and infrastructure (such as bus services, normally provided or funded in Australian cities by state governments) and the only capital city capable of effectively partnering the state government in planning urban development. This has meant the strategy has required a process involving the state government in a continuous dialogue with Brisbane City and other local governments of the South East Queensland Region, including the second biggest Australian local government, Gold Coast. Even so, the strategy has a significant level of spatial detail. It specifies a finely delineated urban footprint boundary that is being adhered to in recent planning decisions. It shows the location of over 50 existing and potential regional activity centres of varying types and sizes. It also shows road, rail and 'quality' public transport routes where improvements and planning investigations are proposed, though a number of these routes are indicative rather than spatially precise. Population and housing targets are set for each local government area.

The Perth plan relies on short statements, annotated maps, research into transport–land use relationships, public forums for debate on policy alternatives, and a process of continuing development relying heavily on existing state planning structures as mediating and implementing instruments. The plan is still evolving, with the development of the plan in detail (including detailing the metropolitan structure and determining local population, housing and job targets) being identified as a priority task (Government of Western Australia and WAPC 2006). The plan's current form is thus more generalized than the other Australian metropolitan strategies, showing activity centres, networks and non-development areas at a coarser level than the Sydney and Melbourne plans, but is intended to evolve to a finer grain. Adelaide's 2006 plan was a holding operation and was replaced in 2010 (Government of South Australia 2010) by a plan which follows the traditional Australian paradigm of detailed land use delineation.

A changing European paradigm

Early post-war metropolitan spatial strategies in Europe generated blueprint plans similar to those for Australian cities. This reflected not only the prevailing ethos of planning as urban design (see Taylor 1998), but also the variable influence of similar drivers as those argued above for Australian strategies that in turn echoed the post-war rise of the welfare state rather than the longer-standing colonial influences evident in Australia. The 1944 London plan and the associated post-war new town plans were prepared under central government auspices and powers and had new (sub)urban development in the new towns as a central goal. The

Stockholm plan of 1952 was based on development around a new county-owned underground rail system (Hall 2002). While blueprint strategic planning did not entirely disappear from European cities thereafter (for example, Senatsverwaltung für Stadtenwicklung und Umweltshutz Berlin 1994), spatial strategic planning gradually moved away from the deterministic blueprint model. The *Schéma Directeur* for Paris of 1965 was an early example of a new style of plan that showed axes of development and new urban areas in broad-brush fashion (Hall 2002). In subsequent years, metropolitan strategies across much of Western Europe became increasingly more conceptual and less deterministic, with an emphasis on nodes and flows, to the extent that this new style could be considered a new paradigm of metropolitan strategic planning. We now overview influences that have contributed to the emergence of this paradigm, and examples of its expression.

A major underlying driving force has been the increasing assimilation occurring in the European Union (EU) and the need for nation-states to shape their policies to recognize and support this. A powerful source of ideas in this has been Ulrich Beck with his concepts of reflexive modernization and the cosmopolitan democracy. Reflexive modernization involves a reconstitution of the institutions of society with those political and ethical ideals of the Enlightenment (Beck and Willms 2004). Stimulated by the development and enlargement of the EU, Beck has projected these into the idea of a cosmopolitan democracy where 'the recognition of multiplicity' (Beck and Willms 2004: 183) involves a change to the nation-state which becomes a microcosm of world society. In doing this, space is confirmed but in fluid and dynamic dimensions: space-bound realms interweaving with and conditioned by flows and currents of people, capital, ideas, information and goods.

In similar manner, new ideas have developed about strategic spatial planning in Europe (Healey 2004). For example, Richardson and Jensen (2003) construct a 'theoretical and analytical framework for the discourse analysis of socio-spatial relations' in order to understand the 'new set of spatial practices which shape European space at the same time as it creates a new system of meaning about that space based on the language and ideas of polycentricity and hypermobility' (Richardson and Jensen 2003: 7).

Other writers in this vein include Albrechts *et al.* (2003), Harding (2007), Healey (2004, 2006, 2007), Madanipour *et al.* (2001), Neuman (2007) and Salet *et al.* (2003). Hillier (2007) develops a multiplanar theory whose distinctive properties emphasize multiple small narratives – 'a focus on parts while still trying to see a constantly changing impossible "whole"', where space is not a hierarchy ranging from the local to the global but 'a rhizome of multiple connectivities' (Hillier 2007:12) in a process of creative experimentation in space. This dynamic complexity is essential to the 'understanding and working with contingencies of place, time and actant behaviours' (Hillier 2007: 319).

Healey (2004, 2006, 2007) deals more with the spatial planning of large cities while still using a similar '"relational planning" situated within the complex socio-spatial interactions through which life in urban areas is experienced' (2007: 11). This does not emphasize Euclidean space so much as action spaces constituted by a 'complex layering of multiple social relations each with its own space-time dynamics and scalar reach' (Healey 2007: 224). Thus 'the work of [spatial] strategy formation becomes an effort to create a nodal force in the ongoing flow of relational complexity' (Healey 2007: 228).

Each of these planning statements involves the construction of a particular process and methodology at a particular time and place. As Healey (2007: 175,177) reminds us, there are multiple experiences of urban strategy-making and each must be understood as a situated practice deeply structured by the specificities of time and place. But while the more fluid nature of contemporary strategic plans generates much greater variety in individual examples, a number of features are central. Collectively they can be loosely regarded as forming a paradigm.

No contemporary feature is more central than the construction of metropolitan strategies as frames of reference to shape a variety of discourses (Healey 2007). The necessity for such framing arises from two developments in particular. The first is the emergence of sustainability as a meta-narrative or master signifier (Khakee 1997). The economic, social and environmental dimensions of sustainability have their distinctive and often competing discourses. These call forth spatial strategies that are generalized to tie the three sustainability dimensions together to avoid contradictions. The second arises from a new landscape of governance involving a wide range of active participants, rather than just the public sector. This has been driven by a number of factors including new financing imperatives arising from government budget reductions, restructuring of welfare state organization (Albrechts *et al.* 2003), and demands by citizens for much greater involvement in plan-making and governance in general. The emergence of public–private partnerships of various kinds, for example, has required a certain amount of flexibility to be built into plans to allow for a range of project proposals and bargaining outcomes. Thus contemporary (European) spatial strategies attempt to integrate actor–actant networks at various levels, focusing on critical juxtapositions and connectivities (see Healey 2007). Strategic planning is seen as taking place within and by large institutional networks involving multiple scales of governance (Neuman 2007).

A related feature of contemporary strategic plans in Europe is the incorporation of Healey's geographies of relational complexity. This encourages strategies to selectively focus on particular urban places and value-driven critical relationships together with their location in a 'governance landscape' and imagining their effects through time (Healey 2007: 230,232). Dynamics and fluidities are emphasized in contrast to the static geography of modernist strategic plans.

A third central feature is the need for strategic plans to act as vehicles to organize inter-city competition, driven by increased global mobility of capital and labour. This also links to the economic dimension of sustainability. It is the most important aspect of a wider feature, an action orientation that Albrechts (2006) sees as characteristic of contemporary European planning. The need for strategic plans to produce a competitive city necessitates a more fluid, generalized spatial structure to allow the insertion of major private sector initiatives. In particular inter-urban competition is essentially denoted by competition for major events or urban development projects (Swyngedow *et al.* 2003) that have to be accommodated in a manner that is difficult or impossible for long-range plans to anticipate.

This kind of spatial planning encompasses complex negotiations between those who wish to join and leave the dialogue, and those with responsibility for taking decisions about their city. They are much affected by the drive to achieve success in the organization of global capital and labour. The action space is abstract, but qualified by location, spatial relationships and urban conditions. The metropolitan strategies that result have a more fluid and generalized structure and are shorter term and more indicative/less definitive than the previous modernist plans. They are supported by other plans, proposals and positions: for example, infrastructure details are contained in the plans of infrastructure providers (whereas such details can be put into metropolitan plans in Australia because the state governments own or closely regulate and direct the metropolitan infrastructure agencies).

How have these new ideas and imperatives been represented in practice? The European Spatial Development Perspective (Committee for Spatial Development 1999) provides a framework 'towards balanced and sustainable development of the territory of the European Union'. It does not contain any policy maps, as space is treated as an abstract concept which while covering defined units of territory, deals with the relationships and flows which result from the spatial impacts of the policies hammered out by the member states. It does not have any legally binding status but acts as a framework for negotiation about current and emerging issues where space is both a consequence and a determinant of the characteristics of those phenomena. Hillier comments that it challenges 'traditional Euclidean geographies of space as a container for places and events, of distance and time; to linear notions of development pathways in time and essence; and to command-control methods of governance' (Hillier 2006: 277–8), replacing it with 'a creative opportunity for networking, communication and negotiation; for flexibility and fluidity across space and time contingent on context' (Hillier 2006: 283).

The governance of European cities characteristically involves different levels and organs of government in a way that does not occur in the Australian scene, which reinforces the need for planning flexibility in contrast to the general Australian situation. A similar imperative arises from heterogeneity in metropolitan infrastructure provision at a variety of governance scales: individual metropolitan areas have a typical pattern of provision by a mixture of city, province/state and national governments. Beyond the formal structures of government, the inter-connections between regions and nation-states in the EU bring much more fluid-ity, dynamics and players onto the stage of any large city.

In more concrete terms, Albrechts (2004, 2006) has both tried to define the principles of new spatial planning and examined a number of city plans (all but one being European) to see how far they might resonate with those themes. Central to his argument is the association of land use policy with a number of policy instruments to provide a spatial framework for the shaping of social, economic and environmental conditions. In Albrecht's terms:

> a more coherent and coordinated long-term spatial logic for land use regula-tion, for resource protection, for action orientation, for a more open multi-level type of governance, for introducing sustainability, and for investments in regeneration and infrastructure.
>
> (Albrechts 2006: 1153)

Albrechts concludes that while most of the city plans 'demonstrate a shift from traditional technocratic statutory planning (away from regulation of land use) towards a more collaborative and (albeit selective) actor-based approach' (Albrechts 2006: 1166), they still had a long way to go. Judicious selectivity is a particular failing, and in only three cases does the level of government equate with the planning level. Thus, on the evidence of Albrechts, the new paradigm of European spatial strategic planning is not yet hegemonic. Nevertheless, it seems to represent a meaningful frame for characterizing contemporary European met-ropolitan plans. We now briefly indicate how this frame is exemplified in particu-lar city strategies.

Paris regional plans were, as noted, in the vanguard of the move toward more generalized and flexible spatial strategies. The Paris regional plan of 1990 (Hall 2002) is an early illustration of the current European paradigm. The plan map is drawn at a very schematic level, with the principal elements being several types of key nodes, and connections between them. The latter include 'orbital' connec-tions shown as broad arrows that do not specify exact routes, and public transport guideways also shown as indicative arrows. Open space areas, effectively denoting the outer limit of urban development, are also indicated at a very generalized level (too broad to show the Bois de Boulogne or Vincennes, for example).

The Key Diagram of *The London Plan* (Mayor of London 2004: 50) exem-plifies the spatial outcomes of the paradigm. Spatial boundaries of regeneration areas and of open land and the green belt are represented semi-schematically with enough detail to guide some major concerns of the strategy – actions to foster development in the former areas and prevention of development threatening the latter areas. Other components of the strategy are shown in a much more general-ized fashion, though the location of metropolitan centres and opportunity areas are specified. A notable feature is that only two transport routes – the Channel Tunnel Rail Link and Crossrail 1 – are indicated, both in highly indicative fashion.

The spatial strategy diagram of the Hanover city region (Kommunalverband Großraum Hannover 2001) has a similar overall level of generalization as the London Plan, though it also includes differentiation of employment zones and maps the location of motor routes and existing and new rail lines. But the new S-bahn routes are very conceptual, shown only as directional arrows. A broad urban containment ring is depicted, exemplifying the contrast with current Aus-tralian metropolitan strategies with their detailed growth boundaries and urban footprints. Overall, all three examples emphasize critical nodes and flows, the cornerstones of the new Euro plans, in varying degrees.

In summary, while echoes of the three drivers that apply in Australia helped to produce similar blueprint planning in Europe in the post-war period, the replacement of these by other influences/drivers has caused European strategic metropolitan planning to become more generalized within a new paradigm.

Metropolitan planning in the United States

Is there comparable debate and discussion in America, and the development of a distinctive paradigm? Is the Australian experience substantively different to that in America? In reality there is such a diversity of arrangements for metropolitan planning that no clear picture emerges. But there does appear to be some reso-nance with Australian experience where differences in the structure of govern-ance and infrastructure provision can on the one hand be reflected in prescriptive and detailed plans, or on the other in more abstract, broad-brush and indicative intentions.

Unlike Australia where the state government clearly produces and owns the plan for its metropolitan city, there are a variety of sources shaping metropolitan plans of varying authority and effectiveness, with some limited to specific if impor-tant issues such as the integration of land use planning with transportation system development and operation.

Some of the most noteworthy plans have been produced by independent organizations. A recent example of this kind is the New York Regional Plan Asso-ciation's regional plan of 1996 covering three states. In Chicago, the Commercial Club of Chicago, an organization of civic and business leaders, produced a plan in 1999 to prepare the metropolis for the twenty-first century.

The most universal US regional planning organizations are of two kinds. The first are those set up through the Federal Highway Administration to provide local input into urban and transportation planning, which then acts as the basis for allocating federal transportation funds to cities of more than 50,000. There are some 400 of these. The resulting transport plans can have a level of detail that is generally lacking from US metropolitan strategies that attempt to deal with development in general. The second are advisory regional planning agencies set up to plan land use, housing, environmental protection and enhancement, and economic development. These act as important forums for exchange and association of proposals and policies between jurisdictions, but plans are broad, conceptual and indicative in nature. It is impossible for planning to be more precise given the strength of local government, and the diverse range of infrastructure providers from school boards to local authorities and companies providing electricity, gas and water supply, sewerage, roads and public transport (Rottblatt 1998). The National Association of Regional Councils estimates there are some 560 of these kinds of agencies.

The most developed metropolitan planning agencies are established by state legislatures with specific and defined roles, responsibilities and resources. These vary in their significance. The Association of Bay Area Governments was established in 1961 in the San Francisco-Oakland region and incorporates various local governments. Its authority and powers are much weaker than those of the Metropolitan Council for Minneapolis-St Paul, created in 1967 'to plan and coordinate the orderly development of the seven county area' (Metropolitan Council 2004, foreword). Implementation in this case involves deploying and using the powers available to the Council and its constituent authorities. So, like Australian practice, 'local communities are required to adopt comprehensive plans that are consistent with the Council's Development Framework and its four metropolitan system plans – for transportation, aviation, wastewater treatment and regional parks' (Metropolitan Council 2004: 30). Likewise capital works programmes implement and support the Regional Development Framework, updated usually every 10 years, and this gives the Framework more detail and articulation than the broad-brush advisory plans.

A final example of this kind, and perhaps the most well known, is that of Portland, Oregon, the only directly elected metropolitan planning organization in the United States:

In 1992, the region's voters adopted a Metro charter giving Metro jurisdiction over matters of metropolitan concern and requiring the adoption of a Regional Framework Plan. The charter directs Metro to address the following subjects in the plan:

- regional transportation and mass transit systems;
- management and amendment of the urban growth boundary;
- protection of lands outside the urban growth boundary for natural resource, future urban or other uses;
- housing densities.

- urban design and settlement patterns;
- parks, open spaces and recreational facilities;
- water sources and storage;
- coordination, to the extent feasible, of Metro growth management and land use planning policies with those of Clark County, Washington;
- planning responsibilities mandated by state law.

(Metro 2005: 1)

Like the Metropolitan Council for Minneapolis-St Paul, the Regional Framework Plan of 2005 contains considerable detail and substance. The plan illustrates the effect of the governance structure and the infrastructure provision drivers, with a fine-grained, spatially detailed strategy being possible because of a single metropolitan government with extensive powers that also provides critical metro-level infrastructure such as mass transit (Berke *et al.* 2006; Hopkins 2001).

Table 9.1 illustrates the variety of arrangements which exist in the formulation of metropolitan plans. Overall, this variety does not suggest any singular American paradigm. Further, within this heterogeneous landscape there are examples that resonate with the Australian paradigm in terms of the influence of governance structure and infrastructure agency on their level of spatial detail.

Emerging imperatives behind Australian metropolitan plans

While the three drivers posited here as distinctively shaping Australian metropolitan plans appear to remain important (at least in terms of the 'blueprint' character of the recent Sydney and Melbourne strategies), emerging strategic imperatives potentially challenge the picture sketched above. There is little published about the imperatives behind the metropolitan plans in Australia, or in reading between their lines. Notable exceptions include those by McLoughlin (1992) for Melbourne, Hillier (2000) for Perth and Searle (2004) for Sydney. In a more theoretical vein, Gleeson (2006) and Gleeson and Low (2000) have provided a powerful analysis

Table 9.1 A typology of US metropolitan planning authorities and examples

Type of metropolitan planning authority	Examples
Advisory, interest/community-based	Regional Plan for New York, Chicago Metropolis 2020, Envision Utah
Advisory, conceptual, broad brush, mainly government-based	Regional Councils of Government
Limited functions related to joint government planning and funding	FHA metropolitan planning organizations
Set up with varying powers by specific state legislation, government-based	Association of Bay Area Governments, Metropolitan Council for Minneapolis-St Paul, Metro Portland

and critique of the kind of political economy reflected in Australian planning, while Gleeson *et al.* (2004) have developed a socio-theoretical analysis of contemporary Australian metropolitan planning shaped around five themes – policy, space, governance, finance and democracy. These reviews reflect both an emerging extension of traditional metropolitan planning into issues of inclusiveness, social justice, sustainability and environmental enhancement, and also attempts to more rigorously connect Australian practice with methodology, concepts and theory.

These imperatives have received more urgency with changes to the three influences that have been the subject of this review. The governance of Australian cities is changing as the federal government inevitably becomes involved in emerging issues to do with climate change, energy provision, water management, and major infrastructure needs. The current global financial crisis has spawned infrastructure and spending programmes on the part of the federal government that involve the cities. In another realm, the private sector is increasingly enlisted in major partnerships about development where negotiated outcomes are sought. An example of this is the influence in Sydney of the Property Council of Australia, an important coalition of property interests which (as noted above) provided many proposals and targets for population and employment growth in nominated centres that became part of the 2005 metropolitan strategy (Property Council of Australia 2002, 2004). Increasingly, Australian metropolitan strategies can be seen as a framework for multi-sector decision-making as well as the arrangement of urban form and structure.

Regarding infrastructure, two trends are apparent. One is the privatization and outsourcing of many infrastructure systems and facilities. This varies from state to state, but obviously makes issues of accountability, coordination and security more difficult. The second trend is the use of public–private partnerships to fund major infrastructure such as toll roads, railway building and tunnels. These have been problematical in their outcomes, with the state having to pick up failed enterprises (such as the Airport rail line in Sydney) or pay penalty provisions when anticipated traffic loads and revenues are not realized on tunnels and roads. These illustrate the themes of 'splintering urbanisation' (Graham and Marvin 2001) and erosion of certainty about provision and level of service. These deficiencies are illustrated in the case of Sydney. The 2005 metropolitan strategy's commitment to extend the heavy rail system through central Sydney and out to new urban sectors to the south-west and north-west, using state funding, was then removed. Instead there was a confusing series of ad hoc proposals for various 'metros' dependent on federal and/or private sector funding.

A third development is a switch from metropolitan growth predominantly through low-density suburban development into the surrounding countryside almost entirely in the form of separate houses, to renewal, redevelopment and infill in the existing urban fabric. This compact city form is seen as achieving many advantages. Over 60 percent of the population growth envisaged in Sydney and

Melbourne to 2030 or 2031 is seen as taking place in this way. However, Australian planning, policies, institutions and instruments have become used to organizing greenfields development and there are obvious changes needed. One is to develop more sophisticated ways of bringing about change than relying on zoning. The necessity for new approaches here has already been recognized by the formation of special development corporations to expedite brownfields development in Sydney, Melbourne and Perth, using a public–private partnership approach but drawing on state legal authority to do so, thus harking back to the first driver of the Australian paradigm argued above. Nevertheless, many problems in bringing about urban intensification remain (for example, see O'Connor 2003).

In summary, the traditional Australian paradigm faces major challenges in:

- new issues that are a consequence of and a determinant of metropolitan development such as climate change, energy production and sourcing, economic change, security, multiculturalism, water management, affordable housing and mode of transport; and
- changes in the configurations of governance, infrastructure provision, and location and intensity of urban development.

A changing Australian paradigm?

The important changes occurring in all the three main drivers of Australian metropolitan strategies have produced a more complex strategic planning environment. This increasing complexity is reflected in the number of different plans that support (or do not support) the metropolitan strategies. Apart from the broad state development plans and infrastructure strategies introduced over the last 15 years or so that parallel the metropolitan plans (Bunker 2008), there are now also housing plans, water plans, energy plans, environment protection and enhancement plans, business development and innovation plans, education plans, health plans and tourism and recreation plans that are produced mainly by state agencies, but sometimes by business interests such as the Property Council of Australia.

All these trends move from providing apparent certainty with long-term metropolitan strategies, to plans which begin to acknowledge the uncertainties unfolding. There are some indications this may be starting to happen in the current cohort of metropolitan plans. That for Perth departs substantially from the traditional paradigm in its present form, while that for South East Queensland draws on the paradigm in selective fashion. Even those for Sydney and Melbourne, which reassert those principles so strongly, could perversely be argued to show some of the indications of a paradigm shift in showing such a strong reversion to it, as if to demonstrate that the state can still order the future of the city in the face of emerging trends that might demand a more flexible and inclusive response. Nevertheless, a central element of the Sydney plan is a Global Arc from the high-tech industries of the northern suburbs through the CBD to the airport: the Arc is a very Euro-style indicative spatial metaphor signifying the priority area for economic growth (Searle 2008).

The changing Australian circumstances are akin to those happening in Europe, where they have led to metropolitan strategies that much more reflect and shape the interaction of space, society and governance in the city regions. While still planning important physical developments, such strategies tend to be more selective and offer a framework and guidelines for decision-making and negotiation between major partners in development and change. The imperatives behind the new European paradigm are becoming more apparent in Australia. But while Australian metropolitan strategies are starting to evolve toward the European model, they generally retain much of their traditional form. The spatial strategy product of the traditional Australian paradigm is showing resilience as developers and the community continue to look to the state governments to provide planning frameworks and infrastructure that offer the prospect of greater certainty in an increasingly unpredictable world.

While this chapter has not been able to find a United States paradigm in the same way to compare with the European and Australian paradigms, the variety of different arrangements for metropolitan planning does provide some resonance with the argument of this chapter concerning the drivers that generate the Australian paradigm. The more indicative, broad-brush and conceptual plans do reflect complex government structures and relationships, and fractured infrastructure provision. By contrast where state legislation gives power, and responsibility to a metropolitan planning authority, usually constituted from governments, then plans have more detail, prescription and instruments of implementation. In short, then, contemporary Australian, European and American experience demonstrates that regional institutional contexts remain critical in determining the fundamental nature of strategic plans. In Australia, the national character of three key institutional dimensions has produced distinctive metropolitan spatial planning whose paradigmatic nature may now be challenged by the way the state governments respond to emerging influences.

References

Albrechts, L. (2004) 'Strategic (spatial) planning re-examined', *Environment and Planning B*, 31: 743–758.

Albrechts, L. (2006) 'Shifts in strategic spatial planning? Some evidence from Europe and Australia', *Environment and Planning A*, 38: 1149–1170.

Albrechts, L., Healey, P. and Kunzmann, K. (2003) 'Strategic spatial planning and regional governance in Europe', *American Planning Association Journal*, 69: 113–129.

Beck, U. and Willms, J. (2004) *Conversations with Ulrich Beck*, Oxford: Polity Press.

Berke, P. R., Godschalk, D. R. and Kaiser, E. J. (2006) *Urban Land Use Planning*, Urbana: University of Illinois Press.

Bunker, R. (2008) *Metropolitan Strategies in Australia*, University of New South Wales City Futures Research Centre, Research Paper No. 9, Sydney: University of New South Wales.

Bunker, R. and Searle, G. (2007) 'Seeking certainty: recent planning for Sydney and Melbourne', *Town Planning Review*, 78: 619–642.

Butlin, N. (1982) *Government and Capitalism: Public and Private Choice in Twentieth Century Australia*, Sydney: George Allen and Unwin.

Committee for Spatial Development (1999) 'European Spatial Development Perspective: Towards Balanced and Sustainable Development of the Territory of the EU', presented at the Informal Meeting of Ministers Responsible for Spatial Planning of the Member States of the European Union, Potsdam.

Cumberland County Council (1948) *County of Cumberland Planning Scheme Report*, Sydney: Cumberland County Council.

Department of Infrastructure, Victoria (2002) *Melbourne 2030: Planning for Sustainable Growth*, Melbourne: Department of Infrastructure.

Department of Planning, NSW (2005) *City of Cities: A Plan for Sydney's Future*, Sydney: Department of Planning.

Gleeson, B. (2006) *Australian Heartlands: Making Space for Hope in the Suburbs*, St Leonards, NSW: Allen and Unwin.

Gleeson, B., Darbas, T. and Lawson S. (2004) 'Governance, sustainability and recent Australian metropolitan strategies: a socio-theoretic analysis', *Urban Policy and Research*, 22: 345–366.

Gleeson, B. and Low, N. (2000) *Australian Urban Planning: New Challenges, New Agendas*, Sydney: Allen and Unwin.

Government of South Australia (2006) *Planning Strategy for Metropolitan Adelaide*, Adelaide: Government of South Australia.

Government of South Australia (2010) *The 30 Year Plan for Greater Adelaide*, Adelaide: Government of South Australia.

Government of Western Australia and Western Australian Planning Commission (2006) *Statement of Planning Policy: Network City (Draft)*, Perth: Government of Western Australia.

Graham, S. and Marvin, S. (2001) *Splintering Urbanism: Networked Infrastructures, Technological Mobilities and the Urban Condition*, London: Routledge.

Hall, P. (2002) *Urban and Regional Planning*, 4th edn, London: Routledge.

Harding, A. (2007) 'Taking city-regions seriously? Response to debate on "city-regions: new geographies of governance, democracy and social reproduction"', *International Journal of Urban and Regional Research*, 31: 443–458.

Healey, P. (2004) 'The treatment of space and place in the new strategic spatial planning in Europe', *International Journal of Urban and Regional Research*, 28: 45–67.

Healey, P. (2006) 'Transforming governance: Challenges of institutional adaptation and a new politics of space', *European Planning Studies*, 14: 299–319.

Healey, P. (2007) *Urban Complexity and Spatial Strategies: Towards a Relational Planning for Our Times*, London and New York: Routledge.

Hillier, J. (2000) 'Going round the back? Complex networks and informal action in local planning processes', *Environment and Planning A*, 32: 33–54.

Hillier, J. (2007) *Stretching Beyond the Horizon: A Multiplanar Theory of Spatial Planning and Governance*, Aldershot: Ashgate.

Hopkins, L. D. (2001) *Urban Development: The Logic of Making Plans*, Washington, DC: Island Press.

Khakee, A. (1997) 'Agenda-setting in European spatial planning', in Healey, P., Khakee, A., Motte, A. and Needham, B. (eds), *Making Strategic Spatial Plans: Innovation in Europe*, pp. 255–268, London: University College London Press.

Kommunalverband Großraum Hannover (2001) Der Zukunftsdialog Profil für die Region Hannover. *Beiträge zur Regionalen Entwicklung* 93.

Madanipour, A., Hull, A. and Healey, P. (eds) (2001) *The Governance of Place*, Aldershot: Ashgate.

Mayor of London (2004) *The London Plan: Spatial Development Strategy for Greater London*, London: Greater London Authority.

McLoughlin, B. (1992) *Shaping Melbourne's Future: Town Planning, the State and Civil Society*, Melbourne: Cambridge University Press.

Melbourne and Metropolitan Board of Works (1954) *Melbourne Metropolitan Planning Scheme*, Melbourne: Melbourne and Metropolitan Board of Works.

Metro (2005) *Regional Framework Plan*, Portland, OR: Metro.

Metropolitan Council (2004) *2030 Regional Development Framework*, St Paul, MN: Metropolitan Council.

Neuman, M. (2007) 'Multi-scalar large institutional networks in regional planning', *Planning Theory and Practice*, 8: 319–344.

Newman, P. and Thornley, A. (2005) *Planning World Cities: Globalization and Urban Politics*, London: Palgrave Macmillan.

O'Connor, K. (2003) 'Melbourne 2030: a response', *Urban Policy and Research*, 21: 211–215.

Office of Urban Management (2005) *South East Queensland Regional Plan 2005–2026*, Brisbane: Office of Urban Management.

Premier of South Australia (1994) *Planning Strategy for Metropolitan Adelaide*, Adelaide: Premier's Department.

Property Council of Australia (2002) *Initiatives for Sydney: The Voice of Leadership*, Sydney: Property Council of Australia.

Property Council of Australia (2004) *Metro Strategy: A Property Industry Perspective*, Sydney: Property Council of Australia.

Richardson, T. and Jensen, O. (2003) 'Linking discourse and space: towards a cultural sociology of space in analyzing spatial policy discourses', *Urban Studies* 40: 7–22.

Rottblatt, D. (1998) 'North American Metropolitan Planning Re-examined', Paper presented to the Association of Collegiate Schools of Planning Conference, Pasadena, California.

Salet, W., Thornley, A. and Kreukels, A. (eds) (2003) *Metropolitan Governance and Spatial Planning: Comparative Studies of European City-Regions*, London: E and FN Spon.

Searle, G. (2004) 'Planning discourses and Sydney's recent metropolitan strategies', *Urban Policy and Research* 22: 367–391.

Searle, G. (2008) 'The construction of a spatial planning metaphor: Sydney's Global Arc', Paper presented to the Planning Institute of Australia 2008 Congress, Sydney.

Senatsverwaltung für Stadtenwicklung und Umweltshutz Berlin (1994). *Flächennützungsplan Berlin – FNP94*, Berlin: SSU.

Stilwell, F. and Troy, P. (2000) 'Multilevel governance and urban development in Australia', *Urban Studies* 37: 909–930.

State Planning Authority of New South Wales (1968) *Sydney Region Outline Plan*, Sydney: State Planning Authority.

Swyngedow, E., Moulaert, F. and Rodriguez, A. (2003) '"The world is a grain of sand": Large-scale urban development projects and the dynamics of "glocal" transformations', in Moulaert, F., Rodriguez, A. and Swyngedow, E. (eds) *The Globalized City*, pp. 9–28, Oxford: Oxford University Press.

Taylor, N. (1998) *Urban Planning Theory since 1945*, London: SAGE.

Taylor, P. (2006) 'Book review of Newman, P. and Thornley, A. (2005). *Planning World Cities: Globalization and Urban Politics*', *European Planning Studies*, 14: 1163–1164.

Town Planning Committee of South Australia (1962) *Report on the Metropolitan Area of Adelaide*, Adelaide: Government Printer.

Western Australian Government Printer (1955) *Plan for the Metropolitan Region, Perth and Fremantle*, Perth: Western Australian Government Printer.

Western Australian Planning Commission and Department of Planning and Infrastructure (2004) *Network City: Community Planning Strategy for Perth and Peel*, Perth: Western Australian Planning Commission and Department of Planning and Infrastructure.

Chapter 10

Reorienting urban development?
Structural obstruction to new urban forms

Pierre Filion

There is a growing sense that the North American low-density and automobile-dependent urban form is unsustainable from the perspective of quality
of life, the economy and the environment. Yet for all the calls for a transformation of development patterns, trends inherited from the post-Second
World War period die hard. This chapter employs structuration theory to
generate a conceptual framework that identifies factors of structural stability
and transformation. It then uses this framework to account for difficulties
in achieving a transition to higher density and less automobile-dependent
forms of development. The evidence originates from Toronto and points
to an uncertain transition in urban development. The chapter closes with a
consideration of society-wide implications of the difficulties in redirecting
structural tendencies when an environmental crisis looms on the horizon.

Introduction

There is a growing gap between urban discourses and the urban development
reality. Rising energy costs, climate change and other forms of environmental
degradation, together with depleting quality of life (largely a function of worsening traffic congestion) provide a context supportive of alternatives to present
forms of urban development. In North America, where the trend towards low-
density automobile-dependant development is most advanced, the latest wave
of attempts at formulating and implementing models that depart from prevailing urban forms are subsumed under the Smart Growth label. However, while

Submitted by the Association of Canadian University Planning Programs (ACUPP).

Originally published in *International Journal of Urban and Regional Research*, 34.1: 1–19 (2010).
© 2010 By kind permission of Wiley-Blackwell Publishing.

much attention is currently given to Smart Growth by planners and politicians, urban development maintains for the most part in its low-density and automobile-dependent trajectory.

The focus is on obstacles to a shift in urban development. In a recent article, Downs (2005) demonstrated how redistributions of costs and benefits detrimental to a majority of citizens, incapacity to meet transit modal share objectives and the need to rely on unpopular authoritarian forms of planning impede a transition to Smart Growth-type development. While not in disagreement with the obstructions identified by Downs, this chapter attempts to broaden the range of obstacles considered, by focusing on public sector investment potential, production and consumption trends, interests vested in present urban forms, the organization of municipal government and urban dynamics. It frames the discussion within a conceptual approach, derived from structuration theory, which explores conditions associated with structural stability and transformation.

The empirical substance originates from Toronto metropolitan development since the 1950s. Toronto stands out among large North American metropolitan regions by its vigorous attempts, at different times over the past six decades, to alter development models dominant across the continent. Its planning capacity was exceptional, thanks to the early creation of a metropolitan government and direct provincial government intervention in matters of land use. Moreover, with persistent demographic and economic growth, metropolitan development and urban form issues were repeatedly given prominence on the political agenda. If there was one metropolitan region with the potential to depart from the dominant low-density and dispersed North American urban form, it was Toronto. Yet, notwithstanding distinctive urban form features, much of Toronto development from the 1950s onwards has conformed to the continental norm.

The Toronto narrative identifies a range of factors assuring the perpetuation of existing forms of development despite rising aspiration for alternative models. It ends on a note of uncertainty by assessing the urban transition potential of emerging planning policies, residential density trends and critical attitudes towards present urban forms. While some factors appear to promote a shift away from styles of development that have dominated the last 60 years, others favor their perpetuation.

The implications of the chapter reach much further than its thematic and geographic specificity. Beyond its focus on urban transition and Toronto, this chapter is about the capacity of making required societal changes in the face of rising crises.

Reactions against prevailing forms of urban development

Especially since the 1970s, the North American dispersed urban development model, pieced together over the period following the Second World War, has come under severe criticism from the media, the literature, planners, politicians and the public.[1] Its low-density, rigid land use specialization and automobile dependence have been depicted as a source of social segregation and anomie, a strain on public resources, a waste of depleting sources of energy, and above all, a foremost cause of environmental damage (Putman 2000; Richardson and Bae 2004; Soule 2006). Critiques of this model first rallied under the energy conservation banner in vogue in the 1970s and early 1980s, and then joined the sustainable development movement from its inception in the mid 1980s (Romanos, Hatmaker and Prastacos 1981; Roseland 1992).

Widespread adherence to Smart Growth principles is the latest among manifestations of mounting dissatisfaction with prevailing development patterns. Smart Growth originated in the 1990s – a period of resentment against urban expansion – when in many places growth became synonymous with rising pollution and property taxes, crowded services and traffic congestion. It was introduced as an approach capable of easing the adverse impacts of growth and thereby making it more palatable (Sierra Club 2000). Interest in Smart Growth is also fuelled by an increased sense that present forms of urban development cannot be replicated forever.

More so than sustainable development, where the focus is mostly on the environment, Smart Growth combines multiple objectives. Environmental considerations are certainly prominent, but so are attempts at abating adverse financial and quality of life sequels of urban development (Ye, Mandpe and Meyer 2005). Equity, especially as it pertains to housing affordability, also figures among Smart Growth objectives (Bullard 2007). As a result, Smart Growth embraces a broad range of measures: urban growth boundaries, green belts, different forms of urban intensification, public transit improvement, the creation of pedestrian-hospitable environments, multifunctional and mixed-income developments (Smart Growth Network 2002; 2003).

References to Smart Growth have quickly found their way into the media, the political discourse and planning documents, to the extent that it is now customary for plans to claim adherence to the Smart Growth credo. The problem confronting Smart Growth is that, after nearly ten years of existence (and at least two more decades of other attempts at transforming urban development), urban indicators still point in the wrong direction. Notwithstanding substantial achievements such as downtown residential intensification, pockets of high residential density in the suburbs, new urbanism and transit-oriented developments, and the success of many light rail transit lines across North America, the tendency is for car reliance to increase, land uses to remain highly segregated and overall density (though not always residential density) to continue falling. Despite all the talk about Smart Growth and some realizations, the main thrust of urban development is still towards a further advancement of the automobile–land use relationship (Downs 2005; Song 2005; Tomalty and Alexander 2005).

The chapter is preoccupied with reasons for the incapacity – until now – for Smart Growth and earlier movements advancing urban development alternatives to launch an urban transition of the scale of the one that changed the course of development 60 years ago. Just as we have fully adapted most urban development to automobile use after the Second World War, will we be able to create an urban form that conforms to Smart Growth principles in the near future?

Structural reproduction and transition

To understand resistance to change, but without excluding possibilities of transition, I draw from elements of structuration theory as formulated by Anthony Giddens. Structuration theory attempts a middle course between structure and agency (Bryant and Jary 1991; Giddens 1976; 1979; 1984). In its view, structures are maintained and recursively reproduced by the actions of agents. In the words of Giddens, "the structural properties of social systems are both medium and outcome of practices they organize" (Giddens 1984: 24; see also Cohen 1989: 46). In the structuration theory perspective, agents are more than passive supports of deterministic structures; they are rather portrayed as imbued with practical consciousness, which allows them to handle structural constraints and resources, with various degrees of competence, in the pursuit of personal objectives. It is through their actions, informed by reflective knowledge, that social practices are ordered and structures reproduced (Giddens 1984: 3 and 22).[2]

Structures can be perceived as consisting of regular patterns of behavior. Agents encounter structures as rules and resources that are inherited from the past (as in the case of language) and resulting from collective behavior (as for the operation of markets). Structures coalesce into social systems, which Giddens defines as 'patterned social relations across time' (Giddens 1984: 377; see also 1979: 65–66). Social systems are distinguished from each other by their scope and the characteristics of the agents and the types of behavior they embody (Giddens 1989: 300; Sewell 1992: 22–26). If societies in their entirety qualify as social systems, they are at the same time themselves constituted of manifold social systems (Giddens 1984: 164). The concept of social system thus allows for a diversity of types of structures, each with its own rules and resources.

In this perspective, the patterning of behavior that constitutes structures can be the outcome of formal or informal, conscious or unconscious rules along with mechanisms controlling access to resources. Means of patterning also include habits, mutual expectations and the exercise of power. Cities, when perceived as structures, include all of the above patterning devices, along with the effect on behavior of their built environment, spatial distribution of activities and transportation system. The chapter's empirical focus on conditions for urban form reproduction and transformation is thus associated with the stability and change of behavior patterns within cities.

A key implication of structuration theory is that with their understanding of specific structures, however partial and clouded by ideology and preconceptions, agents take a utilitarian approach to structures. A range of possible actions ensue; agents can decide to maintain their current behavior within a structural context, thus contributing to its perpetuation; or they can opt for alternative behaviors and strategies. Structural contexts can be profoundly modified by these actions and strategies, provided that individual and collective agents involved muster enough political and economic power, and that alternative behaviors rally a sufficient number of individuals. When possible, agents can exit structures by shifting from one structure to another or simply abandoning certain behaviors. And of course, effects on structures can be the aggregate outcome of decisions taken independently from each other, a common occurrence in the marketplace, or of orchestrated initiatives, as frequently found within the political process.

Attitudes towards structures are shaped by features that encourage their reproduction and, occasionally, their demise or transformation. The relative endurance of structures – consider language, institutions, market relations and, indeed, urban form – suggests a bias in favor of stability over change. I identify five conditions (Figure 10.1) that assure a reproduction of structures and, thereby, a perpetuation of urban forms (Barley and Tolbert 1997 for an institutional perspective; Giddens 1979: 198–233; Gregson 1989; Jary 1991; Wright 1989):

1   Structures must enjoy *legitimacy*, that is, their existence should be justified by dominant discourses, and their existence should appear to be in agreement with major societal objectives. In sum, structures should mirror world views of the time.
2   Existing structures benefit from a *familiarity* effect. They constitute a habitus, which provides a zone of comfort by shielding agents from the unknown and attendant risks.
3   Structures must appear to be *efficient* and thus capable of meeting performance expectations. There must also be an absence of major conflicts and contradictions between structures, a source of disruptions.
4   Structures must be perceived by agents as conducive to the pursuit of their *interests*, by allowing them to carry out their self-defined (although not without influence from structural contexts) strategies and objectives.
5   Finally, the ongoing reproduction of structures and social systems is facilitated by the *absence of credible alternatives*, capable of fulfilling the above conditions.

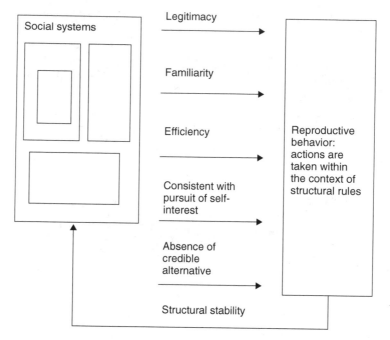

10.1 Structural reproduction

Inversely, when their capacity to act as instruments of structural reproduction is compromised, these same conditions (with the exception of familiarity – a universal factor of stability which transformative forces need to overcome) can convert into factors of change. For example, shifts in values can challenge the legitimacy of existing structures and the emergence of contradictions among structures challenges their efficacy. The resulting debilitation of structures triggers searches for alternatives.[3] However, the replacement of structures requires capacity and can be ensconced in power struggles between competing constituencies, some of which favor the status quo while others champion a transformative agenda. It also helps to have a vision of possible alternatives (see Figure 10.2).

These conditions manifest themselves in different aspects of societal life, shaping the production of the city. I now briefly describe how investment potential, production and consumption tendencies, vested interests, administrative arrangements and urban dynamics have generally contributed to keep prevailing development patterns in place over the last decades.

Government investment capacity can work both as a force of inertia and change. It procures conditions for the efficient operation of the urban system and thereby favors the pursuit of agents' interests therein. However, at the same time,

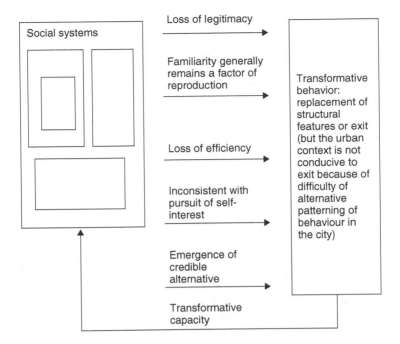

10.2 Structural transformation

depending on political priorities, the availability of financial resources can lead to the creation of alternative infrastructure systems and a resulting transformation of journey patterns and land use. When reduced investment curtails infrastructure development and maintenance, cities lose some of their efficiency – a possible incentive for a search for new urban models. But these same financial conditions curb the possibility of setting up the alternative infrastructure networks needed to induce a change in urban development.

Production and consumption tendencies have resulted in rising reliance on the truck and the automobile along with growing space consumption, accentuating the transportation and land use trends that originated in the post-war period. In production, increasing value was given to automobile and truck accessibility as the labor force settled in the suburbs and just-in-time delivery arrangements proliferated (Lang 2003; Noijiri 2002). Meanwhile, the last decades have witnessed ever growing per capita residential space, made necessary by a mounting accumulation of goods and demand for more personal space in a culture that values individualism (Miles 1998; NAHB 2007; Rorty 1997; Rosenblatt 1999). The pursuit of flexibility, which derives from individualism, favors higher rates of automobile ownership and use, as solo driving gains in popularity.

Interests vested in extant urban forms further contribute to a perpetuation of the dispersed urban model. Constituencies with such interests comprise different businesses involved in the production of the dispersed environment or which derive benefits from this urban form. For example, urban historians have documented the role of the automobile industry in influencing the early decisions that have led to the automobile orientation of US cities (e.g. Bottles 1987; Yago 1984). Households also have an interest in low-density urban forms, motivated by preference for single-family homes, social segregation, tranquillity and the aesthetics of the suburban subdivision. Fear of the unknown and of its possible impact on home values represents an impetus for NIMBY-type reactions against departures from the dispersed pattern (Schively 2007).

The dependence of municipal administrations on property tax revenues explains the readiness with which they accommodate market preference for dispersed types of development. In theory, an elevation in density is fiscally rewarding to a municipality, but this is only the case if developers and consumers are won to intensified development. Otherwise, land zoned for high density will remain fallow and become a fiscal liability. Better, then, to conform to prevailing preferences and allow this land to be developed at a lower density. Inter-municipal competition, where suburban administrations use their low-density configuration and automobile orientation to lure development, is especially favorable to the spread of dispersed urbanization (Lewis 1996).

Urban dynamics – the interaction between different aspects of the city – are another factor of structural stability as these interactions involve mutual adjustments while generally ruling out radical transitions. At the core of urban dynamics is the relationship between transportation and land use, which – over the last 60 years – has involved an ever-advancing mutual adaptation between low-density land use and a transportation system based around automobiles and trucks. Relationships between supply and demand within different markets with direct land use implications – housing, retailing, employment, etc. – also figure among urban dynamics (Filion and Bunting 2006). Urban dynamics point to the difficulty of changing one aspect of the city while leaving its other dimensions untouched. One example is the failed attempt of new urbanism to bring retailing back to neighborhoods. The competition of large establishments, which rely on their customers having access to an automobile (to capture wide catchment areas), proved to be impossible to overcome (see Hernandez, Erguden and Bermingham 2006). Once engaged in interactive patterns, the different elements of urban systems acquire a great deal of stability. It thus follows that only large-scale interventions, addressing simultaneously different aspects of the urban phenomenon, have the potential of bringing about a deep structural transformation of the city.

From a structural point of view, the above mentioned factors owe their role in perpetuating the dispersed urbanization model to their association with conditions for structural reproduction. Each of these factors contributes legitimacy, familiarity, efficiency and opportunities for the advancement of self-interest, to the prevailing style of urban development. However, there are obviously differences in the way in which the factors promote stability. Investment capacity and urban dynamics mostly contribute efficiency, while the advancement of self-interest plays a prominent role in the other factors.

As is the case for structures in general, urban form transformation ensues when conditions for structural reproduction are disrupted. The case study will illustrate conditions that led either to the reproduction or transformation of urban environments.

Metropolitan-scale planning in Toronto

This section applies the structuration theory model to the Toronto case study, which focuses on metropolitan planning and development since the mid 1950s. The period is divided into three phases: (1) the adoption of an attenuated form of dispersion under the auspices of Metro Toronto (1954–1970); (2) all-out dispersion in newly developed suburbs (1971–2002); and (3) provincial government attempts at Smart Growth. For the two first phases, the model plays an interpretive role by identifying circumstances responsible for observed urban form outcomes. However, for the last phase, the model becomes predictive by relying on the five conditions associated with structural reproduction and change to forecast the outcome of the provincial Smart Growth strategy.

1954–1970: adoption and entrenchment of an attenuated dispersed model

Pressures were strong over the high-Fordist period to adapt the urban environment to the symmetric relationship between escalating industrial capacity and rising aspiration for a consumerist lifestyle, largely centered on the automobile (Glennie and Thrift 1992). Traditional urban forms were too dense to provide the space needed to accumulate durable goods and, above all, ill-suited to a generalized reliance on the automobile. It was as if these urban forms held back the economic and technological promises of the time. Conditions were thus ripe for a profound transformation of development patterns. Apart from widespread public adherence to modern urban models and an economic requirement for urban transition, the combination of post-war prosperity and an interventionist Keynesian state secured resources needed for the development of infrastructure systems compatible with new urban forms (Brenner 2004: 114–171; Sewell 1993). These consisted primarily of expressways and arterial road networks. Moreover, during the 10 or 15 years that followed the Second World War, the dispersed suburban model rapidly took shape and was readily accepted as an alternative to the traditional city.

Toronto experienced these pressures for urban change as much as any other North American metropolitan region, but responded to them in a fashion that produced distinct outcomes. In the post-war years, Toronto was facing a severe housing shortage – responsible for high rates of doubling-up. One contributing factor was the inability of suburban municipalities to finance infrastructures needed to sustain rapid external development. The provincial government (which is constitutionally responsible for municipal and land use planning matters) responded by establishing a second tier administration, Metro Toronto, with planning jurisdiction over the entire metropolitan region. Metro was successful in pooling resources from established municipalities to finance the expansion of suburban jurisdictions.

Under the planning helm of Metro Toronto, the metropolitan region was adapted to the space consumption and transportation norms of the time. New suburban areas included abundant green space, a large proportion of single-family homes, and were suited to high levels of car use. Moreover, as elsewhere in North America, the 1950s and 1960s witnessed the construction of an expressway network. However, this network was of a smaller scale than that of most other North American metropolitan regions, a consequence of Metro Toronto's dual transportation strategy. As it was proceeding with expressway development, Toronto was also building a subway system – the first new subway construction to take place in North America since the Second World War. Likewise, Metro Toronto pursued a mixed-density policy across its entire territory, which resulted in numerous pockets of high-density residential development in suburban municipalities. Both its dual transportation and its mixed density strategies singularized Toronto among most of its North American counterparts (Filion 2000; Frisken et al. 1997). These distinctive features are largely a consequence of the presence of a metropolitan-wide agency with the capacity to formulate and implement its own planning strategy. While in other metropolitan regions, multiple suburban municipalities competed to provide low-density and automobile-oriented options to potential developers, Metro Toronto enjoyed the planning coordination and financial capacity needed to pursue an attenuated version of urban dispersion (Frisken 1993; 2007; Rose 1972).

Opposition from low-density neighborhood residents was kept to a minimum by concentrating high residential density along arterial roads and buffering it from single-family subdivisions. Overall, with highway construction keeping pace with urban growth and development adopting automobile-friendly configurations, the period witnessed a mutual adjustment between the new transportation systems and land use patterns. The expansion of highways fostered increased automobile use and accessibility in a relatively congestion-free environment.

Notwithstanding higher density and more reliance on public transit than in most other North American urban areas, by 1970 the dispersed model had taken root in Toronto. Massive funding was sunk into highway networks, land use was adapting to the near ubiquity of the automobile, dispersion conformed to consumerism, and constituencies (notably developers and single-family homeowners) coalesced around features of this urban form (see Figure 10.3).

1971–2002: low-density outer-suburban development

The second period is characterized by a further advancement of the dispersion model, fuelled in part by the creation in the early 1970s of five new regional governments around Metro Toronto. These administrations, which assumed de facto responsibility over the portions of the metropolitan region experiencing growth

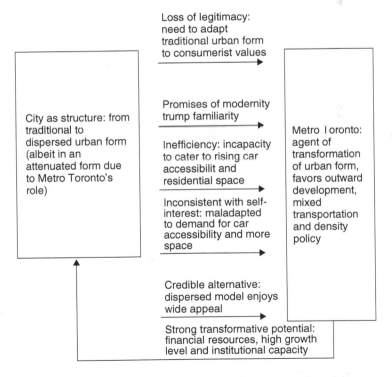

10.3 1954–1970, adoption and entrenchment of the dispersed model

over this period, generally failed to share Metro Toronto's commitment to mixed density and balanced transportation. Responding to market trends, they readily adhered to the low-density automobile-oriented model. Over these years, more than during the first period, development in Toronto resembled that of its North American counterparts, as residential density fell with the rise of single-family homes as a proportion of all new units (Filion et al. 2004).

The event that opens – and possibly best defines – the period is the rejection by the provincial government of a plan that proposed a linear metropolitan form paralleling Lake Ontario so as to minimize water and sewer infrastructures and take full advantage of east–west transportation corridors (Ontario 1970). Beyond this urbanized perimeter, development was to be concentrated in a few satellite communities designed to be largely self-sufficient. In the end, in blatant disregard for the vision of the plan, the provincial government, under pressure from landowners, developers and municipalities north of Toronto, constructed a sewer system servicing the area to the north of the built perimeter, thus opening it to conventional-style suburban development. A comparable situation surfaced in the early 1990s. The ambitious intensification and public transit proposals of the Office for the Greater Toronto Area (a provincial advisory agency) met with a similar fate, this time because of reduced government intervention capacity in the midst of a recession (OGTA 1991; 1992).

The 1971–2002 period is characterized by mounting tension between, on the one hand, factors making dispersion ever more ingrained and, on the other hand, factors responsible for a decline in the efficiency and legitimacy of this urban form, and in its compatibility with the self-interest of agents. Dispersion remained associated with consumerist values. If anything, one could argue that this was ever more the case, with the rise of individualism and attendant demand for more personal space and flexibility in transportation. Increasing familiarity with the dispersed model also marked the second period. As a rising proportion of the public was raised in this type of environment, it often (though not always, as gentrification demonstrates) aspired to the suburban lifestyle. The ongoing expansion of the dispersed model bolstered the economic and political power of the interests associated with this form of development, a major source of influence on suburban municipal administrations (MacDermid 2006; 2007). Additionally, the mutual adjustment of land use with growing reliance on the automobile progressed over these years, thanks to the expansion of the dispersed realm and the mushrooming of new automobile-oriented land use formulas such as big box stores, power malls, mega supermarkets and suburban multiplex cinemas.

Meanwhile, over this period, transportation infrastructures (both public transit and highways) failed to keep pace with rapid urban expansion. Infrastructure development became a victim of a deteriorating fiscal climate, the outcome of globalization, which reduced the capacity of governments to raise taxes from increasingly footloose corporations, and of the introduction of successive income tax cuts (Genschel 2005; Hays 2003; McBride 2005). In Canada and in the province of Ontario, the effect of these circumstances was compounded from the mid 1990s onwards by a firm commitment on the part of governments to trim expenditures in order to eliminate deficits (Lewis 2003; McQuaig 1995). Also responsible for insufficient infrastructure investment was the falling priority given to this sector of expenditure within provincial budgets, as other items, health care in particular, gained in importance (Robson 2001).[4] As expected, the shortfall in infrastructure investment in a context of ongoing dispersed growth translated into worsening traffic congestion and a resulting loss in the metropolitan region's efficiency and capacity to advance the interests of its collective and individual agents. In the 1980s, these interests were also adversely affected by rapidly escalating housing costs, a phenomenon that was not unrelated to deficient provision of infrastructure responsible for an insufficient supply of serviced land.

The legitimacy of the dispersed model was challenged over this period by ascending awareness of the environmental and quality of life sequels of this form of urbanization. Critical attitudes towards urban dispersion were echoed by the planning and political discourse. Planning documents became vehicles for anti-dispersion sentiments. However, this did not prevent them from simultaneously proposing measures supportive of this form of development. There were some planning-induced departures from dispersion. This was notably the case of the development of three large suburban mixed-use nodes, two of which were strategically located on rail transit lines (Metro Toronto 1981). Likewise, despite a decline in its relative importance within the metropolitan region, downtown Toronto enjoyed substantial diversification. Overall, however, dispersion remained the dominant pattern of development at a regional scale.

Besides the role vested interests and consumer values played in propelling ongoing development trends, two additional factors explain the limited impact of attempts at alternative development. First was the absence of an alternative model of development that would be both credible and acceptable to the public and the development industry. The implementation of nodes was too modest to have a significant metropolitan-wide impact (and there are indeed doubts about the extent to which existing nodes have departed from the dispersed urbanization model) (Filion 2007). The same could be said of the few new urbanism communities. Not only was there an absence of alternative model, but had such a model been available, the organizational and financial capacity to implement it was absent. As noted, the loss of the metropolitan-wide planning capacity marked the opening of this period. In addition, the deteriorating fiscal climate compromised the possibility of creating new infrastructure systems compatible with alternative forms of development. Recall that the implementation of alternative patterns

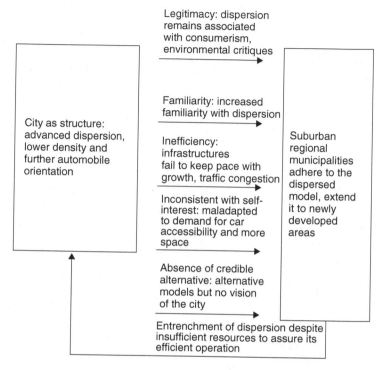

Legitimacy: dispersion remains associated with consumerism, environmental critiques

City as structure: advanced dispersion, lower density and further automobile orientation

Familiarity: increased familiarity with dispersion

Inefficiency: infrastructures fail to keep pace with growth, traffic congestion

Inconsistent with self-interest: maladapted to demand for car accessibility and more space

Suburban regional municipalities adhere to the dispersed model, extend it to newly developed areas

Absence of credible alternative: alternative models but no vision of the city

Entrenchment of dispersion despite insufficient resources to assure its efficient operation

10.4 1971–2002, low-density outer-suburban development

of development requires interventions of a scale sufficient to counter prevailing urban dynamics (see Figure 10.4).

2003 onwards: attempts at Smart Growth development

Two issues prompted the involvement of the provincial government in the planning of metropolitan development in Toronto. There was first wide-scale public protest over the impending urban invasion of the Oak Ridges Moraine, a hilly geological formation, well-provided with lakes and the headwater of all rivers crossing the metropolitan region. Beyond the specificity of the Oak Ridges Moraine conflict, it was the nature and extent of suburban development in the Toronto region that was targeted. The second issue was the anticipation of persistent demographic and economic growth, the expected source of crippling congestion and severe environmental degradation. High public sector expenditure needed to assure the expansion of the dispersed urban form was an additional source of concern (IBI Group 2002; see also Burchell et al. 2005).

It is a Conservative provincial government (a hitherto champion of laissez-faire economics and deregulation) that launched a Smart Growth-inspired metropolitan planning process in 2003. An immediate trigger for the Conservative government's conversion to Smart Growth was rising resentment on the part of its outer-suburban electoral base towards the environmental and congestion downsides of rapidly advancing dispersed-type urban development. The example of the New Jersey Republican administration, an early adherent to Smart Growth with which Ontario Conservatives shared advisors, also played a role. The subsequent Liberal government took over the process and adopted a plan and attendant legislation which: (1) designated a 728,430 hectare green belt that included the Oak Ridges Moraine; (2) required that 40 per cent of new development take place within the built perimeter; (3) adopted a nodal and corridor configuration at the scale of the entire metropolitan region, intended to create land use conditions favorable to public transit ridership (Ontario 2005; 2006). What is more, the province committed to an ambitious public transit investment program (*Toronto Star* 16 June 2007). Together, these measures amount to a particularly bold attempt at modifying the urban development trajectory, especially given the rapidity with which the Province adopted enabling legislation.

While dispersed development is still advancing, as evidenced by the proliferation of new single-family home subdivisions, the ongoing predilection of office space for suburban business parks, and the mushrooming, even in the inner city, of automobile-focused big box stores and power centers, there is evidence of change (e.g. *Globe and Mail*, 3 May 2008; 23 May 2008). Since 2000, the region has experienced a boom in high-rise condominiums, causing the number of new multiple housing units to approach – and at times exceed – that of new single-family homes, a situation unseen since the 1970s. The impact of this boom is particularly visible on the waterfront, in the downtown, in nodes, in redeveloped brownfield sites and in other locations where public transit accessibility is advantageous. The distribution of high-rise condominium structures comes close to the intensification plans of the provincial government.

Continuity or transition?

In light of recent circumstances in Toronto, it is appropriate to reflect on whether we are entering an age of structural inflexion, comparable in intensity to the 1950s, or if urban development is still driven by forces of structural reproduction set in motion over this period. The transition of the 1950s has exposed circumstances that are favorable to structural change. The traditional urban form was inconsistent with the rising appeal of durable goods and thus obstructed the achievement of the self-interest – whose definition was influenced by mounting consumerism – of individuals and businesses. Likewise, rising reliance on automobiles and trucks impaired the efficiency of traditional urban patterns. Moreover, modernism, and the resulting rejection of traditionalism, encouraged individuals to forego familiarity for the promises of newness. Not only was there dissatisfaction with the traditional urban form, but adherence to the alternative suburban model was widespread and ample resources were available to assure its realization. The narrative of the first period has also demonstrated the impact conditions specific to a metropolitan region – in this case the metropolitan planning capacity of Metro Toronto – can have on the form urban transition takes within this region.

There are similarities between the present structural context and the circumstances that have accompanied the urban transition of the 1950s, notwithstanding the change of status of dispersion over the intervening period from source of innovation to obstacle to urban transition. Urban dispersion appears to be losing ground from the perspective of legitimacy, in the present case because it is in conflict with ascending environmental values (Carter 2001; Dobson 2007). Dispersion is also confronted with efficiency problems due to the incapacity to provide sufficient road space to prevent traffic congestion.[5] Efficiency deficits translate into impediments to the achievement of self-interest, such as difficulties in reaching desired activities due to excessive traffic or in finding an affordable home with an acceptable commuting time.

However, one should not be too quick to assume that the dispersed model is fully discredited. While dissatisfaction with the functioning of dispersed metropolitan regions is widespread, there remains a great deal of attachment to the low-density suburban residential formula. Much of the middle class still adheres to conventional suburban models, despite the presence of a segment that expresses a preference for medium- and high-density environments rich in urban amenities.[6] Divisions between the two types of consumers are echoed by the development industry. Lobbies representing developers engaged in low-density projects are particularly vocal opponents of initiatives intended to limit dispersion. For example, the Toronto homebuilding industry objects to urban growth boundaries and the green belt because they cause a reduction in the land available for development. The industry cloaks itself as the defender of the public interest against the housing affordability crisis that would ensue from restrictions on available developable land. It stresses the need for plentiful land availability to respond to widespread preference for ground-related (as opposed to multistorey) housing (see Parson 2004 and Auciello 2006 for statements from the Greater Toronto Home Builders' Association).[7]

It is not only mixed attitudes towards prevailing urban forms that raise doubts about the possibility of a profound urban transition; so do possibilities of achieving an alternative development pattern. A variety of measures are proposed to intensify the metropolitan region and increase non-automobile journey shares: green belt, urban containment, public transit networks and nodal patterning. However, unlike the suburban model of the 1950s, they do not amount to an alternative vision of the city associated with a generalized aspiration for an emerging lifestyle. One problem with present development alternatives is their lack of integration in a fashion that would involve all levels of planning, from the broad metropolitan scale to the configuration and design of local developments. Deficient cohesion is exemplified by the imperfect connection between residential density and reliance on non-automobile modes. Numerous pockets of high residential density are sprouting up but, apart from those located in the core, their automobile modal shares remain sufficiently high to cause severe traffic congestion in surrounding areas (*Toronto Star* 17 May 2008). If they persist, high levels of automobile dependence will limit the potential for future density increments – a further example of the need to intervene on different fronts to alter urban dynamics.

Turning to the capacity of effecting deep changes in development, the provincial government has demonstrated its commitment to shifting patterns of development in Toronto. The will to change has been expressed in planning legislation and in pledges for generous public transit infrastructure funding. These engagements were made at a time when provincial finances were favorable, following successful efforts at deficit control and nearly 15 years of prosperity. The example of Metro Toronto in the 1950s and 1960s has shown that a dedicated administration can have a substantial impact on development patterns. But for all the sums committed to public transit, these are unlikely to counter the effect that decades of priority to highway construction have had on modal shares. In addition, the province alone will not be able to achieve a development transition of the scope that is proposed in its plans. Local and regional governments will need to rally around the alternative development approach and take drastic measures in order to achieve the desired turnaround. There is a surprisingly broad municipal and regional support for the provincial strategy. Only two administrations (one municipal and one regional) have manifested open hostility to the provincial government plan. However, we have yet to reach the stage when numerous suburban sectors will be forced to accommodate intensified development. This is when local administrations, which are much more electorally sensitive to local constituencies than the comparatively distant provincial administration, are likely to show their vulnerability to enduring attachment on the part of their residents to prevailing suburban configurations (Clingermayer 1994; Heilig and Mundt 1984).

What is the likely outcome of all these circumstances? As factors of structural change are not as clearly aligned as they were over the first urban transition, we cannot expect as profound a transformation in urban development. Most likely, the coming transition will be of a partial nature, which is not surprising given the absence of vision of the form a fully transformed metropolitan region would take.

Intensification will happen in some areas, public transit will be improved, but, over-all, dispersion will remain the dominant form of urbanization. Still, positive impacts of such a partial transformation are not trivial, for they afford lifestyle options to people wishing for a non-suburban automobile-centered lifestyle and improve the environmental performance of the metropolitan region. However, particularly in a rapidly growing region, they would fall well short of the level of transformation needed to substantially improve air quality, reduce the emission of greenhouse gases, lower reliance on fossil fuels and fully achieve the quality of life and financial rewards stemming from Smart Growth. Perhaps the best illustration of a partial transition is offered by Portland, Oregon, which is heralded for its muscular Smart Growth interventions and resulting improvements on many environmental indices, but whose development remains primarily low-density, functionally specialized and automobile dependent (Song and Knaap 2004; 2007) (see Figure 10.5).

Of course, in present circumstances one cannot ignore the impact a crisis can have on factors of structural reproduction and change. If it persists, the current economic downturn will alter most of these factors. With declining purchasing

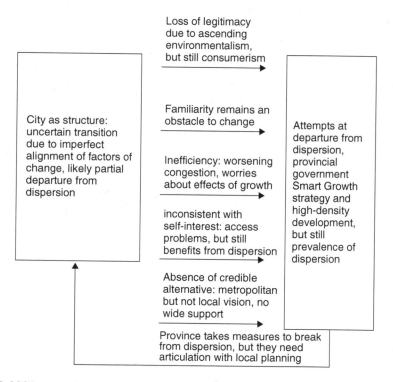

10.5 2003 onwards, attempts at Smart Growth development

power, the legitimacy of an urban form conducive to consumption will be challenged as it becomes increasingly at odds with the economic reality of households. In the same vein, the self-interest of the growing cohort of the economically marginalized, who are already confronted with shrinking economic opportunities, will be further set back by accessibility problems in automobile-focused suburban environments. What is more, the efficiency of urban systems could be compromised by insufficient resources to maintain and operate infrastructures. The economic crisis will then represent an additional factor of transition, alongside concern for the environment and for the overall functioning of the dispersed urban form.

If at first glance this additional factor appears to elevate the likelihood of urban form change, deeper considerations paint a more hesitant picture. A lasting economic crisis will clearly alter behavior, but this emergent behavior will have to contend with a built environment and infrastructure network that are ill-suited to crisis circumstances. The extent of behavioral change will be limited by the prevailing urban form. Will it be possible to transform the urban environment to make it more compatible with emerging values and economic circumstances? While the Obama administration has committed to use its economic stimulus package to advance an environmental and energy conservation agenda, the Canadian federal government has shown far less conviction in this matter (Leonhardt 2009; *Toronto Star* 29 January 2009).

Economic recovery interventions could set the ground for a more environmentally benign and energy efficient urban future. But harking back once more to the previous urban transition, it appears that despite strenuous public sector efforts, relatively little urban development, and thus change in urban form, took place in either the US or Canada over the Great Depression (see Trout 1977). Large-scale public and private sector urban development had to wait until the 1950s, when prosperity fuelled government budgets, the urban growth industry and consumer spending and thereby provided conditions for a profound urban transition. If there is a lesson of relevance to the contemporary context, it is that any major urban structure transformation will have to wait until the recovery. Still, changes of values and in political and economic power distribution among actors with an interest in urban matters will take place between now and then, and will ultimately have a determining impact on the future urban form trajectory.

The chapter has relied on concepts central to structuration theory (that is, structures perceived as patterned behavior and agents as reflective actors with the collective capacity to maintain or transform structures) to interpret conditions associated with the reproduction or modification of urban form. It has identified conditions underlying decisions and behavior on the part of agents, which have either a reproductive or transformative effect on urban form. The chapter has demonstrated the capacity of the approach to interpret past trends in urban development and possibly forecast the respective role of inertia and change in the future evolution of urban form.

Findings from the Toronto case study, as interpreted by the structuration theory model, have implications at three levels. First, they have highlighted the role various factors specific to Toronto – institutions, plans, priority changes, etc. – have played in forging an urban form that conforms largely to the North American norm, while departing from this model in some important respects. Second, the case yields findings of relevance to efforts undertaken across North America to move away from the urban development trajectory of the past 60 years. It identifies conditions that must be met to assure the success of these initiatives, as well as difficulties in aligning these conditions.

Finally, reaching beyond its urban specificity, the structural approach used in this study has broad societal applications. It underlines difficulties in making radical changes, even in the face of widespread awareness of the need for such transformations, in societies that are at once democratic, capitalist and consumerist. The predominance of mechanisms of reproduction and the obstacles transition must overcome account for an overall tendency towards structural stability. The message is not an optimistic one from an environmental perspective. The Toronto case can be seen as an illustration of the difficulties inherent in making profound and timely transformations to avoid climate change or other forms of irreversible environmental damage.[8]

Conclusion

The chapter has investigated reasons for a widening gap between loud calls for a shift in the trajectory of urban development and the rigidity of present growth patterns. Urban development has failed to adapt to mounting dissatisfaction with the quality of life, environmental and financial sequels of dispersion. The entrenchment of present patterns of development has been interpreted as a consequence of the presence of powerful factors of structural reproduction. While the post-Second World War period has exposed conditions that led to a profound transformation of urban patterns, the contemporary context points to only a partial presence of such circumstances. There are, to be sure, criticisms of prevailing patterns, but these are not as widespread as they were over postwar years, due to persistent attachment of large constituencies to dispersion for lifestyle and economic motives. Perhaps most importantly, insufficient financial and institutional capacity and the absence of a cohesive model enjoying broad support jeopardize the possibility of implementing an alternative form of urban development.

Acknowledgments

I acknowledge financial support from the Social Science and Humanities Research Council of Canada (standard research grant 39878).

Notes

1 Throughout the chapter, the term 'dispersion' connotes a form of urban development characterized by relatively low density, functional specialization, reliance on the automobile and a scattering, as opposed to concentration, of activities.
2 Structuration theory is not the only perspective that lends itself to interpretations of structural change (see Hamnet, McDowell and Sarre 1989; Wright 1978). It is, however, less restrictive in its definition of structure, and thus allows for a broader approach to structural change than other perspectives.
3 A parallel can be drawn here with the abnormal science phase of the paradigm shift process (Kuhn 1962).
4 In 1967–8, transportation capital expenses accounted for 9.7 per cent of Ontario government expenditures. By 2007–8, this share had regressed to 3 per cent (Ontario 1968; 2007).
5 Traffic congestion is of course a consequence of insufficient highway investment. But it can be argued that in large metropolitan regions congestion is an inevitable consequence of dispersion, irrespective of the amount of highway space that is provided (see Marshall 2000).
6 Ley (1996) divides members of the so-called new middle class into two categories: those employed in the arts and social sciences, who tend to opt for central residential locations, and those in technical fields, who show an enduring predilection for the suburban lifestyle.
7 However, of late, the Greater Toronto Home Builders' Association has rescinded its opposition to the green belt, likely because its members with green belt land have been generously compensated. Still, the GTHBA remains opposed to any expansion of the green belt.
8 Because of the structural inertia depicted in this chapter, our civilization could figure among those identified by Diamond (2005), which have collapsed owing to an inability to respond sufficiently rapidly to signs of environmental stress.

References

Auciello, D. (2006) "Land supply: cost concerns builders," Greater Toronto Home Builders' Association, Toronto. www.newhomes.org/articles_sun.asp?id=416& SearchType=ExactPhrase&terms=places%20to20grow (accessed May 2008).
Barley, S. R. and Tolbert, P. S. (1997) "Institutionalization and structuration: structuring the links between action and institution," *Organization Studies,* 18: 93–117.
Bottles, S. L. (1987) *Los Angeles and the Automobile: The Making of the Modern City,* Los Angeles, CA: University of California Press.

Brenner, N. (2004) *New State Spaces: Urban Governance and the Rescaling of Statehood*, Oxford: Oxford University Press.

Bryant, C. G. A. and Jary, D. (1991) "Introduction: coming to terms with Anthony Giddens," in Bryant, C. G. A. and Jary, D. (eds), *Giddens' Theory of Structuration: A Critical Appreciation*, pp. 1–31, London: Routledge.

Bullard, R. D. (ed.) (2007) *Growing Smarter: Achieving Livable Communities, Environmental Justice, and Regional Equity*, Cambridge, MA: MIT Press.

Burchell, R. D., Downs, A., McCann, B. and Mukherji, S. (2005) *Sprawl Costs: Economic Impacts of Unchecked Development*, Washington, DC: Island Press.

Carter, N. (2001) *The Politics of the Environment: Ideas, Activism, Policy*, Cambridge: Cambridge University Press.

Clingermayer, J. C. (1994) "Electoral representation, zoning politics, and the exclusion of group homes," *Political Research Quarterly*, 47, 969–984.

Cohen, I. J. (1989) *Structuration Theory: Anthony Giddens and the Construction of Social Life*, Houndmills, Basingstoke: Macmillan.

Diamond, J. (2005) *Collapse: How Societies Choose to Fail or Succeed*, New York: Penguin Books.

Dobson, A. (2007) *Green Political Thought*, London: Routledge.

Downs, A. (2005) "Smart growth: why we discuss it more than we do it," *Journal of the American Planning Association*, 71, 367–378.

Filion, P. (2000) "Balancing concentration and dispersion? Public policy and urban structure in Toronto," *Environment and Planning C*, 18: 163–189.

Filion, P. (2007) "The urban growth centres strategy in the Greater Golden Horseshoe: lessons from downtowns, nodes, and corridors," Toronto: Neptis Foundation.

Filion, P. and Bunting, T. (2006) "Understanding twenty-first-century urban structure: sustainability, unevenness, and uncertainty," in Bunting, T. and Filion, P. (eds), *Canadian Cities in Transition: Local through Global Perspectives* (3rd edition), pp. 1–23, Toronto: Oxford University Press.

Filion, P., McSpurren, K. and Tse, A. (2004) "Canada–U.S. metropolitan density patterns: zonal convergence and divergence," *Urban Geography*, 25: 42–65.

Frisken, F. (1993) "Planning and servicing the GTA: the interplay of provincial and municipal interests," in Rothblatt, D. and Sancton, A. (eds), *Metropolitan Governance: American/Canadian Intergovernmental Perspectives*, pp. 153–204, Berkeley, CA: Institute of Governmental Studies Press.

Frisken, F. (2007) *The Public Metropolis: The Political Dynamics of Urban Expansion in the Toronto Region 1924–2003*, Toronto: Canadian Scholars' Press Inc.

Frisken, F., Bourne, L. S., Gad, G. and Murdie, R. A. (1997) "Governance and social well-being in the Toronto area: past achievements and future challenges," Department of Geography, University of Toronto, Toronto (Research paper No. 193).

Genschel, P. (2005) "Globalization and the transformation of the tax state," *European Review* 13, Suppl. 1: 53–71.

Giddens, A. (1976) *New Rules of Sociological Method*, London: Hutchinson.

Giddens, A. (1979) *Central Problems in Social Theory: Action, Structure and Contradiction in Social Analysis*, London: Macmillan.

Giddens, A. (1984) *The Constitution of Society: Outline of the Theory of Structuration*, Oxford: Polity Press.

Glennie, P. D. and Thrift, N. J. (1992) "Modernity, urbanism, and modern consumption," *Environment and Planning D,* 10: 423–443.

Globe and Mail (3 May 2008) "Class warfare: for urbanites, the mall represents everything they hate about the suburb," M6, D. Gilmor.

Globe and Mail (23 May 2008) "Opponents of Leslieville mall question quality of new jobs," A13, J. Rusk.

Gregson, N. (1989) "On the (ir)relevance of structuration theory to empirical research," in Held, D. and Thompson, J. B. (eds), *Social Theory of Modern Societies: Anthony Giddens and His Critics,* pp. 235–248, Cambridge: Cambridge University Press.

Hamnet, C., McDowell, L. and Sarre, P. (eds) (1989) *The Changing Social Structure,* London: Sage Publications.

Hays, J. C. (2003) "Globalization and capital taxation in consensus and majoritarian democracies," *World Politics,* 56: 79–113.

Heilig, P. and Mundt, R. J. (1984) *Your Voice at City Hall,* Albany, NY: SUNY Press.

Hernandez, T., Erguden, T. and Bermingham, P. (2006) "Power retail growth in Canada and the GTA: 2005." Toronto: Centre for the Study of Commercial Activity, Ryerson University.

IBI Group (in association with Metropole Consultants and Dillon Consulting Ltd) (2002) "Toronto-related region futures study: sketch modeling of four alternative development concepts." Toronto: Neptis and Ontario Smart Growth.

Jary, D. (1991) "Society as a time-traveler: Giddens on historical change, historical materialism and the nation-state in world society," in Bryant, C. G. A. and Jary, D. (eds), *Giddens' Theory of Structuration: A Critical Appreciation,* pp. 116–159, London: Routledge.

Kuhn, T. S. (1962) *The Structure of Scientific Revolutions,* Chicago, IL: University of Chicago Press.

Lang, R. E. (2003) *Edgeless Cities: Exploring the Elusive Metropolis,* Washington, DC: Brookings Institution Press.

Leonhardt, D. (2009) "The big fix," *The New York Times Magazine,* 1 February, pp. 22–29, 48, 50–51.

Lewis, P. G. (1996) *Shaping Suburbia: How Political Institutions Organize Urban Development,* Pittsburgh, PA: University of Pittsburgh Press.

Lewis, T. (2003) *In the Long Run We're All Dead: The Canadian Turn to Fiscal Restraint,* Vancouver: UBC Press.

Ley, D. (1996) *The New Middle Class and the Remaking of the Central City,* Oxford: Oxford University Press.

MacDermid, R. (2006) "Funding municipal elections in the Toronto region." Paper presented at the Annual General Meeting of the Canadian Political Science Association, Toronto. www.cpsa-acsp.ca/papers-2006/MacDermid.pdf (accessed May 2008).

MacDermid, R. (2007) "Campaign finance and campaign success in municipal elections in the Toronto region." Paper presented at the Annual General Meeting of the Canadian Political Science Association, Saskatoon. www.cpsa-acsp.ca/papers-2007/MacDermid.pdf (accessed May 2008).

Marshall, A. (2000) *How Cities Work: Suburbs, Sprawl, and the Roads Not Taken*, Austin, TX: University of Texas Press.

McBride, S. (2005) *Paradigm Shift: Globalization and the Canadian State*, Halifax, NS: Fernwood.

McQuaig, L. (1995) *Shooting the Hippo: Death by Deficit and Other Canadian Myths*, Toronto: Viking.

Metro Toronto (1981) *Official Plan for the Urban Structure*, Toronto: Metro Toronto.

Miles, S. (1998) *Consumerism: As a Way of Life*, London: Sage.

NAHB (National Association of Home Builders) (2007) *Housing Facts, Figures and Trends*, Washington, DC: NAHB.

Noijiri, W. (2002) "'Just-in-time,' the regulation approach and the new industrial geography," *Human Geography* 54: 471–492.

OGTA (Office for the Greater Toronto Area) (1991) *Growing Together: Towards an Urban Consensus in the GTA*, Toronto: OGTA.

OGTA (1992) "GTA 2021: the challenge of our future – a working document," Toronto: OGTA.

Ontario (Government of) (1968) *Budget 1968*, Toronto: Queen's Printer.

Ontario (1970) *Design for Development: the Toronto-Centred Region*, Toronto: Queen's Printer and Publisher.

Ontario (2005) *Greenbelt Act (Bill 135), Royal Assent 13 June 2005*.

Ontario (Ministry of Public Infrastructure Renewal) (2006) *Growth Plan for the Greater Golden Horseshoe*, Toronto: Government of Ontario.

Ontario (Ministry of Finance) (2007) *Ontario Budget 2007*, Toronto: Queen's Printer.

Parson, M. (2004) "Affordable housing depends on land supply," Greater Toronto Home Builders' Association, Toronto. www.newhomes.org/NewHomes/uploadedFiles/Articles/2004/July_31_04.pdf (accessed May 2008).

Putman, R. D. (2000) *Bowling Alone: The Collapse and Revival of American Community*, New York: Simon and Schuster.

Richardson, H. W. and Bae, C-H. (2004) *Urban Sprawl in Western Europe and the United States*, Aldershot: Ashgate.

Robson, W. B. P. (2001) "Will the baby boom bust the health budget? Demographic change and health care financing reform," Toronto: C.D. Howe Institute.

Romanos, M. C., Hatmaker, M. and Prastacos, P. (1981) "Transportation, energy conservation and urban growth," *Transportation Research, Part A: General*, 15A: 215–222.

Rorty, R. (1997) *Truth, Politics and "Post-Modernism,"* Assen: Van Gorcum.

Rose, A. (1972) *Governing Metropolitan Toronto: A Social and Political Analysis 1953–1971*, Berkeley, CA: University of California Press.

Roseland, M. (1992) *Toward Sustainable Communities: A Resource Book for Municipal and Local Governments*, Ottawa, Ont.: National Round Table on the Environment and the Economy.

Rosenblatt, R. (1999) *Consuming Desires: Consumption, Culture, and the Pursuit of Happiness*, Washington, DC: Island Press.

Schively, C. (2007) "Understanding the NIMBY and LULU phenomena: reassessing our knowledge base and informing future research," *Journal of Planning Research,* 21: 255–266.

Sewell, J. (1993) *The Shape of the City: Toronto Struggles with Modern Planning,* Toronto: University of Toronto Press.

Sewell, W. H. Jr. (1992) "A theory of structure: duality, agency and transformation," *American Journal of Sociology,* 98: 1–29.

Sierra Club (2000) "Sprawl costs us all." www.sierraclub.org/sprawl/report00/ (accessed May 2008).

Smart Growth Network (2002) "Getting to Smart Growth: 100 policies for implementation," Washington, DC: International City/Country Management Association. www.smartgrowth.org/pdf/gettosg.pdf (accessed May 2008).

Smart Growth Network (2003) "Getting to Smart Growth II: 100 more policies for implementation," Washington, DC: International City/Country Management Association. www.smartgrowth.org/pdf/gettosg2.pdf (accessed May 2008).

Song, Y. (2005) "Smart growth and urban development pattern: a comparative study," *International Regional Science Review,* 28: 239–265.

Song, Y. and Knaap, G.-J. (2004) "Measuring urban form: is Portland winning the war on sprawl?," *Journal of the American Planning Association,* 70: 210–225.

Song, Y. and Knaap G.-J. (2007) "Quantitative classification of neighborhoods: the neighborhoods of new single-family homes in the Portland metropolitan area," *Journal of Urban Design,* 12: 1–24.

Soule, D. C. (ed.) (2006) *Urban sprawl: A Comprehensive Reference Guide,* Westport, CO: Greenwood Press.

Tomalty, R. and Alexander, D. (2005) "Smart growth in Canada: implementation of a planning concept," Ottawa, Ont.: Canada Mortgage and Housing Corporation, Policy and Research Division.

Toronto Star (16 June 2007) "A 17.5 billion transit promise; Premier's pre-election pledge will create jobs, ease congestion, reduce greenhouse emissions," A06, T. Kalinowski.

Toronto Star (17 May 2008) "Fast growth, slow traffic: condos and townhouses continue to spring up around Yonge St. and Sheppard Ave., plunging residents into congestion chaos," A08, V. Lu.

Toronto Star (29 January 2009) "This sea of red ink could use a splash of vision," A07, J. Travers.

Toronto Star (1 February 2009) "On the wrong side of the fence," IN03, P. Gorrie.

Trout, C. H. (1977) *Boston, the Great Depression, and the New Deal,* New York: Oxford University Press.

Wright, E. O. (1989) "Models of historical trajectory: an assessment of Giddens's critique of Marxism," in Held, D. and Thompson, J. B. (eds), *Social Theory of Modern Societies: Anthony Giddens and His Critics,* pp. 77–102, Cambridge: Cambridge University Press.

Yago, G. (1984) *The Decline of Transit: Urban Transportation in German and U.S. Cities 1900–1970,* Cambridge: Cambridge University Press.

Ye, L., Mandpe, S. and Meyer P. B. (2005) "What is 'Smart Growth' – really?" *Journal of Planning Literature* 19: 301–315.

Designing cities

Chapter 11

Safe urban form

Revisiting the relationship between community design and
traffic safety

Eric Dumbaugh and Robert Rae

While concerns about traffic safety were central to the development of conventional community design practice, there has been little empirical examination into the relationship between community design and the incidence
of traffic-related crashes, injuries, and deaths. This study examines the relationship between community design and crash incidence. It presents a brief
historical review of the safety considerations that helped shape conventional
community design practice, followed by the results of negative binomial
models developed from a GIS-based database of crash incidence and urban
form.

   The authors find that many of the safety assumptions embedded in
contemporary community design practice are not substantiated by the
empirical evidence. While it may be true that disconnecting local street
networks and relocating non-residential uses to arterial thoroughfares can
reduce neighborhood traffic volumes, these community design configu
rations appear to substitute one set of safety problems for another. Surface arterial thoroughfares, arterial-oriented commercial uses, and big box
stores were all found to be associated with an increased incidence of traffic-
related crashes and injuries, while higher-density communities with more
traditional, pedestrian-oriented retail configurations were found to be associated with fewer crashes. Intersections were found to have a mixed effect
on crash incidence. We conclude by discussing the likely reasons for these
findings – vehicle operating speeds and systematic design error – and outline three general design considerations that may help address them.

Submitted by the Association of Collegiate Schools of Planning (ACSP).

Originally published in the *Journal of the American Planning Association*, 75:3: 309–329 (2009).
Used here with kind permission of Taylor and Francis and the American Planning Association.

> Community design is strongly associated with crash incidence. The speed and operating characteristics of arterial thoroughfares, as well as the design and configuration of commercial and retail uses, appear to be particularly important. The results of this study suggest that access should be strictly managed along arterial thoroughfares, and that commercial and retail uses should be located away from these roadways, or at least oriented towards lower-speed access lanes that limit their connections to the arterial system. Encouraging communities to adopt higher-density, more "urban" design configurations would generally appear to help reduce crash incidence, although designers should be cautious about the potential traffic hazards associated with the use of 4-leg intersections.

Introduction

Traffic fatalities are currently the sixth leading cause of preventable death in the United States (Mokdad et. al 2004). In 2006, there were more than 38,600 fatal traffic crashes in the United States, resulting in the death of almost 43,000 people. Of these, 45 per cent occurred in urban environments, and fully 22 per cent occurred on urban arterials alone (See Table 11.1). These numbers have remained relatively constant for more than a decade (National Highway Traffic Safety Administration (NHTSA) 2006), leading many in the transportation community to begin calling for a more "safety-conscious" approach to transportation

Table 11.1 Fatal crashes in the United States

	Road class	Fatal crashes	Pct. urban	Pct. total
Urban	Freeway	3,974	22.9%	10.3%
	Arterial	8,550	49.3%	22.1%
	Collector	1,416	8.2%	3.7%
	Local	3,405	19.6%	8.8%
	Total urban	**17,345**	**100.0%**	**44.9%**
			Pct. rural	*Pct. total*
Rural	Freeway	2,433	11.7%	6.3%
	Arterial	7,722	37.2%	20.0%
	Collector	6,620	31.9%	17.1%
	Local	4,004	19.3%	10.4%
	Total rural	**20,779**	**100.0%**	**53.8%**
Unknown		524		1.4%
Total: urban and rural		**38,648**		**100%**

Source: Fatality Analysis Reporting System

system planning and design (Federal Highway Administration 2003; Transportation Research Board 2001).

To date, there has been little consideration given to the role of community design in the incidence of traffic-related deaths and injuries. Planners and urban designers typically relegate traffic safety concerns to other transportation-related professions, entering the safety discussion, if they do so at all, principally to advocate for the specific safety needs of pedestrians and bicyclists. Yet the planning profession has not historically taken such a narrow view of traffic safety. Many of the policies and practices that have become embedded in contemporary community design practice, such as the functional classification of roadways, the development of disconnected residential subdivisions, and the location of retail uses along arterial thoroughfares, are all products of early planning efforts to address traffic safety.

This chapter revisits the relationship between community design and traffic safety. We begin by identifying the historical safety assumptions that led to contemporary community design practice, and proceed to formally analyze the empirical relationships between crash incidence and urban form. We find that while community designers during the early twentieth century were correct in their concern about the safety effects of gridiron street networks, the design practices that have emerged in response have often had negative and unintended safety effects. We conclude by discussing the implications of these findings for planning practice, and identifying areas where additional research is needed.

Community design and traffic safety: a twentieth-century perspective

Any attempt to understand the historical relationship between planning and traffic safety must begin with an understanding of the "gridiron" street network. While often romanticized by contemporary proponents of neo-traditional development, the urban grid was popular in the nineteenth century not as a means for promoting pedestrianism, but instead as a tool for encouraging the rapid development of unsettled land. Streets were typically designed at uniform widths, and spaced apart at equal intervals, with the net effect of maximizing the number of "premium" corner lots and making each street as accommodating to development as the next. While some cities, like Savannah, incorporated public spaces and street termini into their designs, more common was the application of the concept in cities like New York and Chicago, which simply expanded the grid ad infinitum toward the horizon, permitting interruptions only where required by physical necessity (Kostof 1991).

The profession's early interest in traffic safety is conjoined with its critique of the grid. With the widespread adoption of the personal automobile during the early decades of the twentieth century, gridded street networks had the side effect of making each street as accommodating to automobile traffic as the next, creating unwanted conflicts between motor vehicle traffic and residential and recreational uses. In an early critique of the grid, no less than Frederick Law Olmsted (Jr.) complained that "it has been the tendency of street planners, whether acting for the city or for landowners, to give quite inadequate attention to the need of the public for main thoroughfares laid out with sole regard for the problems of transportation" (1916: 8). In Olmsted's view, safety, aesthetics, and operational efficiency could all be enhanced if community designers moved away from the standardized street arrangements embodied by the grid, and towards streets and street networks designed to accommodate specific and distinct traffic functions.

The precedent for designing street networks according to their traffic function was found in the design of New York's Central Park, where the senior Olmsted installed grade-separated, limited-access thoroughfares to separate through-moving traffic crossing the park from the lower-speed, more recreational activities for which the park was intended (See Figure 11.1). As Olmsted senior described it, "by this means it was made possible, even for the most timid and nervous, to go on foot to any district... without crossing a line of wheels on the same level,

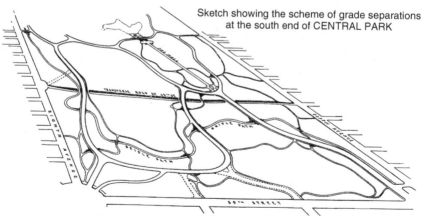

Sketch showing the scheme of grade separations at the south end of CENTRAL PARK

The value of these grade separations lies not so much in the greater safety to pedestrians, and still less in the speeding up or continuity of flow of traffic attainable, but chiefly in the freedom from distraction and in the greater comfort for people who have come to the park for its enjoyment.

11.1 Olmsted's use of functionally designed thoroughfares in Central Park and consequently, without occasion for anxiety or hesitation" (Olmsted 1872, in Olmsted Jr., and Kimball 1922: 47).

This idea was carried forward into what is perhaps the first automobile-era manual on street design, entitled *Width and Arrangement of Streets* (Robinson 1911). Applying the lessons of Central Park, Robinson recommended that streets be designed to serve specific functional purposes, including "main traffic channels that in location and arrangement shall be so nearly ideal that traffic will naturally concentrate upon them, to the end that streets which we do not design for traffic highways shall not be unduly used for traffic" (48). Neighborhood streets, by contrast, should be designed with "some permanent physical handicap, such as indirection, heavy grades or a break in continuity" (10) to encourage automobile traffic to use the main traffic channels.

Beyond simply preserving the residential character of urban neighborhoods, this approach was also presumed to enhance traffic safety. As enumerated in the early professional guidebook *City Planning*:

> If there were a pronounced differentiation between main thoroughfares intended for traffic carriers and secondary or intermediate ones intended for local development, the necessity of very frequent crossings would not exist. Wide traffic streets would afford a better view of vehicles approaching from intersecting streets and good speed could be safely maintained where stopping points were a considerable distance apart.
>
> (Haldeman 1916: 288)

There are two safety ideas embedded here. The first was that by widening and straightening these new automobile thoroughfares, designers would enhance a motorist's ability to see a forthcoming hazard in the right-of-way well before they physically encountered it, providing the driver with ample time to decelerate or change course. Stated in contemporary terms, these thoroughfares would be designed to enhance *sight distance*. The second idea was that safety would be enhanced by eliminating intersections. Logically, fewer intersections would result in fewer intersection conflict points, thereby reducing the number of opportunities for crossing vehicles and pedestrians to be involved in traffic crashes.

Perry, Stein, and the advent of conventional community design

While Olmsted provided the conceptual framework for the design and configuration of contemporary urban streets and street networks, it was Clarence Perry and Clarence Stein who translated these ideas into their conventional form. For Perry, who summarized his life's work in *Housing for the Machine Age* (1939), the problem with turn-of-the-century development practices was that they failed to address community design in a comprehensive, integrated manner. To remedy this, Perry promoted the idea of the neighborhood unit, designed as a self-contained community large enough to populate a local elementary school (6,000–10,000 people).

In addition to the use of functionally designed streets, Perry further proposed reconfiguring land uses to re-enforce the separation of traffic functions. Only residential and neighborhood-supporting civic uses, such as schools and churches, were to be included within a community's confines. All other uses, such as retail and commercial, would be relocated onto the arterial thoroughfares that bounded the community. While acknowledging that this would result in more auto-oriented retail configurations, Perry viewed this configuration as benefiting residents and retailers alike. For residents, moving retail and commercial uses to arterial locations would remove cut-through traffic from their neighborhoods, while still allowing them to conveniently access household-supporting retail as part of their journey-to-work trips. It would further benefit retailers by allowing them to capture both shopping trips from the adjacent neighborhood, as well as new pass-by trips from traffic traveling along the arterial thoroughfare.

Clarence Stein, in partnership with Perry, gave these design concepts their archetypal form in the design of Radburn. While much attention is given to Radburn's larger influence on contemporary community design (see Garvin 1995; Lee and Ahn 2003; Parsons 1994; Schaffer 1982), its intended role as a means for enhancing traffic safety is often overlooked. In describing "The Need for Radburn," Stein writes:

> American cities were certainly not places of security in the twenties. The automobile was a disrupting menace to city life . . . pedestrians risked a dangerous street crossing 20 times a mile. The roadbed was the children's main play space. Every year, there were more Americans killed and injured in automobile accidents than the total American war causalities in any year. The checkerboard pattern made all streets equally inviting to through traffic. Quiet and peaceful repose disappeared along with safety . . . It was in answer to such conditions that the Radburn plan was evolved.
>
> (Stein 1957: 41)

To address these problems, Stein designed a disconnected residential subdivision, characterized by functionally defined roadways "planned and built for one use, instead of for all uses," recognizing that this design entailed "a radical revision of relation of houses, roads, paths, gardens, parks, blocks, and local neighborhoods" (41). Residential uses, if they were to be designed with traffic safety in mind, should be separated from all others, and located along disconnected streets and culs-de-sac. All other uses should be relocated to arterial thoroughfares.

The three safety premises of conventional community design

The safety benefits of conventional community design are premised on three related ideas. The first is that safety could be enhanced by developing a new class of roadway – the arterial thoroughfare – designed to address the specific safety needs of motorists. These roadways should be designed to be wide and straight to increase sight distances, thereby allowing motorists to readily observe and respond to potential traffic conflicts in the right-of-way.

The second safety premise was that street networks should be reconfigured to both prevent vehicle traffic from entering residential areas as well as to reduce the number of conflicts between opposing streams of traffic. This was to be achieved through the development of disconnected street networks that replaced 4-way intersections with T-intersections and culs-de-sac, and which limited the number of access points to the surrounding arterials, with the optimal configuration being a community with exactly one connection to the external street network.

The third safety idea is that land uses should be designed to reinforce the functional separation of traffic. Residential uses should be located principally on disconnected local street networks, while traffic-generating uses, such as neighborhood retail and commercial uses, should be moved to arterial thoroughfares designed to carry heavier traffic loads. These three design concepts were subsequently institutionalized into contemporary land development practices (Southworth and Ben-Joseph 1995), and their lasting influence can be seen in the form and configuration of communities built during the latter half of the twentieth century (see Figure 11.2).

The (limited) empirical basis for conventional community design

But what of the empirical evidence used to justify these practices? Are the communities that emerged in fact safer than the development forms they sought to replace? The guiding study on this subject is a 1957 publication by Harold Marks entitled "Subdividing for Traffic Safety." In this work, Marks sought to examine whether the "limited-access" communities – specifically, disconnected residential subdivisions – were safer than gridiron configurations. In comparing the

11.2 Conventional community design

two types of communities, Marks found that gridiron neighborhoods had fully seven times the number of crashes as the limited-access communities. He further looked at safety performance of 3- and 4-leg intersections. 3-leg ("T") intersections, whether located in gridiron or limited-access communities, reported fewer crashes than 4-leg intersections. Based on these results, Marks concluded that "our new subdivisions have built-in traffic safety" (324).

While Marks' results are compelling, three issues bear noting. First, the study did not control for differences in traffic volumes. Local streets in gridiron networks typically carry heavier traffic loads than those in disconnected subdivisions, which is what led planners to disconnect residential street networks in the first place. Differences in traffic volumes are likely to explain at least some of the differences in crash incidence across the different network types. Second, the safety effects of rearranging neighborhood land uses were not explicitly considered. Instead, Marks simply examined differences in crash incidence for residential areas with different network configurations. Finally, Marks only examined neighborhood streets, and did not consider the safety effects of relocating local traffic and non-residential uses onto arterial thoroughfares. While such network and land use configurations may reduce traffic volumes and crash frequency within a neighborhood's boundaries, it is possible that these localized crash reductions may be offset by substantial increases in the frequency and severity of crashes occurring on the adjacent arterial roads.

During the last 50 years, the only study to formally revisit Mark's findings is a 1995 study by Eran Ben-Joseph, which examined crash frequency for 9 lower-density (6 DU/acre or less) suburban communities in the San Francisco Bay area, all of which consisted of local streets carrying low traffic volumes (1,500 ADT or less). The communities were divided into three types based on their street network configurations: cul-de-sac, loop, or grid. The three gridded communities reported 68 injurious crashes over a 5-year period, compared to 34 for loop communities, and 18 for cul-de-sac communities. This study also considered the effects of traffic volumes on crash incidence. On average, neighborhoods with gridded street networks reported 3 crashes per 100,000 vehicle trips, loop communities experienced 2.8, while cul-de-sac communities reported 2 crashes per 100,000 vehicle trips. In general, the study confirms Marks' earlier findings, although like Marks' study, it did not examine the safety effects of rearranging land uses or re-routing local traffic onto arterial thoroughfares.

Several recent studies have sought to develop crash forecasting models for use in long-range transportation planning applications. While these models were neither designed nor intended to examine the effects of urban form on crash incidence, they include various measures of the built environment at the census tract or traffic analysis zone (TAZ) level, and their results are worth noting. Where density was included in the models, areas with more people, households, or higher population densities generally had higher crash rates (Hadeyeghi, Shalaby, and Persaud 2003; Hadeyeghi et al. 2006; Lovegrove, Gordon, and Sayed 2006; Ladron, Washington, and Oh 2004). Where intersection density was examined, it was found to have mixed effects, with two studies finding higher intersection densities to be associated with increases in crash incidence (Hadeyeghi et al. 2003; 2006), and one study finding them to be associated with significantly fewer crashes (Ladron et al. 2004). Nevertheless, one should be cautious about inferring specific design relationships from these results, as their units of analysis – traffic analysis zones (TAZs) or census tracts – are all too large to draw specific inferences on crash incidence at the level of the individual community.

Revisiting the relationship between traffic safety and urban form

Given the widespread adoption of the community design practices espoused by Perry and Stein, and the limited information on their actual safety effects, we sought to re-examine the relationship between urban form and traffic safety. To do so, we developed a GIS-based database of crash incidence and urban form for the City of San Antonio.

San Antonio was selected for both theoretical and practical reasons. From a theoretical perspective, San Antonio is a high-growth sunbelt city containing a diverse array of design environments, ranging from an historic urban core, street-car suburbs in the areas adjacent to downtown, and more conventionally designed communities located on the City's periphery. As such, its overall form is diverse enough to provide the requisite level of design variation needed to meaningfully model the relationship between community design and crash incidence.

From a practical perspective, the selection of San Antonio allowed us to bypass a major practical barrier to examining traffic safety for large geographic areas – the difficulty of acquiring consistent and reliable information for crashes occurring on local streets. While State Departments of Transportation compile crash information for state-operated roadways, crash data for local streets are typically maintained by local police departments. There is often a great deal of variation in the way individual police departments collect and compile individual police accident reports (PARs), and many jurisdictions do not systematically compile this information at all.

Because the majority of the San Antonio metropolitan region is consolidated within the city limits (roughly 90 per cent of the region's 1.4 million residents reside within the City of San Antonio itself), we were able to acquire crash data for the majority of the region from a single agency – the San Antonio Police Department (SAPD) – in a consistent and reliable form. A second benefit is that this data contains information for *all* crashes to which the SAPD responded, including those occurring on private properties. Thus, unlike most safety analyses, this study includes information for crashes occurring not only along public streets, but also those occurring in residential driveways and the parking lots of shopping centers.

While these data allow us to overcome many of the practical barriers to conducting such an analysis, their principal shortcoming is that they were recorded in a very simplified form. The data only provide information on crash location and severity, and do not identify the specific users involved in a crash (i.e. motorists, pedestrians, or bicyclists), nor do they provide information on crash type (e.g. sideswipe, angle, head-on, run-off-road, etc). As such, this analysis is limited to crash frequency and severity, rather than examining the incidence of specific crash types involving different user groups.

Measuring neighborhood-level crash incidence

To develop our database of crash incidence and urban form, we integrated the crash data supplied by the SAPD with parcel-level land use data supplied by the

Bexar County Appraisal District, street and road network information supplied by the San Antonio-Bexar County MPO, information on traffic volumes, acquired from the Texas Department of Transportation and the City of San Antonio, as well as demographic information acquired from the US Census.

Analyzing this data at the neighborhood level required several methodological assumptions. The first was the identification of the appropriate measures of traffic safety. In addition to examining total crash incidence, we also examined the specific incidence of injurious and fatal crashes, as the design factors that lead to these crashes may have unique characteristics. For this study, total crash incidence is simply the sum of all of the crashes occurring within a community. A fatal crash is defined as a crash leading to the death of one or more persons, while an injurious crash is a crash that resulted in a serious but non-fatal injury. To account for annual fluctuations in crash incidence associated with regression-to-the-mean, we used three years of data (2004–2006) for each of these three safety measures. Descriptive statistics for crashes in the City of San Antonio, by crash type and location, are presented in Table 11.2, below.

A second issue relates to the operational definition of our unit of analysis – crash incidence at the neighborhood level. While the definition of an individual neighborhood can be the subject of debate, we opted to rely on census block group definitions. The decision to do so was based on data availability. From a practical perspective, we needed geographic units for which we could obtain reliable information on a neighborhood's population characteristics, forcing us to rely on either census geography or TAZs. Because larger geographic units mask internal community design variation, we used census block groups, which provide accurate population information but are nevertheless small enough to be relatively homogeneous in their design characteristics. Block groups that overlapped jurisdictions for which we lacked data were excluded, as were block groups with missing or incomplete data, resulting in a total of 747 block groups included in the analysis.

Two additional issues pertained to how we were to address micro-level spatial variation associated with the use of different GIS layers, as well as the related

Table 11.2 Crash incidence for the City of San Antonio, by crash type and location 2004–2006

	Fatal		Injurious		Total	
	Num.	Pct.	Num.	Pct.	Num.	Pct.
Freeway	104	24.0%	5,349	20.4%	29,843	19.8%
Arterial	128	29.6%	9,799	37.4%	51,523	34.2%
Collector	27	6.2%	1,916	7.3%	10,001	6.6%
Local	103	23.8%	6,615	25.2%	39,619	26.3%
Private/Off-Network	71	16.4%	2,555	9.7%	19,640	13.0%
Total	433	100.0%	26,234	100.0%	150,626	100.0%

problem of how to meaningfully assign information occurring along a neighborhood's boundaries. Previous researchers have sought to avoid these attribution problems by eliminating information occurring on the boundaries of their units of analysis (Ladron et al. 2004). The problem with this approach, however, is that both TAZs and census geography often use arterial thoroughfares as geographic boundaries. Since arterial roadways often carry an overwhelming share of the traffic generated by a community, and are thus likely to experience a large share of the crashes occurring in a community, the effective result of eliminating boundary information is that neighborhood crash incidence is likely to be underestimated.

To resolve this problem, we defined our neighborhoods as comprising the block group itself, plus the streets along its edges. To capture the relevant information, we developed 200 ft buffers around each block group (roughly the right-of-way width of a fully designed principal arterial), and assigned the roadway and crash information occurring within the buffer area to the block group it adjoined. This approach thus regards streets located on the edges of neighborhoods as being part of the neighborhood itself, an approach that roughly corresponds with the way individuals define the boundaries of their neighborhood (Lynch 1960). While such an approach does result in some streets and crashes being assigned to more than one neighborhood, it is important to reiterate that the unit of the analysis is the *neighborhood*, not the individual street or crash location. This operational decision provides a consistent framework for addressing problems associated with differences in the spatial definition of individual GIS layers, while also ensuring that essential information on crash incidence was not lost due to the means by which the unit of analysis was operationalized.

Nevertheless, two remaining operational problems remain unresolved. The first is the modifiable area unit problem (MAUP), which relates to the effects that different geographic aggregations may have on the observed values for a variable of interest (Openshaw and Taylor 1979). Because values for a specific variable, such as population or median income, will vary based on the manner in which an area is bounded, the specific values observed are not truly random, but are instead a product of the means by which a geographic area is defined. A second and related problem is the issue of spatial autocorrelation, which relates to the independence of our observations. Because the characteristics of a geographically defined area are likely to be similar to, and perhaps influenced by, the characteristics of adjacent areas, the specific observations cannot be said to be truly independent.

While several studies have sought to examine how MAUP and spatial autocorrelation may affect the results of geographically based analyses of the relationship between transportation and urban form (Buldoc 1992; Horner and Miller 2002; Miller and Shaw 2001; Zhang and Kukadia 2005), none of these studies have examined how these issues may relate to crash incidence, nor on how to adjust negative binomial regression models to address these issues. The resolution to the problems of MAUP and spatial autocorrelation is beyond the scope of this study. Nevertheless, as they may potentially influence our safety estimates, they should be openly acknowledged when considering the results.

Independent and control variables

To examine the relationship between urban form and crash incidence, the following variables were included in this analysis:

- **Block group acreage.** Under conventional practices applied by the US Census, block groups vary in size, with larger block groups located at the periphery of the metropolitan area, in areas that are likely to export much of their traffic to other, more central locations. To account for whatever statistical effects block group definitions might have on our results, we included block group acreage as a control variable.
- **VMT (millions).** Because heavier traffic volumes have been shown to increase crash incidence, this analysis controls for the effects of VMT within a community. We acquired traffic volume information for all freeway, arterial, and collector roadways in the region from the Texas Department of Transportation and the City of San Antonio, and aggregated this data at the block group level.[1] To ensure that the variable was scaled so that model coefficients could be meaningfully interpreted, we converted VMT to million vehicle miles traveled (MVMT), or VMT/1,000,000.
- **Median income (thousands).** Households with higher incomes are likely to be able to afford newer and more crashworthy vehicles, which would be expected to lead to reductions in injurious and fatal crashes, although not a reduction in total crash frequency. This variable controls for the effect of income on crash incidence, and is reported in thousands of dollars ($1,000).
- **Persons aged 18–24.** Young drivers are disproportionately more likely to be involved in total, injurious, or fatal crashes than other age groups. This variable controls for the number of young adults residing in the block group.
- **Persons aged 75 and older.** While crash rates decline for persons between the ages of 25 and 74, there is an increase in crash incidence among individuals aged 75 or older. This variable controls for the presence of adults aged 75 or older residing within a block group.
- **Net population density.** Several recent studies have identified higher population densities as being a crash risk factor. To understand the effects that

population density might have on crash incidence, we calculated the *net* population density of each block group. Net population density is measured as the total population of the block group, divided by the acreage of land dedicated to residential use.

- **Intersection counts.** Because 3-way intersections have been found to have different safety effects from other intersection types, we modeled 3-leg intersections and 4 or more-leg intersections as separate variables, rather than developing a single measure of intersection density that combines their effects. These variables are simply the count of the number of intersections of each type within the block group.
- **Miles of freeways and arterials.** The development of freeways and arterial thoroughfares was presumed to enhance safety by providing roadways appropriate for automobile use. Because this study already controls for VMT, this variable seeks to identify what safety effects, if any, are associated with the presence of freeway and arterial facilities within urban neighborhoods. These variables are the sum of the centerline miles of roadways classified as either freeways or arterials.
- **Arterial-oriented commercial uses.** Relocating commercial and retail uses to arterial thoroughfares was presumed to enhance safety by eliminating non-residential traffic from neighborhoods. Nonetheless, there has been little formal examination of how the presence of these land uses on arterial thoroughfares may affect crash incidence. To calculate this variable, we used GIS to identify each commercial or retail use that is located adjacent to an arterial thoroughfare, and summed the number of these uses for each block group.
- **Big box stores.** An outcome of encouraging arterial-oriented retail, perhaps unforeseen by Perry and Stein, was the advent of the "big box" store. Big box stores draw traffic from a large geographic area, and must typically accommodate a good deal of off-street traffic as vehicles circulate through the site in search of parking (which potentially leads to off-street crashes). For this study, a "big box" store is identified as a retail use comprising 50,000 square feet or more, and having a floor–area ratio (FAR) of 0.4 or less, and this variable is the sum of these uses within a community's boundaries.
- **Pedestrian-scaled retail uses.** A second outcome of the design configurations promoted by Perry and Stein was the decline of pedestrian-scaled retail uses. Pedestrian-scaled retail uses are defined in this study as a commercial or retail use of 20,000 square feet or less, but developed at FARs of 1 or greater (i.e. buildings that front the street or otherwise have little undeveloped surface space). The resulting variable is the count of such uses in a neighborhood. This measure serves as a rough indicator of a neighborhood's "urbanism," and examines what safety effects, if any, that the presence of more traditionally scaled retail and commercial uses may have on crash incidence.

Model specification and reporting

Because the dependent variables are count data that are overdispersed (i.e. the variance is greater than the mean), negative binomial regression models were used for this analysis. Negative binomial regression models have been widely applied in the recent traffic safety literature, and are regarded as the preferred statistical model for analyzing crash frequency and severity (Ladron et al. 2004). The model coefficients report the percentage change of the dependent variable that occurs with each unit of change in the independent variable.

Total crash incidence

As expected, areas with more VMT experience more crashes, with crash incidence increasing by roughly 0.75 per cent with every million miles of vehicle travel. Income did not prove to be related to total crash incidence, while the numbers of both young and older drivers were associated with higher numbers of total crashes. Of the two intersection variables, 3-leg intersections had a slightly positive, but statistically insignificant effect on crash incidence, while 4-leg intersections were associated with a small but significant (0.5 per cent) increase in total crashes. Freeways were not associated with total crash incidence, although arterial thoroughfares were, with each additional mile of arterial thoroughfare being significantly associated with a 15 per cent increase in total crashes. Of the land use variables, the number of arterial-oriented commercial uses and big box stores within a community were associated with significant increases in crash frequency, with each additional arterial-oriented commercial use increasing total crashes by 1.3 per cent, and each additional big box store increasing total crashes by 6.6 per cent. Neighborhood retail uses, by contrast, were associated with a 2.2 per cent *reduction* in crash incidence. Population density was also significantly associated with fewer crashes, with each additional person per net residential acre decreasing crash incidence by 0.05 per cent.

Injurious crash incidence

The factors associated with the incidence of injurious crashes are largely similar to those influencing injurious crashes, with two exceptions. The presence of older adults ceases to be a significant predictor of injurious crashes, while income is associated with a significant reduction in injurious crashes. Arterial thoroughfares again have a profoundly negative effect on traffic safety, with each additional centerline mile of arterial roadway increasing injurious crashes by roughly 17 per cent. 4-leg intersections, which are locations where conflicting traffic streams cross, are likewise associated with significant increases in injurious crashes. This was not true, however, for 3-leg intersections, which were associated with fewer injurious crashes, although not at statistically significant levels. Each additional arterial-oriented commercial use was found to increase injurious crashes by 1.1 per cent, and each additional big box store was found to increase injurious crashes by an additional 4 per cent. Conversely, pedestrian-scaled retail uses were again associated with significantly fewer injurious crashes, with each additional pedestrian-scaled retail use corresponding to a 3.4 per cent injurious crash reduction. Population density was again associated with a significant reduction in injurious crashes, with each additional person per net residential acre being associated with a 0.06 per cent decrease in injurious crashes.

Fatal crash incidence

The factors influencing the incidence of fatal crashes differ notably from those affecting total and injurious crashes. While median income was associated with decreases in fatal crash incidence, none of the remaining demographic or land use variables entered the model at significant levels. Instead, fatal crashes appear to be principally influenced by the effects that roadway and street network design have on vehicle speeds. Each additional mile of freeway within a community was associated with a 5 per cent increase in fatal crashes, although only at a 75 per cent level of statistical confidence. Arterial thoroughfares, by contrast, are associated with significant increases in fatal crashes, with each additional arterial mile associated with a 20 per cent increase in fatal crashes. Conversely, both 3- and 4-leg intersections were associated with significant reductions in fatal crashes, with each additional 3-leg intersection reducing fatal crash incidence by 0.7 per cent, and each additional 4-leg intersection reducing fatal crashes by 1 per cent. This reduction in fatal crashes is likely attributable to the fact that intersections force one or more streams to decelerate or come to a stop, which reduces vehicle speeds and thus crash severity.

Holistic discussion

Conventional community design practice attempts to enhance safety by channeling traffic onto arterial thoroughfares, eliminating intersections on local street networks, and relocating traffic-generating uses away from residential areas. While such strategies may appreciably reduce neighborhood traffic volumes, it is not clear that they also improve traffic safety. Two related factors, unseen and unaccounted for during the early half of the twentieth century, appear to be involved.

The first and perhaps most obvious factor is the moderating effect of *speed* on traffic safety. The original safety assumption regarding arterial roadways, espoused in *City Planning* (1916) and other early professional works, was that widening and straightening these thoroughfares would reduce crash incidence by enhancing sight distances, thus increasing a driver's preparedness to identify and respond to hazards in the right-of-way. Such an assumption was undoubtedly true in 1916, when vehicle speeds were constrained by the performance capabilities of automobiles, rather than a roadway's design characteristics. The Model T, for instance, could travel at maximum speeds of no more than 40–45 mph, and likely traveled at much lower speeds given the pavement conditions of the time. With these performance constraints, widening and straightening arterial thoroughfares would likely have increased sight distances without also increasing operating speeds.

In the intervening century however, advancements in automotive engineering now allow vehicles to readily travel at speeds in excess of 100 mph, making a roadway's geometric design characteristics the primary constraint on operating speeds. Under these conditions, wider and straighter roadways lead motorists to travel at higher speeds, which in turn consumes whatever safety benefits may have been associated with increasing sight distances (Wilde 1994; Aschenbrenner and Biehl 1994). These effects, predicted under risk homeostasis theory,[2] are readily evidenced by the safety performance of freeways, which have no statistically meaningful effect on crash incidence, either positive or negative. Instead, their principal effect is simply to enable vehicles to travel at higher speeds (see Table 11.3).

A second and related factor is one that can best be described as *systematic design error*. Systematic design error deals with human behavior, and occurs

Table 11.3 Crash incidence and urban form: holistic results

	Total	*Injurious*	*Fatal*
Block group acreage	−0.0006***	−0.0004**	0.0005
MVMT	0.0075***	0.0064***	0.0056***
Median income (000)	0.0003	−0.0044***	−0.0083**
Pop.18–24	0.0010***	0.0008***	0.0004
Pop. 75 and older	0.0006*	0.0003	−0.0002
Net population density	−0.0005*	−0.0006**	−0.0005
No. 3-leg intersections	0.0008	−0.0009	−0.0073*
No. 4 or more leg intersections	0.0050**	0.0068***	−0.0099*
Freeway miles	−0.0181	−0.0028	0.0488
Arterial miles	0.1495***	0.1714***	0.1998**
No. of commercial arterial uses	0.0131***	0.0111***	0.0053
No. of big box stores	0.0658***	0.0401***	−0.0367
No. of neighborhood retail uses	**−0.0218***	**−0.0335***	**−0.0120**

* Significant at the 0.1 level
** Significant at the 0.05 level
*** Significant at the 0.01 level

when the real-world use of a designed environment differs from its intended use in a predictable, non-random manner. The resulting errors that occur, and the crashes, injuries, and deaths that result, are thus *systematic outcomes* of the design itself, and an indicator of faulty design (Dumbaugh 2005b; 2006b).

Surface arterial thoroughfares

The safety problem with urban arterials can be best understood as a product of systematic error. Widening and straightening these roadways to increase sight distances has the unintended side effect of also enabling higher operating speeds, which in turn increases *stopping sight distance*, or the distance a vehicle travels from the point at which a driver initially observes a hazard, to the point at which he or she can ultimately bring their vehicle to a complete stop. Higher stopping sight distances pose no particular problem when vehicles are traveling at relatively uniform speeds, with little or no interruption to the through-moving traffic stream that would require immediate braking. When these operating conditions can be met, as they are on grade-separated freeways, higher operating speeds would be expected to have little or no effect on crash incidence.

But the problem is that the operating conditions for which arterials are intended typically *cannot* be met on urban surface streets, where pedestrians, bicyclists, and crossing vehicles are all embedded into the traffic mix. Avoiding crashes under these conditions often requires motorists to brake quickly, an action that higher operating speeds and stopping sight distances make them less able to successfully do (see Figure 11.3).[3] The result is a systematic pattern of error – an inability to quickly respond to other roadway users entering the travelway – that leads to increased crash incidence. This is confirmed by the results of our study, with each additional mile of arterial thoroughfare being associated with a 15 per cent increase in total crashes, a 17 per cent increase in injurious crashes, and a 20 per cent increase in fatal crashes.

Arterial-oriented commercial uses and big box stores

The land use changes encouraged by Perry and Stein exacerbate the systematic error problem. While they were correct in that relocating these uses would channel traffic away from neighborhoods, they failed to consider how the location of these uses on arterial roadways would affect arterial operating conditions. Commercial and retail uses, whether placed within neighborhoods or along arterials, require access to the streets on which they are located. In the case of arterials, this leads to the placement of driveways along the arterial thoroughfare, which in turn

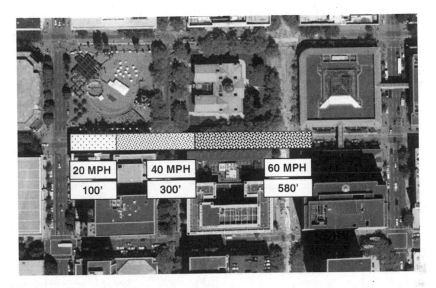

11.3 Average stopping sight distances for different vehicle speeds, superimposed over Portland's street grid for illustrative purposes

results in lower-speed, access-related traffic being introduced into the arterial traffic stream. Where through-moving traffic is traveling at speeds greater than that of vehicles turning into and out of driveways, the result is an increased incidence of rear-end collisions. At arterial locations that also lack a raised median, this can further introduce cross-traffic conflicts as vehicles attempt left-turning maneuvers, leading to an increased incidence of side-impact (angle) collisions (Dumbaugh 2005b). Each additional arterial-oriented commercial use in a community increases both total and injurious crashes by slightly more than 1 per cent, while each big box store increases total crashes by 6.6 per cent, and injurious crashes by 4 per cent.

Interestingly, these land uses were not associated with significant increases in fatal crashes. We suspect this is probably attributable to the heavier levels of traffic congestion occurring when retail and commercial uses are present on arterials, which reduces operating speeds and may in turn reduce crash severity, if not crash frequency. Nevertheless, future research is needed to bear this assertion out.

While the data used in this analysis did not permit pedestrian and bicyclist crashes to be explicitly examined, it is likely that arterials with concentrations of big box stores and commercial uses are particularly problematic for these groups. A study by Miles-Doan and Thompson (1999) examined pedestrian crash incidence in Orlando, FL, finding that the majority of pedestrian crashes occurred along arterial thoroughfares lined with strip commercial uses. The concentration of commercial and retail uses near each other, or near adjacent residential areas, can reduce trip distances to levels that make pedestrian travel viable, thereby adding pedestrians into the overall traffic mix. As described above, vehicle speeds occurring along arterials are often too high to permit vehicles to quickly stop in the event that a pedestrian unexpectedly enters the travelway.

To Perry's (1939) credit, he acknowledged that placing commercial uses on arterial thoroughfares likely created a pedestrian safety problem, but it was a problem he ultimately left unsolved.[4] In practice, the solution to this problem in the United States has been to continue to locate such uses on arterial thoroughfares, but to reduce posted speed limits. In the absence of aggressive police enforcement however, such practices have been uniformly unsuccessful at reducing vehicle operating speeds (Armour 1986; Beenstock, Gafni, and Goldin 2001; Zaal 1994). The principal alternative, adopted by European designers, is to simply design urban surface streets to reduce vehicle speeds to safe levels.

Pedestrian-scaled retail

Pedestrian-scaled retail – the type of retail that was largely abandoned during the postwar period – was found to be associated with reductions in all types of crashes, and at significant levels for both total and injurious crashes. This is consistent with recent research on the subject, which finds that the pedestrian-scaled nature of these environments communicate to motorists that greater caution is warranted, leading to increased driver vigilance, lower operating speeds, and thus a better preparedness to respond to potential crash hazards that may emerge. The effective result is a reduction in crash incidence (Dumbaugh 2005a; 2005b; 2006b; Garder 2004; Naderi 2003; Ossenbruggen, Pendharkar and Ivan 2001).

Intersections, street networks, and traffic control

Intersections are a more complicated matter. Both of the intersection types examined in this study improve safety by reducing the incidence of fatal crashes. Yet, at least in the case of 4-way intersections, these reductions are accompanied by significant increases in total and injurious crashes. 3-leg intersections, by contrast, reported fewer injurious crashes and more total crashes, although not at conventional levels of statistical significance. The seemingly conflicting nature of these results can be understood as the result of the tension between their effects on *traffic conflicts* and *vehicle speeds*. Intersections are locations where conflicting streams of traffic cross, creating locations where crashes are more likely to occur. T-intersections are safer than 4-way intersections in that they produce fewer intersection conflict points and interrupt longer roadway segments, thus preventing vehicles from achieving higher operating speeds. Nonetheless, both intersection types force through-moving vehicles to decelerate, if not stop completely (UK Department for Transport 2007), which in turn reduces impact speeds, and thus the incidence of fatal crashes. Considering intersections exclusive of other exogenous factors, it would appear that hybrid street networks, which contain dense concentrations of T-intersections, would appear to be preferable to either disconnected residential subdivisions or gridiron configurations.

Nevertheless, the safety performance associated with the presence of intersections in a community is a result of their mixed effects on speed and traffic conflicts. Design strategies that target these effects directly are likely to offset the disadvantages associated with any specific network or intersection type. The speed-reducing benefits associated with frequent intersections can likely be achieved in limited-access communities through the use of speed-reducing traffic calming devices such as speed humps, chokers, and chicanes. Likewise, modifications in the type of intersection control used in well-connected street networks may help reduce both traffic conflicts and crash incidence. Roundabouts and traffic circles, which reduce conflict points between opposing streams of traffic, have proven very effective at reducing crash incidence, and would appear to be a promising strategy for balancing traffic safety with network connectivity (Ewing 1999; Zein et al. 1997).

With the exception of the limited installation of speed humps, traffic calming devices are not used by the City of San Antonio, and our study results are limited to the safety effects of the presence or absence of intersections. Thus, while we conclude that hybrid street networks are likely preferable to other network configurations, *ceteris paribus*, we strongly suspect that other network configurations, used in conjunction with an appropriate suite of traffic calming and traffic control devices, may provide safety benefits that are equivalent or even better. On this subject, future research is needed.

Density (and VMT)

Finally, it is important to observe that this study's findings for population density differed notably from those of previous studies, which reported population density to be a significant crash risk factor. Our study, which addresses the built environment in a more comprehensive manner, found population density to be associated with significantly fewer total and injurious crashes. While the safety benefits of higher densities are minor at the level of the individual community, we suspect that their benefits may be agglomerative at larger geographic levels. Individuals living in higher-density environments drive less (Ewing and Cervero 2001), thus reducing their overall exposure to a crash event. When these reductions in VMT are aggregated across a larger population, they can potentially add up to notable reductions in population-level crash incidence. Indeed, this assertion is strongly suggested by a recent study by Ewing, Scheiber, and Zegeer (2003), which found that more sprawling counties, characterized by lower residential densities and larger blocks, experienced significantly higher traffic fatality rates than denser, more compactly developed counties. Differences in VMT are likely responsible for at least part of this difference. As with intersections and network configurations, however, future research is needed to explicitly understand the moderating influence that VMT may have on developmental density and crash incidence.

Implications for planning practice

Considered as a whole, these results have three implications for professional practice. The first is a need to *manage the mobility and access functions of urban arterials.* While urban arterials are typically designed and intended for higher-speed vehicle operations, the presence of commercial and retail uses forces them to accommodate lower-speed, access-related functions as well, resulting in the speed differentials and traffic conflicts that produce traffic crashes.

Two strategies are available for addressing the problem of mixed traffic on urban arterials. The first is *access management,* which attempts to enhance the arterial's mobility function by reducing or eliminating its access function. This is typically achieved by consolidating or eliminating arterial driveways, installing a raised median to restrict left-turning movements, and increasing the spacing between signalized intersections, preferably at distances of ½ mile or more (Florida Department of Transportation 2006; Transportation Research Board 2003). While access management has proven effective at both reducing crashes and increasing vehicle operating speeds (Dumbaugh 2006a), it does so by designing arterials to have the limited-access characteristics of freeways, and is a solution more appropriate to suburban environments, rather than urban ones. The higher operating speeds and greater spacing between signalized intersections found on these roads complicate safe pedestrian crossings and limits developmental access.

In areas where pedestrian activity is present or expected, or where eliminating a roadway's access function is either undesirable or contextually inappropriate, the primary alternative to access management is to reduce operating speeds to levels that are compatible with its access-related functions. This approach, sometimes referred to as the *livable street* approach, incorporates design features intended to encourage or enforce lower operating speeds, such as aligning buildings to the street, incorporating landscaping or street appurtenances along the roadside, enhancing the aesthetics of pavement and signage, and incorporating other elements intended to explicitly reduce operating speeds, such as traffic calming and full intersection control. In short, livable streets emphasize access over higher-speed mobility. When compared against conventional arterial treatments, livable streets report roughly 35–40 per cent fewer crashes per mile traveled, as well as a complete elimination of traffic-related fatalities (Dumbaugh 2005a; Naderi 2003).

Safety may be further enhanced by *orienting retail and commercial uses towards lower-speed thoroughfares.* Two strategies can be readily applied to accomplish this. The first, often promoted by advocates of neo-traditional development, is to simply design arterial thoroughfares in urban areas for *lower operating speeds.* In response to concerns about excessive speeds on urban arterials, the Institute of Transportation Engineers (ITE), in partnership with the Congress for the New Urbanism (CNU), are developing design specifications for three new thoroughfare types – boulevards, avenues, and commercial streets – that are intended to carry arterial traffic volumes at design speeds of 35 mph or less (ITE 2006).

The second approach, often employed by access management proponents, is to require developments located along arterials to incorporate *internal access lanes* that manage their connections to the arterial system. These access lanes are lower-speed, collector-type facilities that consolidate the traffic generated by a development and channel it towards a managed arterial location. Increasingly, developers have begun designing access lanes to look and function like traditional urban commercial streets. Regardless of their specific configuration, the effective result of the provision of access lanes is to relocate access-related traffic from the arterial thoroughfare, and onto lower-speed streets that are better able to safely accommodate on-site circulation and access.

Finally, there is a need for a *network-level approach to land use planning, speed management, and access control.* Traffic safety is a more complicated matter than simply configuring streets and land uses to prevent residential cut-through traffic, and there is no one-size-fits-all solution to the resulting safety problems that occur on urban arterials. Instead, safety is likely better advanced through community-level design solutions that pay attention to how different land use and street network configurations may influence vehicle speed and systematic design error. Residential, commercial and retail uses all require developmental access, and are likely to attract pedestrian traffic as well, particularly when clustered together. Addressing safety in these environments will likely require streets and street networks designed to encourage lower operating speeds and meaningfully accommodate local access and circulation needs. By contrast, safety on higher-speed, mobility-oriented thoroughfares appears to be best enhanced when access to these facilities is strictly managed to reduce the hazards posed by traffic conflicts and speed differentials.

While our results suggest that hybrid street networks are preferable to gridiron or functionally designed networks, the thoughtful use of traffic calming and intersection control at the network level is likely to remedy the specific deficiencies associated with different developmental configurations. Considered broadly, it is the tension between speed and access that leads to crash incidence, and there are likely multiple design configurations that may meaningfully balance them.

Conclusion

Designers during the early half of the twentieth century were rightly concerned about the safety effects of 4-way intersections and gridiron street networks, and this study should not be used to infer that a wholesale return to nineteenth century gridiron configurations is desirable. Instead, our results suggest that many of the community design practices that emerged in response to these concerns, such as the development of functionally designed street networks and the relocation of commercial and retail uses to arterial thoroughfares, are practices that have largely substituted one set of safety problems with another.

This study has sought to identify general trends in the relationship between community design and crash incidence, and has examined the existing safety literature to outline likely causes and potential solutions. Yet we should observe that cross-sectional analyses, such as this one, should not be used to derive explicit inferences on causality. Future research is needed to develop our professional understanding of the specific ways in which community design affects the behaviors that result in traffic crashes, as well as the frequency and severity of the crashes that occur. With these caveats in mind, we conclude by observing that the relationship between community design and traffic safety appears to be an important one, and that the activities of planners may help advance – or hinder – urban traffic safety objectives. It is our hope that this study will serve as a starting point for better understanding the relationship between community design and traffic safety.

Acknowledgements

This research was supported by a grant from the Southwest University Transportation Center. We would also like to thank Cecilio Martinez with the San Antonio-Bexar County Metropolitan Planning Organization and Kurt Menking with the Bexar County Appraisal District, who were instrumental in assembling the data used in this study, as well as Lijing Zhou, Rajanesh Kakumani, Nitin Warrier, and Doug Wunneberger, who provided invaluable assistance in compiling and analyzing this data. We would also like to thank Rachel Askew and the *Journal of the American Planning Association*'s editors and reviewers for their invaluable comments and recommendations on an earlier version of this chapter.

Notes

1 Developing neighborhood-level VMT estimates proved methodologically complex, and warrants a brief description. The Texas Department of Transportation provided average daily traffic volumes (ADT) in linear form for all state highways (freeways and principal arterials) in the metropolitan area. The City of San Antonio supplemented this information with traffic counts at 804 locations off the state system. These streets consist of all principal arterials, minor arterials, and collector roadways in the region. Traffic volumes for select local roadways were also provided, but because of the limited number of roadways for which local ADT was available (roughly 10 per cent of all local road mileage in the City), we omitted this information from our calculation of VMT. As such, our data present ADT for freeways, arterials, and collectors.

A second issue related to the integration of the state and City ADT information. Unlike the state data, the City's ADT data were provided in point form. To make these data compatible, we converted the San Antonio point data into a linear form, and extrapolated the corresponding point value for the length of the roadway. Where multiple ADT values were present on a given roadway, the extrapolated values were extended outwards towards the midpoint between them.

Next, because the individual road segment lengths rarely corresponded to block group boundaries, it was necessary to clip the line segments to correspond with block group definitions. To identify the VMT associated with streets located on the boundaries of neighborhoods, we again used a 200 ft buffer surrounding each block group to clip the individual line segments, thus ensuring all related roadways were included in the analysis. Once the road segments were clipped, the VMT for each road segment was calculated by multiplying the segment ADT by the length of the road segment, and then multiplying this value by 365 (days) and 3 (years). As such, block group VMT is computed as:

$$\sum SegADT * SegLength\ 365 * 3$$

2 Risk homeostasis theory, as detailed by Wilde (1994), asserts that individuals make decisions on whether to engage in specific behaviors or activities by weighing the relative utility of an action against its perceived risk. While all actions involve some risk, risk homeostasis theory asserts that individuals will adjust their behavior to maintain a static level of exposure to perceived hazard or harm. With respect to driving behavior, risk homeostasis theory posits that drivers intuitively balance the relative benefits of traveling at higher speeds or engaging against their individual perceptions of how hazardous engaging in such behavior might be.

3 Stopping sight distance is calculated as the sum of the brake reaction distance, which is the distance traveled between when a driver observes a hazard and applies the brakes, as well as the braking distance itself (AASHTO 2004). The stopping sight distances presented in Table 11.4 represent "average" estimates that assume a reaction time of 2 seconds, and a vehicle deceleration rate of 11.2 ft/s^2. They further presume that drivers are reasonably alert to oncoming hazards and are driving vehicles that provide adequate braking performance and maintain adequate pavement friction during deceleration. Actual stopping sight distances may increase substantially if drivers are not alert to oncoming hazards or are driving a substandard vehicle. Table 11.4 lists stopping sight distances for different combinations of reaction times and operating speeds, assuming a typical vehicle deceleration rate of 11.2 ft/s^2.

Table 11.4 Estimated stopping sight distances, by speed and reaction time

	Reaction time		
	1.5s	2.0s	2.5s
20 mph	478 ft	522 ft	566 ft
20 mph	82 ft	97 ft	112 ft
25 mph	115 ft	134 ft	152 ft
30 mph	153 ft	175 ft	197 ft
35 mph	195 ft	221 ft	246 ft
40 mph	241 ft	271 ft	301 ft
45 mph	294 ft	327 ft	360 ft
50 mph	350 ft	387 ft	424 ft
55 mph	412 ft	452 ft	493 ft
60 mph	478 ft	522 ft	566 ft

4   Perry struggled with the issue of pedestrian safety on arterials, suggesting that "it might be possible to provide an underpass, accessible by stairs or a ramp, on each side of the street" or, in a notable flight of fancy, exceptionally wide bridges could be located over the arterial, resulting in "new store sites thus created 'out of the air' – literally in the air – . . . bring[ing] a revenue that would take care of both construction and maintenance" (1939: 71).

References

American Association of State Highway and Transportation Officials (2004) *A Policy on the Geometric Design of Highways and Streets* (5th ed.), Washington, DC: AASHTO.

Armour, M. (1986) "The effect of police presence on urban driving speeds," *ITE Journal*, Feb: 40–45.

Aschenbrenner, M. and Biehl, B. (1994) "Improved safety through improved technical measures? Empirical studies regarding risk compensation processes in relation to anti-lock braking systems," in Trimpop, R. M. and Wilde, G. J. S. (eds) *Challenges to Accident Prevention: The Issue of Risk Compensation Behaviour*, Groninger: Styx Publications.

Beenstock, M., Gafni, D., and Goldin, E. (2001) "The effect of traffic policing on road safety in Israel," *Accident Analysis and Prevention*, 33(1): 73–80.

Buldoc, D., Laferrier, R., and Santarossa, G. (1992) "Spatial autoregressive error components in travel flow models," *Regional Science and Urban Economics*, 22: 371–385.

Dumbaugh, E. (2005a) "Safe streets, livable streets," *Journal of the American Planning Association*, 71(3): 283–298.

Dumbaugh, E. (2005b) "Safe streets, livable streets: a positive approach to urban roadside design." Doctoral Thesis, Georgia Institute of Technology, Department of Civil and Environmental Engineering.

Dumbaugh, E. (2006a) "Access Management Planning Assistance Program: Enhancing the safety of arterial roadways in the Atlanta Metropolitan Region." Report submitted to the Atlanta Regional Commission.

Dumbaugh, E. (2006b) "The design of safe urban roadsides," *Transportation Research Record: Journal of the Transportation Research Board, 1961*: 74–82.

Ewing, R. (1999) *Traffic Calming: State of the Practice*, Washington, DC: Institute of Transportation Engineers.

Ewing, R. and Cervero, R. (2001) "Travel and the built environment: a synthesis," *Transportation Research Record: Journal of the Transportation Research Board, 1780*: 87–114.

Ewing, R., Schieber, R. and Zegeer, C. (2003) "Urban sprawl as a risk factor in motor vehicle occupant and pedestrian fatalities," *American Journal of Public Health*, 93(9): 1541–1545.

Federal Highway Administration (2003) "Considering safety in the transportation planning process." Available at http://tmifhwa.dot.gov/clearinghouse/docs/safety/descriptions.htm (accessed 27 July 2003).

Florida Department of Transportation (2006) *Median Handbook*, Tallahassee: Florida Department of Transportation, February.

Garder, P. E. (2004) "The impact of speed and other variables on pedestrian safety in Maine," *Accident Analysis and Prevention*, 36(4): 533–42.

Garvin, A. (1995) *The American City: What Works, What Doesn't*, New York: McGraw Hill.

Hadayeghi, A., Shalaby, A. S. and Persaud, B. N. (2003) "Macro level accident prediction models for evaluating safety of urban transportation systems," *Transportation Research Record: Journal of the Transportation Research Board 1840*: 87–95

Hadayeghi, A., Shalaby, A. S. Persaud, B. N. and Cheung, C. (2006) "Temporal transferability and updating of zonal level accident prediction models," *Accident Analysis and Prevention*, 38(3): 579–589.

Haldeman, B. A. (1916) "Transportation and main thoroughfares and street railways," in Nolen, J. (ed.), *City Planning*, New York: Appleton and Co.

Horner, M. W. and Miller, H. T. (2002) "Excess commuting and the modifiable areal unit problem," *Urban Studies*, 39(1): 131–139.

Institute of Traffic Engineers (2006) *Context Sensitive Solutions in Designing Major Thoroughfares for Walkable Communities*, Washington, DC: Institute of Transportation Engineers.

Kostof, S. (1991) *The City Shaped: Urban Patterns and Meanings Through History*, New York: Bullfinch Press.

Ladron de Guevara, F., Washington, S. and Oh, J. (2004) "Forecasting travel crashes at the planning levels: simultaneous negative binomial crash model applied in Tucson, Arizona," *Transportation Research Record: Journal of the Transportation Research Board, 1897*: 191–199.

Lee, C. M and Ahn, K. H. (2003) "Is Kentlands better than Radburn? The American garden city and new urbanist paradigms," *Journal of the American Planning Association*, 69(1): 50–71.

Lovegrove, G. and Sayed, T. (2006) "Macro-level collision prediction models for evaluating neighbourhood traffic safety," *Canadian Journal of Civil Engineering*, 33(5): 609–621.

Lynch, K. (1960) *The Image of the City*, Cambridge: MIT Press.

Marks, H. (1957) "Subdividing for traffic safety," *Traffic Quarterly*, 11 (3): 308–325.

Miles-Doan, R. and Thompson, G. (1999) "The planning profession and pedestrian safety: lessons from Orlando," *Journal of Planning Education and Research*, 18(3): 211–220.

Miller, H. J. and Shaw, S. (2001) *Geographic Information Systems for Transportation: Principles and Applications*, New York: Oxford University Press.

Mokdad, A. H., Marks, J. S., Stroup, D. F. and Gerberding, J. L. (2004) "Actual causes of death in the United States, 2000," *Journal of the American Medical Association*, 291 (10): 1238–1245.

Naderi, J. R. (2003) "Landscape design in the clear zone: the effects of landscape variables on pedestrian health and driver safety," *Transportation Research Board 82nd Annual Conference Proceedings* (CD-ROM), Washington, DC: Transportation Research Board.

National Highway Traffic Safety Administration (NHTSA) (2006) *Traffic Safety Facts 2006* (DOT HS 810 818), Washington, DC: U.S. Department of Transportation.

Olmsted, F. L. (1872) "Central Park," in Olmsted, Jr., F.L. and Kimball, T. (eds) (1970) *Frederick Law Olmsted, Landscape Architect 1822–1903*, New York: B. Blom.

Olmsted, F. L., Jr. (1916) "Introduction," in Nolen, J. (ed.), *City Planning*, pp. 1–18, New York: Appleton & Co.

Openshaw, S. and Taylor, P. (1979) "A million or so correlation coefficients: three experiments on the modifiable area unit problem," in Wrigley, N. (ed.) *Statistical Applications in the Spatial Sciences*, pp. 127–144, London: Pion.

Ossenbruggen, J., Pendharkar, J. and Ivan, J. (2001) "Roadway safety in rural and small urbanized areas," *Accident Analysis and Prevention*, 33(4): 485–498.

Parsons, K. C. (1994) "Collaborative genius: The Regional Planning Association of America," *Journal of the American Planning Association*, 60(4): 462–483.

Perry, C. A. (1939) *Housing for the Machine Age*, New York: Russell Sage Foundation.

Robinson, C. M. (1911) *The Width and Arrangement of Streets*, New York: Engineering News Publishing Co.

Schaffer, D. (1982) *Garden Cities for America: The Radburn Experience*, Philadelphia: Temple University Press.

Southworth, M. and Ben-Joseph, E. (1995) "Street standards and the shaping of suburbia," *Journal of the American Planning Association*, 61(1): 65–81.

Stein, C. (1957) *Towards New Towns for America*, Cambridge: MIT Press.

Transportation Research Board (2001) "Safety conscious planning," Transportation Research Circular Number E-CO25. Available at http://gulliver.trb.org/publications/circulars/ec025.pdf (accessed 27 July 2003).

Transportation Research Board (2003) *Access Management Manual*, Washington, DC: Transportation Research Board.

UK Department for Transport (2007) *Manual for Streets*, London: Thomas Telford.

Wilde, G. J. S. (1994) *Target Risk*, Toronto: PDE Publications.

Zaal, D. (1994) *Traffic Law Enforcement: A Review of the Literature*, Leidschendam, Netherlands: Institute of Road Safety Research.

Zein, S. R., Geddes, E., Hemsing, S. and Johnson, M. (1997) "Safety benefits of traffic calming," *Transportation Research Record: Journal of the Transportation Research Board 1578*: 3–10.

Zhang, M, and Kukadia, N. (2005) "Metrics of urban form and the modifiable areal unit problem," *Transportation Research Record: Journal of the Transportation Research Board 1902*: 71–79.

Chapter 12

Evaluating skywalk systems as a response to high-density living in Hong Kong

Patricia Woo and Lai Choo Malone-Lee

Planners have sought to mitigate the adverse impacts of high-density living through various interventions. This research looks into one of the planning responses for this purpose, namely, the implementation of skywalk systems in Hong Kong. Through a combination of qualitative and quantitative methods, the identified impacts of skywalk systems are validated by the community, stakeholders and experts. The effectiveness of skywalk systems as a strategic response to high density living is evaluated, and recommendations are put forward to planners and decision-makers.

This research shows that skywalk systems have been used successfully in residential areas of Hong Kong. Being integrated with a trunk-and-feeder transportation network, skywalk systems have mitigated the traffic–pedestrian conflict of high-density living in the case of Hong Kong, while contributing to persuading the public to use public transportation. The multiple planning benefits include a seamless integration with transport nodes, support for land use intensification, integration into the lifestyles of community users, and offering a diversity of activity options. They are well accepted by the people, and have improved connectivity, pedestrian convenience and safety. An immediately observable impact is that the ground level of purposely designed skywalk systems is often very user-unfriendly for pedestrians. Furthermore, although the existing surroundings of skywalk systems in Hong Kong have not shown loss of street life, the risk of killing street life by diversion is significant and real. It also shows that skywalk systems do not function as good substitutes for public open space at ground level.

Submitted by the Asian Planning Schools Association (APSA).

This chapter was originally presented at the 11th International Congress of the Asian Planning Schools Association, Tokyo, 19–21 September 2011. Used here with kind permission of the authors.

Introduction

The merits of high-density living and urban compaction have been recognized for their reduced occupancy of land, less costly infrastructure and lower emissions as a result of reduced automobile usage. For example, McLaren (2000) cited several studies in support of the environmental benefits, including a 10 to 15 per cent reduction in carbon dioxide emissions or fuel use. As far back as 1974, a study entitled *The Cost of Sprawl* by the US Council on Environmental Quality, Department of Housing and Urban Development, and the Environmental Protection Agency (cited in Register 2002) found that high-density communities required 50 per cent and 45 per cent less land and investment cost in infrastructure respectively, and the air pollution and water usage were reduced by 45 per cent. It was noted that, depending on the actual designs, the energy savings compared to a low density settlement could range from 14 to 44 per cent. Vancouver, a city widely recognized for its quality of life, saw its City Council adopting the Vancouver Ecodensity Charter in 2008 (2010), in pursuit of strategic use of greater density in land use patterns to make the city more resilient to climate change.

However, higher densities pose two major problems for transport planning: one relates to planning for facilities necessary for easy movement of large numbers of commuters within the city, and the other to the no less daunting task of addressing pedestrian–vehicular conflicts arising from the competition for space in the constrained environment of cities.

Planners and architects in various cities have been concerned with these two problems for a long time. From the early years, various forms of passenger–vehicle separation strategies have been discussed, such as those from Le Corbusier (1925) and Corbett (1913); and "skywalk systems" of one form or another have been an integral part of these strategies. The main objective of such above-ground systems is the creation of separate functional planes for pedestrians and traffic so that both could move with minimal conflict or no interference at all from each other. The concept is best illustrated by one of the early drawings by Corbett of a "multi-transit city" (Figure 12.1), where several skywalk systems crisscrossed at different levels, working in concert with each other.

Skywalk systems

A skywalk system is defined by Robertson (1993) as "a network of elevated interconnecting pedestrian walkways . . . (which) consists of skybridges over the streets, second-storey corridors within buildings and various activity hubs." Thus defined,

12.1 An impression of a multi-level transit city by Corbett (1913)

Source: http://arquinoias.tumblr.com

a skywalk system is an elevated plane of activities with three integral elements, as opposed to the common understanding that it is merely a collection of bridges.

In the 1960s and 1970s, a wave of skywalk construction took place in various parts of North America, particularly in cities with extreme cold weather such as Des Moines, Edmonton, Milwaukee, Minneapolis and St. Paul. They were conceived primarily to provide pedestrians with a weather-proof walking environment. Later, other US cities with a milder climate, such as Charlotte, Dallas and Cincinnati, also took an interest in skywalk systems, viewing them as a useful redevelopment tool to reinvigorate downtowns. However, 40 years on, several US cities have decided to reverse the process, and cities like Cincinnati had 22 of its skywalk bridges around Fountain Square torn down. Other cities, like Des Moines, kept their skywalk systems but have put in extra effort to encourage more street-level activities (Cornwell 2006).

A comparative study by Robertson of the skywalk systems in five mid-western cities in the US (1994) revealed that while the usage rate was relatively high, there were indications that skywalks had brought about negative impacts to street activities, and reduced property values at ground level. With pedestrians removed from the streets on ground, sidewalks became empty, resulting in uncomfortably deserted downtown areas. Robertson noted that skywalks also contributed to the separation of people in the downtown on the basis of economic class. The decline of street activities in the downtowns of these cities had invited adverse media debate such as whether lifting pedestrians off the ground was "suicidal in the long run" (Anderson 1998).

In the Asian city of Hong Kong, however, skywalk systems seem to be thriving. Hong Kong, long noted as an example of extreme high density, has developed many forms of public interventions to mitigate the adverse impacts of high-density living. The city has an average population density of 6,400 persons per sq km, but 50 per cent of its administrative districts have actual densities between 15,000 to 52,000 persons per sq km (Census, HKSAR 2006). Although its GDP per capita exceeds USD 31,000 (Census, HKSAR), which is relatively high in the context of Asian cities, about 90 per cent of all journeys are undertaken by public transport (Secretary for Transport and Housing 2010b). A common question that is asked is, "How does the city persuade its 7.1 million people to choose public transport as their major means of moving around the city?" One contributory reason could be the complementary role played by the pedestrian walkway systems in Hong Kong's transport strategy, a factor that was readily acknowledged by the city's Secretary for Transport and Housing (2010a).

Rotmeyer (2006), in a 2006 study, compared the sustainability of skywalk systems in the US with one in the Central Business District (Central) of Hong Kong, using five indicators, namely temperature control, private versus public, continuous flow, ground floor decay, and high density. She found that while both the Hong Kong and US systems were able to offer pedestrians some relief to inclement weather, the skywalk systems in Hong Kong fared better in the areas of continuous flow, by having high permeability and connectivity, in preventing ground floor decay and having a density high enough to sustain them. Both Hong Kong and the US, however, still faced problems pertaining to barriers against full utilization of skywalk systems by low-income people. Summarizing her observations, Rotmeyer suggested that "a multi-layered city can be sustainable if each layer maintains itself within a balanced functional network of movement and activity."

This is consistent with Robertson's earlier comparative study (1994) which points to similar key elements for success of skywalks: diversity of space, mixed-use and integration with transit, all of which seem to contribute to better walking experience for users. Nonetheless, Robertson also pointed out that the better-designed skywalks would still have similar social and economic problems at ground as their more ordinary counterparts. These problems were illustrated in a preceding survey of some 50 shops by the Liberal Party of Hong Kong in 2002, which revealed that 90 per cent of the shops located on streets near to a recently built skywalk system suffered from a reduction of business, ranging from 10 per cent to 50 per cent, due to a drop of pedestrian flow (Legislative Council 2002).

Up to today, however, there have been very few studies explicitly on the role that pedestrian walkway systems, or skywalk systems, play in the transport management of Hong Kong. There is much scope for studies with regard to their impacts on the built environment and on various aspects of the life of the local community, including people's choice of transportation mode, their willingness to walk, the local economic activities affected, and people's perception of the social and environmental impacts of these systems.

Research and methodology

The research attempts to examine the case of skywalk systems in Hong Kong with a view to providing empirical data and analytical insights on these issues, which may prove useful to planners and transport engineers/architects who contemplate similar interventions to mitigate traffic–pedestrian conflicts in cities.

The study is designed to better understand community perceptions of skywalk systems in Hong Kong on one hand, and to distill the key issues for the successful implementation of such systems in a high-density environment on the other. The nature of the inquiry calls for a combination of qualitative and quantitative methods for a comprehensive analysis of empirical data to answer the directed questions. Out of ten large-scale skywalk systems identified for study, two were specifically chosen as case studies. The first one (referred to as "System A") is a purpose-built skywalk system while the second one (referred to as "System B") is a system constructed post-development in a small town center (figure 12.2).

A three-prong approach was adopted:

- First, the research team conducted a non-participant observation of the intensity, nature and pace of user behavior throughout a study day, with the aim of generating non-communicated information from a spectator's external perspective. The linear observation took place at 13 fixed time points from 8:00am to 10:00pm.
- Second, an on-site questionnaire survey of 120 community users was carried out to find out the perceived social and environmental impacts of skywalk systems. This was supported by in-depth structured interviews with representatives of the property management company concerned and a

12.2 Example of a skywalk system in Hong Kong

first-person account by a long-time resident of the study locality. The latter documented a personal narrative to reveal first-person insights of the day-to-day encounter with such systems.

- Finally, an expert-based study was conducted to elucidate the perceived impacts, with a view to evaluating the effectiveness of skywalk systems as a strategic response to high-density living. A small group of seven experts, including government town planners, an architect, an environmental assessment expert and two academics were invited to this study.

Research findings

Planning, design and implementation

In terms of design and operation, the implementation model of skywalk systems in Hong Kong is found to be significantly different from the US practice in the following ways:

- Most of the large skywalk systems in Hong Kong are spearheaded by the government, and implemented through very detailed specifications in land lease conditions. In new sites, the type and parameters of linkage were determined at the early stage of planning. All the required connections were stipulated in the land lease when the lot was sold.

- Major skywalk systems in Hong Kong are always linked to a mass transit station. The transport system in Hong Kong operates on a trunk-feeder mode (Agriculture, HKSAR 2010) and the city is served by the Mass Transit Railway (MTR) "trunks" while buses, minibuses and taxis "feed" passengers from farther areas to the MTR stations. The bus and minibus interchanges are usually located directly under the rail station, or made accessible via a skywalk system within a fairly short walking distance. Thus the skywalk system has an integral role in the overall operation of the transport network.

- The planned population density in and around such systems is usually amongst the highest in the city. For example, in the two case studies, there are an estimated 27,000 residents and workers living and working within System A, and 12,000 residents and workers living and working within System B. This large population base in the transit shed means that the local community users could access the skywalk systems on foot. Apart from providing a captive user base, this arrangement means that skywalks become deeply entrenched into the lives of the local community.

- Skywalk systems are usually planned to be closely integrated with a variety of activity hubs and residential projects. Hence, apart from the transport-related function, there is a wide variety of attractions as a user moves through the system. For example, in the skywalk system of System A, the linked-up elements include residential complexes, two commercial buildings, two major shopping malls, two public gardens, an MTR station and a bus interchange. In System B, the functional network comprises a town hall for cultural performances, four shopping malls, a hotel-cum-office building, a wet market, landscaped space for relaxation, an MTR station and a bus/minibus interchange.

Observed usage patterns

The time-line observation study affirms the multi-use characteristic of skywalk systems. Observations within the skywalk systems relating to three key areas, namely, use intensity, function/activities, people and atmosphere, were analyzed. General observations were also made of activities on the ground levels around the skywalk systems.

Use intensity

The large local population base resulted in very high intensity of usage of the skywalk systems studied. During the observation period, the rate of user flow at the selected observation points at System A was estimated to be 49.6 persons per minute during peak hours. Even in the quietest period, the rate was recorded at 18.2 persons per minute. The utilization rate of System B ranged from 10.2 to 40 persons per minute.

Function and activities

Although the main function of the skywalk system is that of a pedestrian transit, people use the system for a large variety of activities, including shopping (permitted and unpermitted selling activities take place on the elevated space), having meals, running errands and even conducting educational classes. The designed podium gardens are seen to be frequented by people of all ages, although not always in great numbers. More interestingly, some users chose not to use the dedicated spaces for their intended functions, for example, recreational activities, but would instead make use of "forgotten" spaces in the skywalk system. For example, an elderly person was observed to be performing his morning exercises at the waiting area for elevators; a group of women took up part of a wide passageway for their t'ai chi sword dance practice and parents spontaneously converted spaces decorated for festive celebration into a toddlers' playground.

People and atmosphere

People of all ages use the skywalk systems, although working-age adults form the largest proportion of the users (over 80 per cent). It was observable that elderly persons using the skywalk systems appear to be better dressed than the average elderly in the district.

The general ambience of the place differs at different times of the day. For example, in the mornings, lunch hours and early evenings, most of the users appear to be hurrying to work and home respectively. In the late evenings, there is a distinctly more relaxed atmosphere as people seem to stroll along more leisurely, perhaps for after-dinner gatherings or night shopping.

Ambience of ground areas around skywalk systems

There were distinct signs of activity decline in the general areas around skywalk systems, and the following specific observations were recorded:

- System A was purpose-built to connect communities separated by a major highway. As such, design of the ground space is heavily oriented towards vehicles, and observed to be generally lacking in human activity. The overall ambience was unfriendly for pedestrians.

- As for System B, which was only recently added to an existing town center, negative impacts on life on the back streets at grade were not evident. However, the side of it facing the main road appeared to have become less appealing to pedestrians.

Perception of community users

The face-to-face survey with 120 users revealed an overall favorable disposition towards skywalk systems in Hong Kong. Notwithstanding this, there appears to be a slight ambivalence as the value of vibrant street life was apparently still very much treasured by the local community. This issue would need further investigation.

From the results of the survey, the more specific perceptions of the local community are discussed as follows:

Use of skywalk systems

The responses indicate that between 50 and 60 per cent of the community use the activity hubs of the skywalk system in more or less the same frequency as they would use venues outside the system, i.e. they do not have a special preference for one or the other. The remaining 40–50 per cent were more inclined to shop or socialize outside the skywalk system. This result is clearly testimony to the high level of accessibility provided by public transportation in Hong Kong which facilitates people's decision to shop or socialize at various places by choice. An exception is the non-destination-oriented activities, that is, "hanging around, browsing or just walking for leisure". Such random or passive recreation tends to be conducted within the skywalk system, and this could be further explored and enhanced as potential outlets for "free" or unplanned urban spaces for the surrounding community.

Given the relatively high percentage of educational facilities in System A, it was expected that they would create opportunity for social interactions, for example, interaction between parents while waiting for their children. However, the findings show a rather small chance of such interactions, as most respondents indicate the level of encounters in the skywalk systems as no more than one-off acquaintances. (However, there may be a survey bias here, and the actual numbers higher, as most of the skywalk system users were accompanied by students or young children, and declined to take part in the survey).

Perception of benefits and challenges

The survey shows an overwhelmingly encouraging assessment from users that skywalk systems do bring about favorable lifestyle impacts. In descending order, respondents were positive about the following advantages brought forth by the skywalks, namely, "pedestrian convenience," "pedestrian comfort" and "pedestrian safety." Two other lifestyle impacts are significant. First, respondents reported a propensity to increase consumption while using the skywalk systems ("I have bought more"), which may be considered positive from the viewpoint of real estate performance and welcomed by property owners and shop tenants. Second, another beneficial incidental outcome seems to be an increase in the amount of walking in daily life, which is considered good from a "healthy city" perspective.

Environment-wise, most users seem to have a strong perception that greenery and public seating areas in the skywalk system were inadequate. This could be an area requiring management response if more spontaneous interaction were to be promoted.

Level of acceptability

There is a high level of acceptability of skywalk systems on functional grounds (pedestrian convenience, pedestrian comfort and pedestrian safety). Even when the temperature factor (air conditioning) was discounted, the number of people choosing skywalks was very close to those who choose the streets. Air pollution and poor walkability of the streets around the skywalks could be the contributory reasons.

Community users, however, still value a vibrant street life. Almost 40 per cent of the respondents thought that shop-lined streets were "more comfortable" than skywalk systems. This reflects the still dominant preference of the community for the more diverse and stimulating environment offered by street-level activities, despite their equal acknowledgment of the convenience brought by elevated linkages.

A further informal investigation into this issue of "life on the streets" was carried out through a personal in-depth interview with a long-time resident of the relevant district. In the interview, both benefits and pitfalls were highlighted, demonstrating perhaps a somewhat ambiguous relationship with skywalk systems. For this resident, the biggest unhappiness was that navigating at grade within the district has become very difficult, as the ground level has been mainly given over to accommodate motorized traffic as pedestrians were expected to use the skywalk system. This has resulted in lengthy detours for some areas which were otherwise easily accessible in the past. The issue was such an annoyance for the local community that politicians had to fight hard with the government for more road-crossing facilities. The respondent also observed that the skywalk system has brought about significant increases of people flow into the district, especially tourists and younger people from other districts, and new residential estates built atop the skywalk systems have driven up the average property prices. Other positive responses relate to the addition of recreational spaces by the developers who were well received. In this regard, the resident concluded that the skywalk system has contributed to the modernization of the old and derelict district within which it is located.

Management issues

From the management perspective, the implementation of skywalk systems has created some specific property management issues including coordination and security.

Coordination

The following are some management coordination issues relating to skywalk systems:

- Several management entities are often involved as land parcels developed by different developers are linked up in a skywalk system.
- There are varying degrees of publicness in a skywalk system – some sections can be public facilities built by the government; other sections require public passageways or voluntarily provided privately managed public space created in exchange for concessions; and some are privately owned with fairly accessible semi-public spaces such as shopping malls.
- High connectivity between all the components means that they are not unaffected by each other's functions and activities, for example, a sudden surge of pedestrian flow may be encountered at various parts of the system due to large-scale functions at one of the component venues. Another example is that when one segment is lacking in certain amenities (such as wastebins), users tend to meet their needs at another segment on their walking route, thus overloading the wastebins of the other facility.

To overcome the problems, the management company of System A, for example, has opted for a coordinated approach by setting up a "Development Owners Committee (DOC)." Through meetings of the DOC, members get to understand the needs and concerns of other landlords within the skywalk system and also share information about feedback received from residents and community users.

Accessibility and security

Given the high rate of people flow, the ease of access to the activity hubs and relatively open podiums of skywalk systems which are also connected to residential units above, privacy and security concerns are inevitable. Design solutions are often used to solve the problems, for example, by providing different entrances for residents and other pedestrians. Usually security guards are on duty 24 hours a day at the residential end of the entrances to ensure trespassers are kept away.

Conversely, accessibility from ground level to skywalk systems is not always easy, and this can pose problems with older and disabled persons. This is observed in System B, where a part of the system was designed to be fairly spacious but found to have low utilization rate in practice. Interviews with elderly persons who are the most probable users of this "free" space revealed that they found the malls – which can only be accessed via escalators – to be intimidating.

Evaluation

The case studies presented a rich array of information about the role played by skywalk systems in the lives of the people and the transformation of the districts where they are located. It also highlights the thoughtful concepts that are integrated into the design and management of skywalk systems for their smooth operation in Hong Kong.

The research affirms that the intensity of use of skywalk systems is high in Hong Kong. People think positively of them, and use them for a mix of functional, social and recreational purposes. The improved connectivity improves property prices on one hand, while bringing about more choices to local residents on the other. However, streets are still regarded to be more appealing, and the concerns about diverting a substantial share of pedestrian patrons away from the small shops at ground level are evidently real. An immediate challenge for planners is to address the poor walking environment at ground level as a result of surrendering the space to vehicular transport.

The study surfaced a number of important social and environmental issues, which will set the agenda of more in-depth planning debate and public intervention. These are summarized here:

Skywalk systems and social equity

The study highlights that the managed nature of skywalk systems, especially those by private entities, might have innocuously brought about social differentiation in the use of space. Respondents of the survey viewed that the diminished vibrancy of street life has discouraged certain segments of people from using them. This unintended emphasis of the social divide is manifested in two ways: first, by the conscious effort of property managements to keep out "undesirable" people and behaviors (the "inverse prison" concept described by Voyce (2006)); and second, by the unconscious creation of physical and psychological barriers for the poor and elderly to enter.

Another related social issue which surfaced in our study is that skywalk systems may have inadvertently resulted in an urban form that promotes consumerism, one that induces impulse spending as pedestrians are compelled to pass through mall after mall. Indeed, the survey respondents admitted that they actually bought more because of the skywalk–mall linkages. From a more positive perspective, malls provide variety to the walking experience, thus encouraging pedestrians to use skywalks to avoid the hassles of crossing the roads below. This is consistent with the main objective of creating skywalk systems in the first place: people–vehicular traffic segregation. The problem appears to lie more in the planning process, for where land leases do not specify what is to be built, developers would naturally tend to take the most profitable option.

Skywalk systems and urban street life

The study suggests that more has to be done to prevent the ground level from being relegated to the tyranny of cars and the infrastructure that serves them. From the users' angle, skywalk systems accentuate the efficiency of urban living, by providing a time-saving facility that lowers costs of tasks-handling in the city. They are able to achieve this because they draw out all the functional benefits of dense, compact and multifunctional developments as embodied in the concept of Multiple and Intensive Land Use (MILU) (Lau et al. 2007). Skywalk systems enable individual MILU developments to connect and make available their facilities for easy sharing. They also address the conflict between vehicles and people in a situation where urban compaction pushes the density of both to an untenable point. However, their obvious drawback is that pedestrian needs are now to be met at an elevated level, such that ground level conditions become less amenable for walking. By themselves, skywalk systems do not seem to be able to offer sufficient flexibility or varied experience comparable to the streets on the ground. The present study affirms that vibrancy of street life is still valued by a lot of the local residents in Hong Kong.

Contribution to the district as a whole

The economic and environmental benefits of skywalk systems as a planning strategy for high-density living lie also in its contribution to releasing scarce city land for other uses that might better meet the needs of the community. For example, by locating a bus terminus beneath the podiums, one additional piece of land could be available as public open space. The planning question is whether these public open spaces and the adjoining skywalk systems complement each other functionally. The study affirms that a stronger connectivity between the elevated spaces and those at grade would have induced greater pedestrian flow, and hence maximizes the utility of both.

Careful planning as "space," not merely "bridges"

From the institutional point of view, skywalk systems demonstrate the need for cross-sectoral collaboration, such as the planning, transport, social policy and local economy institutions. In a high-density environment, coordinated planning and sensitive handling of people-related issues is paramount for success (Yeh 2000).

From a social angle, skywalk systems can be harnessed to bring about profound and positive impacts to people's lives. With good planning and design, traditionally unconnected areas can be brought into the networks. At present, private residential estates in Hong Kong are largely vertical gated communities, and skywalk systems can link them with developments across social strata using common facilities such as shopping centers. Such an integrated approach would create opportunities for shared experiences between social classes. Hong Kong offers several examples of such opportunities.

The success factors identified for implementing skywalk systems in high-density living can potentially be applied to other high-density urban environments. However, before introducing skywalks, the social context and value system of the city must first be understood. If the benefits of skywalks, namely, efficiency, convenience and connectivity, are not valued by the community, they would turn out to be urban eyesores rather than assets. The US experience is a case in point.

Conclusion

Our research shows that apart from their planned functions as transport-related infrastructure, skywalk systems in Hong Kong fulfill a social function as well which could potentially be enhanced. As currently implemented, they are well accepted by the community, and recognized to have improved connectivity, pedestrian convenience and safety. Being integrated with a trunk-and-feeder transportation network, skywalk systems have mitigated the traffic–pedestrian conflict of high-density living in the case of Hong Kong, while contributing to persuading the local community to use public transportation systems. The multiple planning benefits include a seamless integration with transport nodes, support for land use intensification, integration into the lifestyles of community users, and offering a diversity of activity options.

The study nonetheless also brought to the fore potential adverse impacts to city life that must be addressed through policy and design interventions. The ground space of purpose-designed skywalk systems in Hong Kong has recognizably become less appealing to pedestrians, and there are observable signs of street life decline as a direct result of the overwhelming dominance of vehicular traffic at grade.

In terms of their ancillary role as social space, skywalk systems are currently not adequate in themselves, from the design perspective, as a substitute for public open space at ground level. For greater social and environmental sustainability, skywalk systems have to be planned with the same attention to design details and usage potential as at-grade spaces. With such goal-directed planning and design, skywalk systems would be well placed in future to realize their multifaceted potentials as surrogate public space, pedestrian corridors, feeder of users to the mass transit nodes, activity hubs, enabler of multiple and intensive land use, and connectors of fragmented communities.

A multidisciplinary team of town planners, architects, transport planners, urban designers, landscape architects, sociologists and experts on the local economy should work together with the affected community from the early stage of planning. Sensitivity to community values and usage patterns will help planners decide how the skywalk–ground permeability relationship could be better addressed, and what kind of interface would work best between a skywalk system and the rest of the locality.

The experience in Hong Kong shows that if these concerns are addressed, the benefits of skywalk systems could outweigh some of the drawbacks in a high-density environment, and such systems can become a valuable tool of planning and design intervention.

References

Agriculture, Fisheries and Conservation Department, Government of HKSAR (2010) *A New Geography of Hong Kong, vol. 2.*

Anderson, K. (1988) "Fast life along the skywalks," *Time*, 1 August 1988.

Census and Statistics Department, Government of HKSAR (2006) *2006 Population By-census* data. Available at www.censtatd.gov.hk (accessed 4 November 2010).

Census and Statistics Department, Government of HKSAR (n.d.), *Hong Kong Statistics*. Available at www.censtatd.gov.hk (accessed 4 November 2010).

Corbett, W. H. (1913) *City of the Future* (drawing). Available at http://arquinoias.tumblr.com (accessed 10 March 2011).

Cornwell, L. (2006) "Cities' enthusiasm for skywalk fades," *Boston Globe*. Available at www.boston.com/news/local/connecticut/articles/2006/01/14/cities_enthusiasm_for_skywalks_fades/?page=full (accessed 3 November 2010).

Lau, Stephen S. Y. and Coorey, B. A. S. (2007) "Hong Kong," in *Multifunctional Intensive Land Use: Principles, Practices, Projects, and Policies*, ed. MILU Network editing team, the Netherlands: Habiforum Foundation.

Le Corbusier (1925) *Voisin Plan for Paris.*

Legislative Council, Hansard of motion debate (2002). Available at www.legco.gov. hk/yr02-03/chinese/counmtg/floor/cm1120ti-confirm-c.pdf (accessed 10 March 2011).

McLaren, D. (2000) "Compact or dispersed: dilution is no solution," *Built Environment*, 18(4): 268–284.

Register, R. (2002) *Ecocities: Building Cities in Balance with Nature*, Vancouver: New Society Publishers.

Robertson, K. A. (1993) "Pedestrians and the American downtown," *Town Planning Review*, 64(3): 273–286.

Robertson, K. A. (1994) *Pedestrian Malls and Skywalks*, Aldershot: Avebury.

Rotmeyer, J. (2006) "Can elevated pedestrian walkways be sustainable?," in *The Sustainable City IV: Urban Regeneration and Sustainability, WIT Transaction on Ecology and the Environment*, 93, Southampton: WIT Press.

Secretary for Transport and Housing, reply to LegCo (2010a) Question raised by Hon Chim Pui-chung on 4 May 2010. Available at www.thb.gov.hk/eng/legislative/transport/replies/land/2011/201105042.pdf (accessed 2 March 2011).

Secretary for Transport and Housing (2010b) Speech at the "Green Transport in Hong Kong, Asia's World City" seminar at Shanghai Exp. (accessed 2 March 2011).

Vancouver Ecodensity Charter (2010). Available at http://vancouver.ca/commsvcs/ecocity/pdf/ecodensity-charter-low.pdf (accessed 3 November 2010).

Voyce, M. (2006) "Shopping malls in Australia," *Journal of Sociology*, 42(3): 269 286.

Yeh, A. G. O. (2000) "The planning and management of a better high density environment," in *Planning for a Better Urban Living Environment in Asia*, ed. by A. Yeh and N. M. Kam. London: Ashgate.

Chapter 13

Experiential planning

A practitioner's account of Vancouver's success

Jill L. Grant

Vancouver, BC has achieved international acclaim for its livability and compact urban form, making it one model of good planning practice. What techniques and strategies contributed to effective planning practice in the city? This chapter shares stories from a prominent practitioner in Vancouver, illuminating some of the techniques and processes planners used to help develop consensus around building a socially responsible and progressive city. This chapter presents results from interviews with Larry Beasley, the former director of current planning with the City of Vancouver. Excerpts from the interviews illustrate his locally situated theory of planning practice.

Beasley's stories of practice are not those of a heroic planner, but affirmations of basic planning principles: good processes, practical ethics, and effective organization. Beasley's model of "experiential planning" pursues good city form and function using socially just and politically responsive participatory processes. Vancouver planners helped build consensus by framing meaningful visions based on the everyday experiences and aspirations of residents. I feel this model holds promise for further development.

As we search practitioners' success stories for strategies and outcomes to emulate, it is important to identify which factors made a difference to outcomes. Those seeking to copy Vancouver's success will find some elements of the Vancouver case unique. However, Vancouver planners also developed engagement techniques and management strategies with wide applicability.

Submitted by the Association of Canadian University Planning Programs (ACUPP).

Originally published in the *Journal of the American Planning Association*, 75: 3: 358–370 (2009). Used here with kind permission of Taylor & Francis and the American Planning Association.

Introduction

In opening *The Prospect of Cities*, John Friedmann (2002) opined:

> The city is dead. As it grew in population and expanded horizontally, many attempted to rescue it, to revive it, to hold back urban sprawl, to recover a sense of urbanity and civic order. But the forces that led to its demise could not be held back, much less reversed. (xi)

Today Friedmann lives in Vancouver, BC, a city that has topped *The Economist's* list of the most livable cities in the world since 2002 (BBC News 2002; CNN 2005; Economist.com 2007 2008). Over the last decade Vancouver has become an internationally acclaimed model of a good city. Unlike so many North American cities, Vancouver resisted the forces of sprawl and decay that threatened it. Harcourt, Cameron, and Rossiter (2007) described Vancouver in the 1960s as low density and lacking urban character. Its municipal authorities began considering urban renewal and downtown expressways to address blight. Ultimately, though, Vancouver's leaders rejected those options. How did Vancouver transform itself from provincial backwater to a world-class city, and what role did planning play in the transformation? Punter (2003) and Harcourt et al. (2007) suggested that a series of good political and planning decisions changed Vancouver's trajectory, resulting in the dynamic and diverse city that now wins awards and entices respected planning academics to relocate.

What lessons can a city seen as socially progressive and urbane offer to planning practice and theory? In recent decades Vancouver has adopted policies to mandate the inclusion of affordable and family housing in new developments, and has established design policies and strategies to permit high-rise development and medium-to-high densities. Consequently, many planners view Vancouver as the exemplar of a modernist city that attempts to be socially inclusionary while adapting new urbanism principles for urban design (Berelowitz 2005).

This chapter presents one account of Vancouver's success. By sharing some stories from the practice of Larry Beasley, one of the Vancouver planners credited with a significant role in the social and urban transformation of the city, I offer some answers to Forester's (1999) question: "How are we to structure and encourage democratic deliberations in which participants can learn so much that their senses of hope and possibility, relevance and significance, interests and priorities shift?" (203). Many accounts of the last three decades of urban history in Vancouver suggest that the planning process helped shift the perspectives of local residents to embrace a new kind of urban form and a new image of the good city (Berelowitz 2005; Punter 2003). Beasley coined the phrase "experiential planning" to describe how planners catalyzed the process of urban transformation in Vancouver.

I begin by describing the changes in Vancouver, and then move to tell some of Beasley's stories and consider his theory of practice. I profile the perspective of one planner rather than testing hypotheses or rigorously interrogating factual claims. I aim to develop insights into the way a high-profile practitioner explains his work, and seek to make sense of a lifetime of practice.

Vancouver: poster city for urbanism

Since 1968 the City of Vancouver and its suburban municipalities have collaborated within a regional authority, the Greater Vancouver Regional District (GVRD), now included within the entity called Metro Vancouver. In 1969, the GVRD hired planner Harry Lash and began a public consultation process that resulted in a plan for a "livable region" (Lash 1977). Livability became a watchword for the region at that time, and has remained so. Despite municipal autonomy, local governments have voluntarily cooperated on planning and infrastructure matters. The provincial government's policy of protecting agricultural land limited the availability of land for development and forced plans throughout the Vancouver region to accommodate this as a constraint (Smith and Haid 2004). This drew planners' attention to the need for urban and suburban densification by the early 1970s.

In 1972, British Columbia politics took a significant step to the left with the election of a New Democrat provincial government and a radical city council in Vancouver under the new alliance of The Electors Action Movement (TEAM) and the Council of Progressive Electors (COPE; Punter 2003).[1] The new city council,[2] committed to the idea of a livable city, hired a dynamic, young director of planning, Ray Spaxman. Under Spaxman's leadership, Vancouver planners began to overturn conventional planning wisdom (Harcourt et al. 2007; Punter 2003). For instance, while scholarly planning authorities in Canada criticized apartment living as leaving children "caged in their high-rise trap" (Gertler and Crowley 1977: 339), Vancouver encouraged families to return to the city through redevelopment projects like False Creek South in the 1970s (Vischer 1984). Larry Beasley and Ann McAfee, who went on to become co-directors of planning in Vancouver's period of most intensive development, joined the planning team during Spaxman's early initiatives.

When Vancouver was chosen to host the World's Fair, Expo '86, provincial and federal financial resources assisted regional authorities in creating a regional rapid transit system to link suburban municipalities to the city (Harcourt et al. 2007). An excellent network of pedestrian and cycling trails was built in the city at the same time. External conditions that spurred increasingly high immigration and investment levels from Asia, beginning in the 1980s, encouraged rapid growth in the city as well as in suburban municipalities. Vancouver adopted its "Living First" policy – to facilitate residential growth downtown – in the 1980s to guide redevelopment of waterfront land made available following the dismantling of the fairgrounds of Expo '86 (Beasley 2000). Through an extensive public consultation process to consider options, residents and authorities came to a new consensus that instead of building freeways and bridges they would find ways to encourage more people to live downtown to reduce commuting.

With its urban revitalization projects, Vancouver specifically opted to create household diversity downtown. Taking the risk of pushing to intensify residential development in the urban core, Vancouver demonstrated the viability of the compact city model and established an enhanced role for urban design in place-making. Dozens of slender glass towers with mixed-use bases now grace the Vancouver skyline, illustrating the recent massive inflow of residents and capital to the urban core (Berelowitz 2005; Hutton 1998). The city has facilitated developing new commercial uses to meet residents' daily needs downtown (Beasley 2000). Over 44,000 people have moved downtown in the last 25 years, increasing the proportion of City of Vancouver residents there from 10.4 per cent to 15.2 per cent. Close to 7,000 children now live in the core. In recent years the city has opened new schools and created parks downtown. Policy requires that developers create a mix of housing that includes row housing and apartments, rental and ownership units, market and non-market units, and small and family-sized units. With a mix of uses throughout the urban core, and buildings that address the street to create an attractive pedestrian realm, Vancouver became a noted example of high-rise new urbanism (Grant 2006).

Urban leaders across the United States have sent delegates to learn from Vancouver's achievements (Baker 2005; O'Connor 2007). Organizations such as Massachusetts Institute of Technology (MIT) (Vale 2007), the Congress for the New Urbanism (CNU 2006), and the Canadian Institute of Planners (2008) have recently presented awards to Vancouver planners. Clearly, the profession recognizes that Vancouver's experience offers important lessons about producing good cities.

Writing 40 years ago, Eversley described the planner as "universally feared and disliked" (1973: 3). Recent analyses of Vancouver's experience show quite the opposite; planners there enjoy considerable respect and receive public credit for facilitating the city's transformation into a vibrant urban showcase (Punter 2003). What stories do these planners tell about the keys to their success in practice? This chapter investigates that question with Larry Beasley, arguably the best known of Vancouver's planners.[3] Beasley merits particular attention because of the role he has played within the new urbanism movement. A vigorous proponent of urbanism and believer in the value of modernist high-rise architecture, Beasley stimulated a debate about urban character and architecture within the CNU that some traditionalists resisted ("Tall-building Controversy," 2008: 7). Thus

Beasley has had considerable influence on the development of community design theory and practice.

Widely honored at home and abroad, Larry Beasley retired as Director of Current Planning in 2006 after 30 years of service with the city. A member of the Order of Canada (the highest national honor in Canada), a co-winner of the Kevin Lynch Prize from MIT, and the recipient of honorary degrees from two universities, Beasley has become an iconic figure in Canada and abroad. Recently a Vancouver developer named a 33-storey mixed-use building after him (VancouverReflections.com 2008). Prolific public speaker, international planning consultant, and author (e.g. Beasley 2000; 2004a; 2004b; 2006), Beasley is a highly successful planning practitioner, and acclaimed internationally as an agent of urban transformation. Yet he has said that when he began practice, planners were often seen as part of the problem for cities (Beasley 2004a). What role did planners play in changing this situation in Vancouver? What skills and strategies did they use?

Analysts like Sandercock (2003: 158) warn that theory about best practices derived from case studies may mislead because local conditions shape places and outcomes. However, case studies of practitioners (e.g. Forester 1999; Grant 1994; Hoch 1994; Krumholz and Forester 1990; Throgmorton 1996) reveal the practical challenges planners face in their work, and the frustrations they experience. Forester (1999) suggests that "the insightful analysis of planning situations can encourage better practice not by producing abstract lessons but by showing what can be done through practitioners' vivid, instructive, and even moving accounts of their successes and failures" (7). Since so much planning practice relies on emulation and the borrowing of standards and methods from other places, understanding the strategies for success in Vancouver and the challenges planners met in trying to achieve its ambitious social agenda may inspire and educate others. Personal insights from practitioners complement the evaluations of plans and policies, and show what is involved in making plans that matter (e.g. see Burby 2003; Dalton 1990; Talen 1996).

Earlier generations of Canadian planning practitioners (Blumenfeld 1987; Carver 1975; Gertler 2005; Lash 1977) established the tradition of reflective thinking and writing about their work. Schön's (1983) call to the profession to learn from practice seems to strike a chord with Beasley as well. In an article published in *Plan Canada* summarizing a conference keynote address, Beasley (2004a) drew on his life lessons to enjoin planners to be visionary guides for communities. Beasley argued that every planner is a role model for others who follow and hence has the obligation to learn and teach the lessons of practice well.

Presenting Beasley's stories acknowledges the authority of his influence in contemporary planning practice and theory. The analysis and excerpts that follow came from a series of telephone interviews I conducted with Beasley on four occasions between October 2 2007 and January 7 2009. I recorded and transcribed these interviews, and directly quote excerpts from them below. The conversations focused on the techniques and processes that Vancouver planners used, the challenges they faced, and the lessons they might offer planners elsewhere. Such a method cannot readily be replicated, but it produces unique evidence vital to interpreting events and drawing meaning from them. Moreover, it reveals valuable insights into the ways in which "deliberative practitioners" (to borrow Forester's 1999 term) use and develop theory to inform and describe their practices.

A practitioner's perspective

In Beasley's stories of practice, Vancouver's planning and urban success originated in political choices that citizens and leaders made in the early 1970s which "changed everything." Planners with new sensibilities reformed planning processes and objectives, inspired by the work of writers such as Jane Jacobs (1961). Like many young planners in that post-urban renewal period, Beasley worked as a neighborhood planner under a nationally funded neighborhood improvement program aimed at regenerating older areas. In discussing the times, he described a political and planning culture of innovation, community engagement, and consensus building. By the mid 1980s when he took responsibility for leading the downtown planning process (see Excerpt 1), collaborative planning had become well entrenched in Vancouver. Planners shaped the institutions and conditions that empowered citizens to participate in achieving their visions (Healey 1997; Innes 1996).

Excerpt 1. ***Consensus on urbanism***

I took over the planning of the downtown a few weeks after the close of Expo '86, in January of 1987. We knew that we were going to face the planning of that huge area. The office market had essentially collapsed, as it had everywhere in North America, roughly in 1982–83. So we were in the position that we needed to rethink the city at that point. What was going to be our economic driver? What was going to be the shape of the city? We had vast areas for redevelopment. So we implemented a dreaming process starting in about 1988. And out of that process of everyone around the city talking, a collective conclusion came to the fore that our destiny could be pursued in a positive way if we could entice people to come back and live in the inner city. It was a community-wide consensus.

In fact, if there's one thing that I have to say about Vancouver – the thing that's been fascinating for me in practice here – is that, in the design community, in the planning community, among many politicians, and among many informed citizens, there has been a strong consensus about the vision of the city, about trying to pursue an urban future, of trying to bring the car into its proper balance with other modes of travel in terms of its influence on the shape of the city, in support of density. There's been an almost intuitive consensus that tall buildings were not bad but that they had to be well designed. This, in my view, was set off and came together because of the TEAM council and the discussions that they brought about in the city.

With an urban population growing ever more diverse as a result of heavy immigration, how did Vancouver planners ensure the representativeness of planning processes? Beasley noted that staff designed processes and techniques to reach out broadly, even to the most disadvantaged in the community (see Excerpt 2). They used open houses, workshops, surveys, informal meetings, newsletters, and various other strategies. The commitment to engagement permeated the municipal organization and became a central organizing principle. Beasley also noted that staff operated with the understanding that they had the responsibility to ensure that all constituencies were represented in presentations to city council.

Excerpt 2. Strategies for broadening engagement

In all the public consultation that we do, we try to design techniques that fit the needs of the people involved rather than use techniques that are comfortable and convenient for the planner. Let's take the disadvantaged community of indigent men who live in small hotels. We put the staff person down in the lobby of the hotel and just sit and talk to the men as they come in and have their coffee. It is a much more informal engagement process. Whereas for single parents with small children, we may go to the childcare centre and talk to them there. For young families, we may start a conversation with the children who then take the message home to the parents; then in most cases a parent will not ignore a subject that a child is dealing with in school. You can take the public consultation into schools and reach parents who then can come back and participate with you.

We use various kinds of techniques. When you go very broadly you move back to polling and surveys and those kinds of techniques. They don't go deep but they go very broad. What you try to do in all cases is layer your findings on top of one another to see what kind of common themes tend to emerge. What unique themes suit certain kinds of groups in the community? On the one hand, we find often that general themes emerge that can become very useful throughout our policymaking. On the other hand, sometimes special needs themes emerge that you want to make sure that you maintain some policy places for. . . .

Most of the time, the outcome is more predictable if you have a very well articulated program that you have worked up with people. For example, when we did all of our plans, before we initiated the actual planning, we actually had a public engagement process about the planning process itself. I'll just use one example: the recent Downtown Transportation Plan, which was a later chapter of the Central Area Plan. We were out four to six months talking to special interest groups and downtown organizations about how this planning should take place. Who should be involved? What should be the kind of issues on the table? When we finally started the process we had an awful lot of people who felt they owned that process and therefore they were more inclined to participate. The more that people are engaged to believe in the process, the more likely that the process is going to be successful.

From 1992 to 1995, planning staff elaborated on the consultation model by involving thousands of residents in the process for creating *CityPlan*, the long-term vision for Vancouver (McAfee 1997, 2008). Staff saw their role as facilitating engagement and documenting its outcomes, allowing city council to identify the consensus and set policy. Equity, diversity, and urban form concerns intertwined in this process, leading staff to develop new techniques to inform their approaches and to develop new ways to describe their successes (see Excerpt 3). Beasley argued that the commitment to urbanism grew naturally out of this process as community members learned with staff. In telling the story, Beasley continually emphasized the teamwork within the city that made innovation possible, acknowledging the leadership of the municipal council and of planners Spaxman and McAfee. He also suggested that developers adopted the new direction quickly, as they realized that higher density development would improve their return on investment.

Critics of *CityPlan* argued that with millions of dollars devoted to the process, participation became an end rather than a means in Vancouver, as planners spent years generating little more than wish lists (Seelig and Seelig 1997). That perspective missed what Beasley said was important about the exercise. Participatory processes that simply "harvest opinion," as Beasley put it, become mired in self-interest. Vancouver avoided that pitfall, Beasley said, by framing its questions carefully to facilitate learning. Encouraging residents to think about their own futures and those of their descendants created a moral framework for envisioning places capable of including a wide cross-section of potential residents. The innovative techniques Vancouver planners used helped residents to connect the ideas of affordability and urban diversity to their own personal experiences of place and their aspirations for the futures of their families and their city. Advancing equity planning concerns in the Vancouver context did not involve battles with political leaders or citizens' groups, as in communities like Cleveland (Krumholz 1996; Krumholz and Forester 1990). Rather, planners aimed to understand what was important to people in their everyday lives and then to help them see how progressive planning choices could address those needs. The process thus facilitated change in participants' values.

Excerpt 3. Engagement techniques

CityPlan in the 1990s was a dramatic moment in the city and I will always give Ann McAfee and her team great credit for what they did. I mean, to have 100,000 people in some way contributing and to have 12,000 people very actively involved, it was a huge thing and it changed consciousness. *CityPlan*, though, was also the continuation of a tradition of engagement that has been going on since the '70s. The deep and institutionalized commitment to public engagement started with the TEAM council in the early '70s and under Ray Spaxman's planning leadership. Literally every day planners and engineers and other civic officials are out in the community talking to people about issues. There's something like half a million letters a year that go out to inform people about developments. And those things have dramatically changed everything in the city and opened up extraordinary opportunities for city building. Because people have, in a sense, been educated by this process, about what's good and what's not so good and what works and what doesn't work. And they have more discerning eyes, as it were, which they express in their response to developments and at the polls and everywhere else. . . .

The *CityPlan* process did start a conversation that I don't think we had before. And that was the conversation about what is going to actually happen to you if nothing changes. What's going to happen when your children need a home, need a place to live? Are they going to find a place in this community? Or must they be pushed somewhere else, either downtown or out to the far suburbs? And what's going to happen to you when you're old and you don't really want or need the house anymore? Are you going to be able to find a place in your community? And do you have enough people in your community to have the kind of community support facilities that you are calling for and you need? In that way we've been able to show people that if you don't have enough people, you don't have library services and all of that.

And that kind of conversation – which is not a conversation saying, "Density's good for you," but a conversation saying, "What do you need your neighborhood to be for you?" – has been much more fruitful. With citizens, I think you have to start with what is good for them on their terms. And then if you can illustrate that some of the things that are good for them come from some intensification in their community and build up enough people with that understanding, at least you get an alternative to that small core group of NIMBYists that are there. And then, if you do it very carefully and very well, and you can illustrate that it's not destroying the community, then a lot of other people who really weren't paying too much attention start to shift to say, "It's been pretty good for us." That approach changes opinions. And I think *CityPlan* started that conversation because the City just approached the whole problem in a different way. That was the great brilliance of *CityPlan* that is not often understood.

Having thought carefully about the implications of this sequence of mutual learning activities within Vancouver's planning processes, Beasley coined the phrase *experiential planning* to embody his theory of what motivates citizens to do the right thing for their communities (see Excerpt 4).

Excerpt 4. Experiential planning

It struck me when I read a book a few years ago called *The Experience Economy* [Pine and Gilmore 1999] that what was missing and what we were doing in Vancouver – but we hadn't put a name to it – was to try to create for people the experiences that they aspire to in their day-to-day lives in the city that houses and accommodates that day-to-day life. If we really put our attention there, in addition to our systemic view of the city that we've put most of our planning attention to, then we would start creating cities that draw an emotional connection to the people who are there. It would be a connection of affection, of loyalty, of all those things, because the city becomes fulfilling for them. And as I've worked through that in my own mind, I have dubbed it "experiential planning" because it is a planning approach based on the experiential expectations of the users of that environment. What I like about it for me is that it brings together many of the things I believe in for planning. It brings together, for example, a very tangible rationale for the level of public involvement and engagement that we have with people. It illustrates the value of explicit urban design and developing policy and regulatory frameworks where urban design has a high imperative. And I hope it will start to deal with what I think, from a sustainability point of view, is one of the fundamental difficulties with moving forward – which is about the sustainable city being dense, being mixed-use, being diverse. In a democracy, we're not going to go very far if citizens' experience is leading them to a very unsustainable consumer pattern.

I think the planning epistemology that comes with this is not about just having focus groups and a superficial level of marketing analysis. It's about a fairly deep and continuous level of engagement with citizens, going beyond even the direct issues of material consumption to a lot more spiritual issues and other social issues and other things which really make up the totality of people's experience of the city. For example, I think that we might find that our country gets a lot more heavily into housing the homeless people of the country if, in fact, we tapped into what the experiential expectations were – not just of homeless people, but of other citizens seeing homeless people in their community. I'm not a cynic who says that most people don't care. I think a lot of people care, but they just don't know what to do about it and there's no way for their concern to be expressed through their actions.

Vancouver used these processes of public engagement to develop innovative policy that requires developers to provide 20 per cent non-market units in new housing downtown. Beasley acknowledged the challenges in achieving that social objective,[4] but argued that the principle has become so accepted that even wealthy neighborhoods expect to live by it (see Excerpt 5). The success of Vancouver's practice in encouraging social responsibility made Beasley a committed optimist who suggested that good planning practices can unleash good citizenship.

Excerpt 5. Affordable housing

Right now, the amount of affordable housing in new developments is sitting at about 16 per cent. The numbers aren't quite at 20 per cent. But in a way, I've always been less worried about the numbers than I have about the philosophy behind the numbers: that the philosophy stays solid, that no neighborhood comes to exist without its component of social housing. The exact amount is never going to be enough. There are always waiting lists. We always want to do more, but now every neighborhood has a fair share. It becomes the way the city is built.

Interestingly enough, we had this great drama a couple of years ago where, in one of the neighborhoods in Coal Harbour [an upscale development area], there was a site and the Housing Office thought, well, maybe this is a site the City could cash in because there had been a lot of pressure from the developer saying "You could use that money elsewhere to do a lot more social housing," so the Housing Office raised that issue. And the neighbors all came out and said, "No. We want our share of social housing, too." These are wealthy people! "We want our social housing. That's a part of what we bought into is the social mix." And the City backed away from that conversion and that housing is now built.

Beasley offered many examples of the way that the process of helping planners to work through issues needing resolution changed people's perspectives (see Excerpt 6). From the siting of halfway houses, to accepting a policy legalizing accessory suites, to developing intensification policies: all the policy changes resulted from extensive consultation processes that changed popular opinion. Taking community concerns seriously and finding ways to negotiate agreements between residents and developers about how to resolve disagreements can, in Beasley's view, result in success.

Excerpt 6. Resolving conflicts

[We were proposing] a halfway house for convicted felons that were coming out of prison. We started this and the community rose up in opposition, based on their spontaneous thinking about what such a place might be and what the impacts might be. We started out with a couple of field locations where there really was no way a politician could make a decision and survive. We then pulled people together and learned about these facilities. We went and visited these facilities. We published the information and had a community discussion about that. We set up some guidelines that were then debated in the community. People started to understand what these facilities really were and the probabilities for difficulties

> really were. They got some security that [we would set conditions on the use]: for example 24/7 supervision and conditions like that. With that in hand, and a set of locational criteria that had been put together by community people, we were then able to determine a site that was quite well received in the community – not unanimously – and is functioning as we speak.

To make public engagement processes work effectively, Beasley said, planners need to develop specific skill sets. He especially emphasized the need for mediation and negotiation skills, and for knowledge about effective urban design. Vancouver established an annual staff training budget to ensure that all its planners would have funds to facilitate upgrading. Since the city is only as effective as its staff, managers need to attend carefully to the effectiveness of their planners (see Excerpt 7). Beasley acknowledged some instances where he had to move and retrain planners who didn't work well with a community.

Excerpt 7. Staff effectiveness

I remember a couple of community-based processes where the planner wasn't paying attention. One process I remember where a very good young planner went out into a neighborhood and was working for a number of months. The community came to me and said, "We like this planner, but could you please remove this person because he's not dealing with the issues we care about." The planner had not been listening carefully enough and had a kind of sense of his own of what issues had to be dealt with. He wasn't listening to their issues, so that process went off the rails.

We had another case where the planner's personality simply did not work with the personality of the community. She was one kind of person and the community was predominantly different kinds of people; that didn't work.

It is very tricky putting planning teams out in the field. There is a feeling that you need a methodical process but you also have to carefully think about the personalities, the inclinations, the styles, the rhythms of the way people work and try to match those with a sense of how that is in the community. You have to do some training with your staff, so that they also get the sense of how to morph themselves into what they need to be for different kinds of community processes.

In Beasley's theory of practice, personality matters. Effective planning is not just about getting the process right: the way that the planner operates affects the prospects for success. Yet he also acknowledged that the political and planning system in Vancouver created the context within which planners could make innovations that prove challenging in other places (see Excerpt 8). Without political leaders willing to give planners the authority to negotiate uses and densities, Vancouver might be a different kind of place. The system, the players that operate within it, and the way it structures influence all play a part in determining outcomes.

Excerpt 8. Discretionary approvals

One of the things that happened – and it related to the social housing policy, it related to the family housing policy, it related to the array of standards of public amenities, whether that was child care or community facilities or parks–was that the Mayor of the day, who really led a lot of this, Gordon Campbell, was a policy-oriented guy. On the one hand, it's just his nature. "I want to know what the rules are." On the other hand, he had come out of the development community and he felt the risk and anxiety that the developer feels when there are no rules. And also he was very dedicated to the public objectives of the city and so wanted to make sure that we did all these things. But he was also sensitive to the needs of the development community and felt we couldn't do them capriciously. So when we started to develop and we knew we were going to face developers of some very huge sites – 250 acres, in the case of False Creek North, 80 acres in the case of Coal Harbour – his drive for us as an organization was to "Put down on paper what it is that we're going to want. Bring analysis to that to justify what it's about, and let it be adopted politically before you go in to negotiate." And wow! That was a very good way to do things because as a negotiator for the city, it meant that I could walk into a negotiation with people – who usually in these negotiations are much stronger than you are – with a lot of clarity and a lot of political support. And I knew that my council was behind me. So I could insist. . . .

No negotiation between parties is a fruitful one both ways unless there's power at both sides of the table. And by virtue of the power that's been invested in the planners and those doing the negotiations on the civic side, we could broker real deals. Because the developers would not move without the planners and, in a sense, the planners could not move without the developers. It set up a dynamic where collaboration and interest-based negotiations could occur and they do occur every day in Vancouver. I think, by and large, most savvy developers now will say, "You know, it's a system I can live with and I can see the benefits of."

The discretionary approval process in Vancouver gave planners considerable authority to broker deals that implement principles of social inclusion and urban design. Critics of such processes may worry about backroom agreements or the potential for corruption, but Beasley explained the checks and balances in place to ensure that public purposes were addressed and planners remained accountable (see Excerpt 9). A Development Permit Board made decisions on developments in the city through a transparent public process that involved considering and presenting a body of evidence. The city council allowed expert city staff to evaluate proposed projects in light of policy the council had adopted. Although political leaders in some communities may safeguard the right to decide whether developments proceed, in Vancouver the council delegated that task to staff. Beasley suggested that council members wanted to avoid any perception that they might be inappropriately influenced in making development decisions, but more importantly, they trusted senior staff to make the right decisions (and could fire them if they did not).

Excerpt 9. Development approvals

Several factors protect the integrity of the discretionary regulatory process that we have in Vancouver. One is that it's totally transparent. No matter who has a meeting with whom, ultimately the results of that come into the public realm as a matter of public knowledge and discussion. For example, if there is a meeting between the developer and a staff person on some aspects of project qualities or an amenity, no decision is made there. They come to a working premise that then goes to a public process: either a council process or in our case a Development Permit Board process. The Development Permit Board sits in public and never makes decisions in camera [in closed session]. It is open to anyone to come and talk to it. The point is, it is very transparent and it's very accessible.

The second aspect that protects it is that it takes advice from a number of different perspectives and overlays that advice to create the choices that need to be made by the decision-makers. There is an urban design perspective from the urban design panel. We are not looking for the citizen's view of things there; we are looking for design quality discussions. There is the community's view. There is the special needs view. There is the heritage view. Each one has their own process to engage the public and that is overlaid and put into the public report.

By the time something gets to a decision it's pretty clear who has been saying what. No one is in an omnipotent or power position to just do whatever they want. The Development Permit Board itself is a dynamic decision-making body and it makes most of the decisions on development.

How do planners avoid being pressured about decisions? Beasley noted that ensuring that the process is open and transparent protects the planner. Documenting all of the factors and all of the perspectives helps the participants understand the chain of reasoning that staff used to produce an outcome. Ultimately, though, Beasley insisted that good practice requires ethical practitioners (see Excerpt 10). Process cannot substitute for moral fiber.

Excerpt 10. *Ethical practice*

Over the many years I was on the Development Board and involved in individual decisions having authority as Director of Planning, I never experienced a pressure of a personal type. The reason for that is that I was well paid [and] honored and I didn't care about any of that. I could make the decisions I needed to make based on the process.

Secondly, the process is so transparent it protected me from pressure. That's what I loved about the process. It had so many checks and balances that no one, not even politicians, could impress me. The way I made the judgment was so clear to everybody who's looking that A-B-C led to D. Not just "I felt like it that day": not just me, but also the people with power to make decisions.

Thirdly, was the overlay of different advice coming in. The way choices were made, you could see the advice, and how we reconciled the advice and made the decision. A skeptical person living in a society where there is a tendency towards a strong power base to form is going to say "that's subject to abuse." Yes it is subject to abuse. It takes ethical people to make it function. It is an aspect that often gets lightly dealt with when we talk about these processes. No process works if unethical people are working in that process. So you have to make sure that you choose ethical people and test them in that regard.

Beasley suggested that planners have a professional responsibility to get their analyses right. The planner maintains professional credibility and public respect by doing analysis fully, dispassionately, and fairly; in this, his view echoed that of Forester (1989). For Beasley the planner simultaneously plays a political and ethical role. The planner lobbies for social justice and good urban design, and innovates where it produces community benefits. Beasley argued that each planner's work matters because it affects the potential success of those who follow. Vancouver's contribution represented more than an idiosyncratic local development for Beasley; it constituted a natural experiment for new urbanism, recognized through awards as a trend-setting example of best practices (CNU 2006).

Lessons for practice and theory

To what extent is Vancouver's experience unique? To what extent can others emulate it? The particular constellation of circumstances that created the "Vancouver achievement," as Punter (2003) described it, may not be readily replicated. However, Beasley insisted that people made significant choices that allowed such innovation (see Excerpt 11). Citizens get the kind of places they deserve. Studying effective practice clarifies the potential effects of particular choices and reveals the gamut of options available to those who want to change their future. Vancouver illustrates that if people elect progressive governments that hire progressive staff who use progressive strategies, they can generate progressive outcomes. For Beasley, planning permits purposive action.

> ### Excerpt 11. Choice
>
> I always say, "You get the city you decide you want." I believe you have to decide what kind of city you aspire to as a community and then courageously change whatever you need to change to get there. Vancouver did have to change its regulatory system. It did have to change its political system. It did have to change its development management system. It did have to change its planning system and the way it engaged its public in order to achieve the things that we've [done]. I don't buy the critique [that Vancouver is unique]: I think, in a way, it allows people to be complacent about what they face in their communities.

History, politics, and geography all contributed to Vancouver's adopting socially progressive planning policies and encouraging forward thinking urban design. Electors' transformation of local politics in 1972 unleashed incredible potential. Political leadership committed to principled, policy-based decision-making gave staff the authority to act. Vancouver's system of electing municipal council members at large instead of on a ward basis gave political leaders the ability to take citywide rather than parochial perspectives. Its experience with hosting a world fair provided resources and redevelopment opportunities other cities envy. The city's location on the temperate Pacific Rim made it a magnet for immigration and investment.

Despite the unique features in the local system, however, Vancouver's experience indicates that planners who practice effectively, employing engagement techniques that help people learn, can exert a transformative influence. Practitioners in Vancouver built a tradition of practice that earned considerable respect (see Excerpt 12) by identifying issues and acknowledging varying perspectives, according to Beasley (see Excerpt 13). They avoided oversimplifying or ignoring differences. The process is not foolproof, but, Beasley implied, such humility increased their odds of success.

Excerpt 12. Respect

In our system, even transportation is now jointly managed by Planning and Engineering, which is unusual in many cities. And that has also helped. The other factors are that planners are vested with real decision-making power in regard to development. And that, over time, just builds up a status and a stature for the planners where they are people to be reckoned with. Because ultimately when it comes down to development, they make the decisions. Through all of those measures – this combination of things – planning has never been marginalized here the way I see it in many places. I get really worried in some cities I go to when I see that the planners are not very important to what's going on. And then I see who is and I'm shocked.

What I found fascinating in Vancouver for at least the last decade with every single municipal election, right across the political parties, one of the first things that happens is that every political party declares its support for planning and the planners. It's something they almost have to do because the planners are very highly respected by the electorate, by the population. And part of that is because we've delivered hundreds of millions of dollars of public goods – and the population knows that – through the negotiated process enabled by discretionary zoning. Partly it's because the planners are out there on the street every day in a highly participatory process and citizens see that and they remember that. And partly, it's because we have been putting forward cutting-edge propositions and bringing them to ground, not just talking about them but bringing them to ground in development after development after development.

In Vancouver, reform-minded politicians used planning processes to initiate an ambitious social agenda. Enabling an engaging collaborative process permitted planners to reinforce the social and urban agenda and keep it active even as political regimes changed. Success bred success to the point where planners came to enjoy the respect of citizens and political leaders alike. Through the process of experiential planning, of framing meaningful visions in the context of the everyday experiences and aspirations of ordinary residents, Vancouver planners influenced the politics and outcomes of daily life in the city. By under-standing and engaging citizens and politicians, the planners built an effective constituency for planning that helped residents to reframe their understanding of urban issues and options.

Excerpt 13. Humility

Sometimes what we do is say "Okay, here is the array of public opinion. In this clustering of opinion you have a significant majority of citizens agree-ing on these items. In this array of opinion you have a significant minority of citizens agreeing on this item. Here you have special interest minorities agreeing on items." You just lay all that out.

What often happens in documentation of public consultation is that the people doing the management of it oversimplify. They turn it into zero-sum documentation, which is to say that "most people said this." Often with the richness of opinion it's important to know that most people said something, some significant minorities might have said something else, and small minorities said something else. Perhaps this helps to identify the minorities that have special needs and special concerns. You document and put all that down so it can be understood.

In fact in our city vision process when we ended up pulling the policies into place, we articulated where a policy was strongly endorsed by that com-munity and where policy was more modestly endorsed by that community – so that when decision-making came later and council was making deci-sions on particular things, they knew where [people stood].

Many cities are now trying to emulate Vancouver's success. Some do that by copying the outcomes: by developing urban design policies that code for skinny towers with townhouses and commercial uses at the base. While imitation may be a sincere form of flattery, in planning it risks creating new forms of conformity or locally inappropriate options. Some cities are adopting discretionary approval processes to give staff the authority to negotiate design improvements with devel-opers. While such processes can enhance the quality of design, they also carry the potential that planners may find themselves being accused of being anti-demo-cratic and open to corruption if safeguards are insufficient.

Beasley's story is not a view of the heroic planner single-handedly reshaping the city, or cleverly identifying the particular outcome or strategy that can change history. Instead his experience reinforces the fundamental significance of plan-ning process and organization: these are the key messages of planning theory as well. Vancouver's success represents an argument for ensuring the effectiveness of the planning team, of taking the time and committing the resources to find solutions that work locally, of committing to mutual learning with political leaders and community members, and of reflecting on and learning from mistakes. Beasley's interpretations give the insider's perspective on what others have called "city making in paradise" (Harcourt et al. 2007) or the "Vancouver achievement" (Punter 2003).

Interviews with practitioners present richly textured and nuanced understandings of the relationship between politics and planning practice in ways that further the development of planning theory. While such methods do not lend themselves to ready replication, the stories generated provide a body of evidence through which planning scholars may articulate and interrogate theories of practice. At the same time, planners' stories reveal the ways that practitioners employ and develop theory to account for their successes and failures. Beasley's comments reflected one planner's locally situated theory of planning practice. He described the planner as a powerful force for social transformation working in a responsive way with community members and political leaders to achieve values of social justice and urbanity reinforced and reproduced through effective planning processes. With power comes the responsibility to behave ethically and professionally. With appropriate practice the planner earns societal admiration and respect. Beasley's concept of experiential planning encapsulated the notion of a socially just and politically responsive participatory process in pursuit of good city form and function. That notion has powerful rhetorical potential for a profession committed to positive social change.

Acknowledgements

I am grateful to Larry Beasley for consenting to almost four hours of interviews and allowing me to publish excerpts here. Thanks to Jeff Haggett for his assistance with transcription, and to the reviewers and editors for helpful comments on the manuscript. This research was supported by the Social Sciences and Humanities Research Council of Canada.

Notes

1   As Higgins (1986: 246) notes, explaining when the reform era began is easier than explaining why the shift occurred. In 1968, Canadians elected Pierre Elliott Trudeau on a platform of participatory democracy. The early 1970s saw progressive city governments elected in several cities, including Toronto. Most of the progressive governments were short-lived, but Vancouver's transformation continued even under subsequent regimes.

2   Several members of the TEAM and COPE parties came either from the academic ranks of local universities or from the community groups that had challenged plans for urban renewal (Harcourt et al. 2007; Punter 2003).
3   Since 1972 Vancouver has developed under three planning regimes. Ray Spaxman was director from 1972 to 1994. From 1994 to 2006 Larry Beasley and Ann McAfee were co-directors of planning. Beasley was in charge of current planning (development activities) while McAfee was in charge of long-range or strategic planning.
4   Critics such as Swanson and Yan (2006) suggest that the city's policies for downtown increase the risk of gentrification and displacement that worsen instead of improve the ability to provide affordable housing.

References

Baker, L. (2005, December 25) "Spurring urban growth in Vancouver, one family at a time," *New York Times*, www.nytimes.com/2005/12/25/realestate/25nati.html, (accessed 2 October 2007).
BBC News (2002, October 4) Vancouver and Melbourne top city league. Http://news.bbc.co.uk/2/hi/business/2299119.stm (accessed 31 May 2008).
Beasley, L. (2000, April) "'Living first' in Vancouver," *Zoning News*, www.city.vancouver.bc.ca/commsvcs/currentplanning/living.htm (accessed 2 October 2007).
Beasley, L. (2004a) "Moving forward in Canadian communities: soliloquy of an urbanist," *Plan Canada*, 44(4): 16–19.
Beasley, L. (2004b) "Working with the modernist legacy: new urbanism Vancouver style." Larry Beasley's address to the Congress for the New Urbanism, Chicago, June 2004, www.cnu.org/node/987 (accessed 2 October 2007).
Beasley, L. (2006, May 23) "Ingredients for a vibrant sustainable city: cues from Vancouver's recent experience in the New Urbanism." Presentation to the "Creating Valuable Cities Conference," Hong Kong, www.uli.org/Content/ContentGroups/Events/Conferences/HongKong2006/beasley_larry.pdf (accessed 2 October 2007).
Berelowitz, L. (2005) *Dream City: Vancouver and the Global Imagination*. Vancouver, BC: Douglas and McIntyre.
Blumenfeld, H. (1987) *Life Begins at 65: The Not Entirely Candid Autobiography of a Drifter*, Montreal, QC: Harvest House.
Burby, R. J. (2003) "Making plans that matter: citizen involvement and government action," *Journal of the American Planning Association*, 69(1), 33–49.
Canadian Institute of Planners (2008) CIP awards for planning excellence, www.cip-icu.ca/web/la/en/pa/3a169ad266da4c68a3af96a53e3f11b2/template.asp (accessed 1 June 2008).
Carver, H. (1975) *Compassionate Landscape*, Toronto, ON: University of Toronto Press.
CNN (2005) "Vancouver is 'best city to live'," www.cnn.com/2005/WORLD/europe/10/04/eui.survey/ (accessed 22 July 2007).

Congress for the New Urbanism (2006, November 9) "Vancouver's 'Living First' strategy," www.cnu.org/node/283 (accessed 1 June 2008).

Dalton, L. (1990) "Emerging knowledge about planning practice," *Journal of Planning Education and Research*, 9(1) 29–44.

Economist.com (2007, August 22) "Where the grass is greener," www.economist.com/markets/rankings/displaystory.cfm?story_id=8908454&CFID=16415879&CFTOKEN=94552766 (accessed 31 May 2008).

Economist.com (2008, April 28) "Urban idylls," www.economist.com/markets/rankings/displaystory.cfm?story_id=11116839 (accessed 31 May 2008).

Eversley, D. (1973) *The Planner in Society: The Changing Role of a Profession*, London: Faber and Faber.

Forester, J. (1989) *Planning in the Face of Power*, Berkeley, CA: University of California Press.

Forester, J. (1999) *The Deliberative Practitioner: Encouraging Participatory Planning Processes*, Cambridge, MA: MIT Press.

Friedmann, J. (2002) *The Prospect of Cities*, Minneapolis: University of Minnesota Press.

Gertler, L. (2005) *Radical Rumblings: Confessions of a Peripatetic Planner*, Waterloo, ON: Department of Geography, University of Waterloo.

Gertler, L. and Crowley, R. (1977) *Changing Canadian Cities: The Next 25 Years*, Toronto, ON: McClelland and Stewart.

Grant, J. (1994) *The Drama of Democracy: Contention and Dispute in Community Planning*, Toronto, ON: University of Toronto Press.

Grant, J. (2006) *Planning the Good Community: New Urbanism in Theory and Practice*, New York: Routledge.

Harcourt, M., Cameron, K. and Rossiter, S. (2007) *City Making in Paradise: Nine Decisions That Saved Vancouver*, Vancouver, BC: Douglas and McIntyre.

Healey, P. (1997) *Collaborative Planning: Shaping Places in Fragmented Societies*, Vancouver, BC: UBC Press.

Higgins, D. J. H. (1986) *Local and Urban Politics in Canada*, Toronto, ON: Gage Publishing.

Hoch, C. (1994) *What Planners Do: Power, Politics and Persuasion*, Chicago: APA Planners Press.

Hutton, T. A. (1998) *The Transformation of Canada's Pacific Metropolis: A Study of Vancouver*, Montreal, QC: Institute for Research on Public Policy.

Innes, J. (1996) "Planning through consensus-building: a new view of the comprehensive planning ideal," *Journal of the American Planning Association*, 62(4): 460–472.

Jacobs, J. (1961) *The Death and Life of Great American Cities*, New York: Vintage Books, Random House.

Krumholz, N. (1996) "A retrospective view of equity planning: Cleveland 1969–1979," in Campbell, S. and Fainstein, S. (eds), *Readings in Planning Theory*, pp. 344–362, Oxford: Blackwell.

Krumholz, N. and Forester, J. (1990) *Making Equity Planning Work: Leadership in the Public Sector*, Philadelphia, PA: Temple University Press.

Lash, H. (1977) *Planning in a Human Way: Personal Reflections on the Regional*

Planning Experience in Greater Vancouver, Ottawa, ON: Ministry of State for Urban Affairs.

McAfee, A. (1997) "When theory meets practice: citizen participation in planning," *Plan Canada*, 37(3): 18–22.

McAfee, A. (2008) "An update on CityPlan," in Grant, J. (ed.), *A Reader in Canadian Planning: Linking Theory and Practice,* p. 116, Scarborough, ON: Thomson Nelson.

O'Connor, E. (May 11, 2007) "Atlanta delegation here to learn from city," *Vancouver Province*, www.mayorsamsullivan.ca/ecodensity (accessed 22 July 2007).

Pine, B. J. and Gilmore, J. (1999) *The Experience Economy: Work Is Theatre and Every Business a Stage*, Boston, MA: Harvard Business School Press.

Punter, J. (2003) *The Vancouver Achievement: Urban Planning and Design*, Vancouver, BC: UBC Press.

Sandercock, L. (2003) *Cosmopolis II: Mongrel Cities of the 21st Century*, London: Continuum.

Schön, D. (1983) *The Reflective Practitioner: How Professionals Think in Action*, New York: Basic Books.

Seelig, M. and Seelig, J. (1997) "Participation or abdication?" *Plan Canada*, 37(3): 18–22.

Smith, B. E. and Haid, S. (2004) "The rural–urban connection: growing together in Greater Vancouver," *Plan Canada*, 44(1): 36–39, www.smartgrowth.bc.ca/Portals/0/Downloads/Growing%20Green%20in%20Greater%20Vancouver.pdf (accessed 1 June 2008).

Swanson, J. and Yan, A. (2006) "Affordable housing policy brief," Carnegie Centre Community Association, Carnegie Community Action Project. Retrieved October 2 2007, from http://intraspec.ca/CCAP_HousingBrief.pdf (accessed 2 October 2007).

Talen, E. (1996) "Do plans get implemented? A review of evaluation in planning," *Journal of Planning Literature*, 10(3): 248–259.

Tall-building controversy (2008) *New Urban News*, 13(1): 7–8.

Throgmorton, J. A. (1996) *Planning as Persuasive Storytelling: The Rhetorical Construction of Chicago's Electric Future,* Chicago: University of Chicago Press.

Vale, L. (2007) "MIT Department of Urban Studies and Planning (DUSP) 2006–07 Report to the President," http://web.mit.edu/annualreports/pres07/03.03.pdf (accessed 21 December 2012).

Vancouver.Reflections.com (2008, March 9) "Who is the Beasley in Yaletown named after?" *Vancouver Reflections*, www.vancouverreflections.com/2008/03/09/who-is-the-beasley-in-yaletown-named-after/ (accessed 24 January 2009).

Vischer, J. (1984) "Community and privacy: planners' intentions and residents' reactions," *Plan Canada* 23(4): 112–122.

Index

pedestrians: high-density living 265;
  pedestrian-scaled retail uses 248, 250,
  254; skywalk systems 10, 264–79; traffic
  safety 240, 253–4, 256, 260n4
peri-urban communes 29
Perry, Clarence 10, 239–40, 243, 248,
  252, 254, 260n4
Perth 188–9, 190–1, 193, 200, 202
Pile, S. 44, 46, 64
Pirotte, A. 170
place 25, 26, 28; "large objects" 159;
  vigilantism 76, 88, 90; women's
  insurgent practices in South Africa 81,
  82–4
planning: Africa 45; Australian
  metropolitan planning 187–206;
  informality 48; insurgent planning
  practices 70–1, 72–4, 77–8, 91;
  residents' participation 77; resistance
  to 46, 47, 51–2; scholarship 1–3;
  skywalk systems 269–70, 276, 277;
  staff effectiveness 292–3; as technology
  of domination 7, 44–5, 63; thinking
  about 3–7; Toronto 208; traffic
  safety 237; Vancouver 280–99;
  Zimbabwe 43–4, 51–2, 62, 63–4; see
  also land use planning; metropolitan
  planning
polarization 5, 155–6
police: apartheid South Africa 71;
  Durban 83, 86, 87, 92; vigilantism as
  challenge to the 74; Zimbabwe 53–4,
  56, 58
policy decisions 293–5, 297, 298
political leadership 282, 287, 288, 293,
  296
pollution 182, 209, 265, 273
polycentrism 9, 163, 164, 166, 171,
  173–4, 181, 182–3, 194
population concentration 147
population growth: Australia 201–2;
  Bangalore 101; Mexico City 103–4,
  116, 121, 126, 127–8; Toronto 208,
  220; Vancouver 282–3, 286

population loss 115, 116, 126–8, 137
portfolio diversification theory 106
Portland, Oregon 199–200, 224
Posel, D. 75, 86–7
postmodernism 46–8, 60, 63
poverty: institutional fragmentation 37;
  Lyon 178; Naucalpan 117; women
  78
power relations 45, 46–7; insurgency 90;
  South Africa 79; Zimbabwe 60, 63,
  64
privatization: Amazonia 148; Australian
  metropolitan planning 189, 201;
  structural adjustment programs 124–5
production 213
progressivism 296, 299n1
Property Council of Australia (PCA) 191
public consultation 282, 283, 286–8,
  291–2, 298
public goods 33
public interest 5, 222
public services 118, 125
public space 277–8
public transport: Australian metropolitan
  planning 191, 193; commuting 173,
  174–5; France 166, 173, 174–5,
  176–8, 179, 182; Hong Kong 264,
  267, 270, 272, 277; inequalities 165;
  metropolitan planning 6;
  Naucalpan 123–4; segregation 176–8;
  social issues 167, 179; Toronto 217,
  221, 223, 224; see also bus transport; rail
  transport
public–private partnerships 195, 201, 202
Punter, J. 281, 296, 299

quadricentrism 171
quality of life 156, 209, 226

Raby, R. 45–6, 47, 60, 64
race 83–4
racism 5
Radburn 240
Rae, Robert 10, 235–63